Cutting-edge Technology and Systems for Water Treatment and Reuse

AF631509

In Focus – Special Book Series

Cutting-edge Technology and Systems for Water Treatment and Reuse

Editors

Peng-Kang Jin, Ying-Xue Sun, Can Wang, Huang Huang and Yin-Hu Wu

IWA PUBLISHING

中建环能 SCIMEE

Published by **IWA Publishing**
Republic – Export Building, 1st Floor
2 Clove Crescent
London E14 2BE, UK
Telephone: +44 (0)20 7654 5500
Fax: +44 (0)20 7654 5555
Email: publications@iwap.co.uk
Web: www.iwapublishing.com

First published 2021
© 2021 the Authors

Apart from any fair dealing for the purposes of research or private study, or criticism or review, as permitted under the UK Copyright, Designs and Patents Act (1998), no part of this publication may be reproduced, stored or transmitted in any form or by any means, without the prior permission in writing of the publisher, or, in the case of photographic reproduction, in accordance with the terms of licenses issued by the Copyright Licensing Agency in the UK, or in accordance with the terms of licenses issued by the appropriate reproduction rights organization outside the UK. Enquiries concerning reproduction outside the terms stated here should be sent to IWA Publishing at the address printed above.

The publisher makes no representation, express or implied, with regard to the accuracy of the information contained in this book and cannot accept any legal responsibility or liability for errors or omissions that may be made.

Disclaimer
The information provided and the opinions given in this publication are not necessarily those of IWA and should not be acted upon without independent consideration and professional advice. IWA and the Editors and Authors will not accept responsibility for any loss or damage suffered by any person acting or refraining from acting upon any material contained in this publication.

British Library Cataloguing in Publication Data
A CIP catalogue record for this book is available from the British Library

ISBN: 9781789062670 (paperback)

Contents

 © 2021 The Authors

Editorial: Cutting-edge technology and systems for water treatment and reuse

Water treatment and reuse is the key strategy to address the problems of water shortages. To facilitate communication and application of technology and research in water reclamation, the National Conference on Water Treatment and Reuse of China has been held annually from 2017 by the Committee of Water Treatment and Reuse, Chinese Society for Environmental Sciences.

The theme of the 4th conference was water eco-nexus and water environment remediation. The conference, held from July 31st to August 2nd in 2020, brought together researchers working in different areas related to water treatment and reuse from China, Singapore, Japan, Ireland and many other countries and regions. More than 15,000 researchers from universities, institutes and companies attended the virtual conference online.

The conference covered 15 main topics:

1. Physical separation theory and technologies
2. Chemical transformation theory and technologies
3. Biological purification theory and technologies
4. Natural treatment theory and technologies
5. Water treatment theory and technology based on microalgae
6. Disinfection and microbial risk control
7. Reclaimed water storage and reuse
8. UV treatment theory and technologies
9. Standards and policy for water quality control and water resource management
10. Urban water system and water recycling
11. Industrial water system and water recycling
12. Rural water system and water recycling
13. Pollution control and utilization of rainwater runoff
14. Yellow River basin water environment remediation
15. Desalination technology and processes

In this special issue of *Water Reuse*, 15 papers that focus on cutting-edge technology and systems for water treatment and reuse, were selected among papers presented at the 4th National Conference on Water Treatment and Reuse of China. This special issue is sponsored by CSCEC Scimee Sci. & Tech. Co., Ltd to cover the article processing charges. The 15 papers can be subdivided into three main topics.

The first five articles are in the fields of desalination technology and processes. Bai *et al.* (2021) analyze the fouling mechanism of ultrafiltration membrane caused by denitrification filter effluent during ozonation. Zhang *et al.* (2021) investigate the fouling characteristic of natural organic matter in capacitive deionization. Yu *et al.* (2021) focus on micro-coagulation as pretreatment to improve the performance of a UF-RO system treating the effluent from a coastal municipal wastewater treatment plant. Tong *et al.* (2021) present a review on advances in capacitive deionization technology for water recycling. Luo *et al.* (2021) reveal the adverse effects of chlorine disinfection on biofouling of a pilot-scale reverse osmosis system and identify the key chlorine-resistant bacteria causing membrane fouling.

Further, six articles are related to physical and chemical treatment technology and process of non-conventional water resources. Qiao *et al.* (2021) study the removal efficiencies and mechanism of aniline degradation by peroxydisulfate activated with magnetic Fe-Mn oxides composite. Schnabel *et al.* (2021) investigate photocatalytic degradation of hydrocarbons and methylene blue using floatable titanium dioxide catalysts in contaminated water. Yao *et al.* (2021) focus on the application of integrated dissolved ozone flotation process in a centralised fracturing wastewater treatment plant. Lin *et al.* (2021) present a study on the adsorption characteristics and influencing factors of antibiotics in wastewater by iron-rich farmland soil for the safe reuse of livestock wastewater in farmland. Wang *et al.* (2021a) provide a feeding preparation strategy for the supercritical water oxidation system disposing liquid hazardous wastes from various industrial manufactures. Lu *et al.* (2021) investigate the degradation of 1,4-dioxane using a pulsed switching peroxi-coagulation process.

Lastly, four studies deal with problems concerning biological and natural treatment technology and process of non-conventional water resources. Xin *et al.* (2021) present a study on the behavioral shifts of constructed wetland when the

This is an Open Access article distributed under the terms of the Creative Commons Attribution Licence (CC BY 4.0), which permits copying, adaptation and redistribution, provided the original work is properly cited (http://creativecommons.org/licenses/by/4.0/).

doi: 10.2166/wrd.2021.001

treated water was changed from domestic wastewater to mariculture wastewater. Wang *et al.* (2021b) investigate copper removal from semiconductor chemical mechanical planarization wastewater in the presence of nano-SiO_2 through biosorption. Liu *et al.* (2021) study *Chlorella* growth, photosynthetic pigments and high-valued products accumulation in coastal saline-alkali leachate. Wang & Tian (2021) investigate seasonal variations of pollutants removal and microbial activity in integrated constructed wetland–microbial fuel cell systems.

This special issue deals with various problems and investigates different treatment technologies concerning water reuse. In response to the challenges of severe water shortages and the deterioration of water quality, the focus on cutting-edge technology and systems for water treatment and reuse is fast-growing worldwide. Progress is needed to provide high quality reclaimed water, minimize the risks to environment and human health, and support use of non-conventional water resources.

Guest Editors

Peng-Kang Jin

School of Environmental and Municipal Engineering, Xi'an University of Architecture and Technology, Shaanxi, China

Ying-Xue Sun

School of Ecology and Environment, Beijing Technology and Business University, Beijing, China

Can Wang

School of Environmental Science and Engineering, Tianjin University, Tianjin, China

Huang Huang

School of Environmental Science and Engineering, Sun Yat-Sen University, Guangdong, China

Yin-Hu Wu IWA

School of Environment, Tsinghua University, Beijing, China

REFERENCES

Bai, Y., Wu, Y.-H., Tong, X., Wang, Y.-H., Ikuno, N., Wang, W., Shi, Y.-L. & Hu, H.-Y. 2021 Revealing the membrane fouling mechanism caused by the denitrification filter effluent during ozonation by model assessment. *Water Reuse* **11** (2), 149–159.

Lin, L., Hu, X., Liang, J., Huang, Z., Yu, G. & Chong, Y. 2021 Adsorption of tylosin in wastewater by iron-rich farmland soil and the effect of iron reduction and common cations. *Water Reuse* **11** (2), 248–256.

Liu, X.-y., Hong, Y. & Gu, W.-p. 2021 Influence of light quality on *Chlorella* growth, photosynthetic pigments and high-valued products accumulation in coastal saline-alkali leachate. *Water Reuse* **11** (2), 301–311.

Lu, Y., Shi, H., Yao, J., Liu, G., Luo, H. & Zhang, R. 2021 1,4-dioxane degradation using a pulsed switching peroxi-coagulation process. *Water Reuse* **11** (2), 268–278.

Luo, L.-W., Wu, Y.-H., Wang, Y.-H., Tong, X., Bai, Y., Chen, G.-Q., Wang, H.-B., Ikuno, N. & Hu, H.-Y. 2021 Aggravated biofouling caused by chlorine disinfection in a pilot-scale reverse osmosis treatment system of municipal wastewater. *Water Reuse* **11** (2), 201–211.

Qiao, L., Shi, Y., Cheng, Q., Liu, B. & Liu, J. 2021 The removal efficiencies and mechanism of aniline degradation by peroxydisulfate activated with magnetic Fe-Mn oxides composite. *Water Reuse* **11** (2), 212–223.

Schnabel, T., Jautzus, N., Mehling, S., Springer, C. & Londong, J. 2021 Photocatalytic degradation of hydrocarbons and methylene blue using floatable titanium dioxide catalysts in contaminated water. *Water Reuse* **11** (2), 224–235.

Tong, Y., Zhou, S., Zhou, J., Zhang, G., Li, X., Zhao, C. & Liu, P. 2021 Advances in efficient desalination technology of capacitive deionization for water recycling. *Water Reuse* **11** (2), 189–200.

Wang, X. & Tian, Y. 2021 Seasonal variations of pollutants removal and microbial activity in integrated constructed wetland–microbial fuel cell systems. *Water Reuse* **11** (2), 312–327.

Wang, B., Lu, M., Zhao, L., Zhang, Z., Zhao, Y. & Qin, S. 2021a Feeding preparation strategy for the supercritical water oxidation (SCWO) system for disposing of liquid hazardous waste. *Water Reuse* **11** (2), 257–267.

Wang, X., Buer, G., Fan, W., Gao, L. & Huo, M. 2021b Copper removal from semiconductor CMP wastewater in the presence of nano-SiO_2 through biosorption. *Water Reuse* **11** (2), 289–300.

Xin, Y., Liu, L., Wei, L., Huang, X. & Liu, C. 2021 Recycling of rural abandoned constructed wetlands: mariculture wastewater treatment. *Water Reuse* **11** (2), 279–288.

Yao, Z., Jin, X., Liang, J., Wang, R., Jin, P., Li, G., Yao, L. & Wang, Z. 2021 Application of an integrated dissolved ozone flotation process in centralised fracturing wastewater treatment plant. *Water Reuse* **11** (2), 236–247.

Yu, T., Xu, C., Chen, F., Yin, H., Sun, H., Cheng, L. & Bi, X. 2021 Microcoagulation improved the performance of the UF–RO system treating the effluent from a coastal municipal wastewater treatment plant: a pilot-scale study. *Water Reuse* **11** (2), 177–188.

Zhang, H., Tian, J., Hao, X., Liu, D. & Cui, F. 2021 Investigations on the fouling characteristic of humic acid and alginate sodium in capacitive deionization. *Water Reuse* **11** (2), 160–176.

© 2021 The Authors

Revealing the membrane fouling mechanism caused by the denitrification filter effluent during ozonation by model assessment

Yuan Bai, Yin-Hu Wu, Xin Tong, Yun-Hong Wang, Nozomu Ikuno, Wei Wang, Yu-Long Shi and Hong-Ying Hu IWA

ABSTRACT

Membrane fouling has been a major obstacle for stable operation of ultrafiltration. In this study, prevailing fouling models were applied to assess the fouling behavior of the denitrification filter (DNF) effluent during ozonation. In order to clarify the fouling mechanism, correlation analysis and redundancy analysis (RDA) were conducted to investigate the correlations among model parameters, fouling potential and water features of the DNF effluent. The combined intermediate-standard model exhibited superior determination coefficients ($R^2 > 0.99$). Based on analytical results, the model parameter of intermediate blocking (K_i) and standard blocking (K_s) was fairly applicable to describe the fouling of higher molecular weight (F1, MW >4,000 Da) and lower molecular weight fractions (F2, MW = 2,000–4,000 Da and F3, MW < 2,000 Da), respectively. In comparison, F1 played a predominant role in the fouling behavior of the DNF effluent. Increased ozone dosage resulted in decreased membrane fouling contribution of F1 and increased fouling contribution of F2 and F3 during ozonation. The change of fouling contributions was attributed to the transformation of high MW fractions into lower MW fractions by ozonation. This study clarified the relationships between model parameters and the membrane fouling process caused by organic fractions with specific molecular weight, thus demonstrating the membrane fouling mechanism of the DNF effluent during ozonation.

Key words | correlation analysis, intermediate-standard model, membrane fouling mechanism, ozonation, redundancy analysis

Yuan Bai
Yin-Hu Wu (corresponding author)
Xin Tong
Yun-Hong Wang
Hong-Ying Hu IWA
Environmental Simulation and Pollution Control State Key Joint Laboratory, State Environmental Protection Key Laboratory of Microorganism Application and Risk Control (SMARC), School of Environment,
Tsinghua University,
Beijing 100084,
China
E-mail: *wuyinhu@mail.tsinghua.edu.cn*

Nozomu Ikuno
Kurita Water Industries Ltd.,
Nakano-ku, Tokyo 164-0001,
Japan

Wei Wang
Yu-Long Shi
Beijing Drainage Group Company Limited,
Beijing,
China

Hong-Ying Hu
Shenzhen Environmental Science and New Energy Technology Engineering Laboratory,
Tsinghua-Berkeley Shenzhen Institute,
Shenzhen 518055,
China

HIGHLIGHTS

- Correlation between different MW fractions and model parameters was established.
- Fouling contribution of high and low MW fractions was assessed by model parameters.
- Change of model parameters illustrated conversion from high to low MW fractions.
- Intermediate blocking showed predominance over standard blocking during ozonation.
- Fouling control by ozonation could be quantitatively evaluated by model assessment.

This is an Open Access article distributed under the terms of the Creative Commons Attribution Licence (CC BY 4.0), which permits copying, adaptation and redistribution, provided the original work is properly cited (http://creativecommons.org/licenses/by/4.0/).

doi: 10.2166/wrd.2021.005

GRAPHICAL ABSTRACT

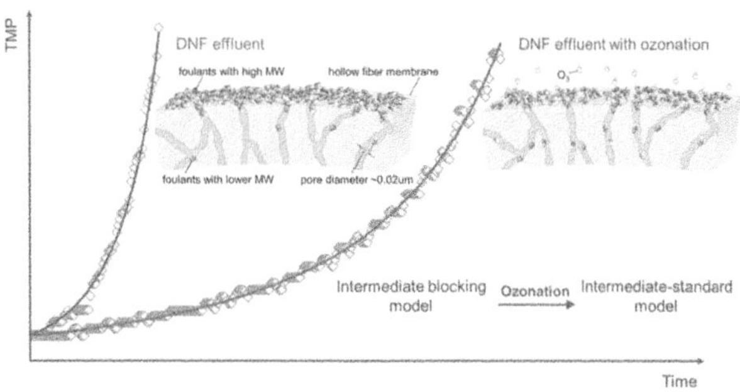

INTRODUCTION

Nowadays, membrane filtration processes such as ultrafiltration (UF) have occupied a pivotal position of in-depth sewage treatment for water reclamation (Shen *et al.* 2012; Song *et al.* 2019). Membrane fouling continues to be one of the most significant problems working with UF (Guo *et al.* 2012; Yoo 2018). The investigation of the membrane fouling mechanism could contribute a great deal to solve the fouling problem and achieve stable operation of the system. Therefore, it is of great significance to identify and reveal the membrane fouling mechanism in the process of wastewater reclamation.

Membrane fouling is mainly related to pore blocking and the formation of the cake layer by deposition of foulants inside or on top of the membrane (Cerón-Vivas *et al.* 2018). Four prevailing mechanisms such as complete blocking, standard blocking, intermediate blocking and cake filtration have been adopted to describe membrane fouling (Ho & Zydney 2002; Iritani 2013). Complete blocking assumed that particles completely plugged membrane pores without superposition. In standard blocking, small particles deposited onto the internal pore walls of membrane, thus leading to a restriction of pore volume. Intermediate blocking was introduced by assuming that particles can settle on other deposited particles or membrane surface. Cake filtration occurred when particles accumulated on the membrane surface by forming a permeable cake of increasing thickness (Bowen *et al.* 1995; Suarez & Veza 2000).

The operation patterns of the UF membrane were based on constant-pressure mode (Aslam *et al.* 2015) and constant-flux mode (Kovalsky *et al.* 2009), the latter being more applied in most full-scale wastewater treatment systems (Kim & DiGiano 2009; Goldrick *et al.* 2017). Based on the filtration system with constant flux, fouling models derived from these four mechanisms have been developed to interpret the transmembrane pressure increase during operation (Hermia 1982; Hlavacek & Bouchet 1993). Table 1 summarizes the four fouling models with corresponding equations, which were related with the operating transmembrane pressure (P), initial transmembrane pressure (P_0), filtration time (t), initial flux (J_0) and the corresponding model parameters (K_b, K_s, K_i and K_c). On this basis, Bolton *et al.* (2006) proposed five new fouling models that accounted for the combined effects of the above-mentioned individual fouling mechanisms derived from Darcy's law. The combined models related to the same parameters (P, P_0, t and J_0) as individual models but considered simultaneously two of the model parameters (K_b, K_s, K_i and K_c) involved. Table 2 exhibits the equations of five combined fouling models for constant-flux operation.

Table 1 | Equations of the individual fouling models for constant-flux operation

Model	Equation	Fitting parameter
Complete	$\frac{P}{P_0} = \frac{1}{1 - K_b t}$	K_b (s^{-1})
Standard	$\frac{P}{P_0} = \left(1 - \frac{K_s J_0 t}{2}\right)^{-2}$	K_s (m^{-1})
Intermediate	$\frac{P}{P_0} = e^{K_i J_0 t}$	K_i (m^{-1})
Cake	$\frac{P}{P_0} = 1 + K_c J_0^2 t$	K_c (s/m^2)

Table 2 | Equations of the combined fouling models for constant-flux operation

Model	Equation	Fitting parameters
Cake-complete	$\frac{P}{P_0}=\frac{1}{(1-K_bt)}\left(1-\frac{K_cJ_0^2}{K_b}\ln(1-K_bt)\right)$	K_c (s/m^2), K_b (s^{-1})
Cake-intermediate	$\frac{P}{P_0}=\exp(K_iJ_0t)\left(1+\frac{K_cJ_0}{K_i}(\exp(K_iJ_0t)-1)\right)$	K_c (s/m^2), K_i (m^{-1})
Complete-standard	$\frac{P}{P_0}=\frac{1}{(1-K_bt)\left(1+\frac{K_sJ_0}{2K_b}\ln(1-K_bt)\right)^2}$	K_b (s^{-1}), K_s (m^{-1})
Intermediate-standard	$\frac{P}{P_0}=\frac{\exp(K_iJ_0t)}{\left(1-\frac{K_s}{2K_i}(\exp(K_iJ_0t)-1)\right)^2}$	K_i (m^{-1}), K_s (m^{-1})
Cake-standard	$\frac{P}{P_0}=\left(\left(1-\frac{K_sJ_0t}{2}\right)^{-2}+K_cJ_0^2t\right)$	K_c (s/m^2), K_s (m^{-1})

Both individual and combined fouling models are widely applied to assess the blocking process of synthetic solutions, surface water or wastewater (Wang *et al.* 2018; Kirschner *et al.* 2019; Tong *et al.* 2020). Xing *et al.* (2019) applied fouling models to evaluate the fouling mitigation by UV and UV/chlorine oxidation, and standard pore blocking was proved to be crucial during the filtration process. Liu & Kim (2008) evaluated the performances of both blocking laws and artificial neural networks (ANNs) model, demonstrating that the combined cake-complete, cake-intermediate and ANNs models showed relative high consistency with experimental data. Liu *et al.* (2019) also used the fouling models on the ceramic membrane filtration for wastewater reclamation, reporting that the combined intermediate-standard model showed the best fit in all cases.

Due to the complexity of wastewater, researchers have revealed that the combined fouling models exhibited better fitting performances compared with single ones (Cerón-Vivas *et al.* 2018; Xiong *et al.* 2019; Li *et al.* 2020).

To illustrate the fouling process, models were proposed with idealized particles and membrane pores based on theoretical assumptions. But in practical engineering, the fouling phenomena caused by wastewater were rather complex and closely related to various water features, such as viscosity, hydrophobicity and molecular weight (MW) distribution (Kang *et al.* 2011; Shen *et al.* 2012). Therefore, to further demonstrate the influence of water features on membrane fouling, it is of importance to analyze the model assessment results based on water features. This is of great practical significance to clarify the membrane fouling mechanism and propose fouling control strategies. However, related researches are rarely reported.

In this study, we compared the fitting performances of different models in the fouling process of the denitrification filter (DNF) effluent during ozonation. Furthermore, correlation analysis and redundancy analysis (RDA) were conducted to investigate the correlations among the model parameters, fouling potential and water features of the DNF effluent, thus endowing the model parameters with practical significance in our system. Based on the fitting results at different ozone dosages, the quantitative relationships among ozone dosage, water features and fouling potential were established. The membrane fouling mechanism of the DNF effluent during ozonation was revealed and this study helped glean valuable insight into membrane fouling control by model assessment in the UF system.

MATERIALS AND METHODS

Ozonation and membrane fouling experiments

Feed water of this study was the DNF effluent collected from the wastewater reclamation plant of Gaobeidian, Beijing. The feed water was transported to the laboratory immediately after sampling for experiments. The major water characteristics including DOC (dissolved organic carbon), COD (chemical oxygen demand), SS (suspended solids),

TN (total nitrogen), pH, and turbidity of the feed sample are listed in Supplementary Material, Table S1.

Ozonation experiments were conducted using a stock solution with maximum ozone concentration of 40 mg/L prepared by an ozone generator (CF-YG20, Shanmeishuimei High Technology Co., Ltd). The volumes of the original DNF effluent and the stock solution of water samples with different ozone dosages were calculated firstly before ozonation, and the final DOC concentration of each sample was set at 4 mg/L. After ozonation treatment, the DOC of each water sample was almost unchanged and no residual ozone could be detected after 6-min storage.

The filtration experiments were conducted under the constant flux of 180 L/m^2h using a dead-end filtration system as shown in Supplementary Material, Figure S1, which was described in detail in our previous work (Bai *et al.* 2020). The membrane module was constructed with a single hollow fiber membrane (PVDF) with a nominal pore diameter of 0.02 μm. A constant flow pump was set to maintain the working flux. The threshold transmembrane pressure (TMP) was fixed at 3 bar, and the pump would be stopped at once when the TMP reached the fixed value. An electronic balance was used to monitor the flow rate, while a pressure sensor was used to monitor the TMP change during the filtration process in real time. Each filtration experiment in this study has been repeated three times.

Water feature characterization

The DOC concentration, fluorescence excitation emission matrix spectra and UV_{254} value of water samples were determined by a total organic carbon (TOC) analyzer (TOC-VCPH, SHIMADZU), a fluorescence spectrophotometer (F-7000, Hitachi) and a UV–Vis spectrophotometer (UV-2450, Shimadzu, Japan), respectively. A rapid digested spectrophotometry method (HJ/T 399-2007) was used to determine COD concentration. A pH meter (Seven2Go, Mettler Toledo, Switzerland) was used to measure the pH value. Chinese States Standard Testing Methods (State Environmental Protection Administration 2002) were used to measure TN. The apparent MW distribution was determined following the reported method (Wang *et al.* 2017) by a high-performance size-exclusion chromatography (HPSEC; LC20-AD, Shimadzu, Japan). The mobile phase, which was composed of 2.4 mM KH_2PO_4, 3.6 mM K_2HPO_4 and 25 mM Na_2SO_4, was injected of 100 μL with a flow rate of 0.5 mL/min. Water samples were filtered by a 0.45 μm membrane, and the filtrates were used for the above-mentioned analyses.

The filtration and analytical results of water samples with ozonation treatment were reported by the authors' previous study. For more detailed information, refer to Bai *et al.* (2020).

Modeling analysis of the membrane fouling process

In this study, to further investigate the membrane fouling mechanism of DNF effluent during ozonation, five fouling models including four individual models and one combined model (intermediate-standard model) in Tables 1 and 2 were applied to fit the experimental data in ORIGIN. The fitting results of models were evaluated based on the sum of squares of residuals (SSR) and the determination coefficient (R^2).

Statistical analysis of the membrane fouling process

A method of RDA was adopted to determine the correlations among the model parameters, fouling potential and water features of the DNF effluent during ozonation with different ozone dosages (0–2 mg O_3/mg C). The water features of membrane influent, including DOC, UV_{254}, EEM, F1, F2 and F3, were set as six explanatory variables. EEM represented the fluorescence intensity of excitation emission matrix spectra. F1, F2 and F3 represented the proportion of high (>4,000 Da), medium (2,000–4,000 Da) and low (<2,000 Da) apparent MW fractions, respectively. Model parameters K_i and K_s of the combined intermediate-standard model and the filtration time t for the TMP reaching 3 bar were set as three different response variables. All explanatory and response variables consisted of the results of three repeated experiments. Logarithmic transformation was performed on the raw data of all response variables and the significance of the explanatory variables for response variables was evaluated by 499 times of Monte-Carlo permutation test. RDA analysis was performed with CANOCO version 5.0 (Microcomputer Power, USA) according to published methods (ter Braak & Smilauer 2012).

To assess the correlations among the model parameters, fouling potential and water features of the DNF effluent at different conditions of ozonation (0–2 mg O_3/mg C), IBM

SPSS Statistics version 25.0 was applied in the correlation analysis of the data.

RESULTS AND DISCUSSION

Filtration tests of DNF effluent with ozonation treatment

Filtration tests of DNF effluent with ozonation is shown in Figure 1. The control sample without ozonation showed the fastest fouling rate with the filtration time of 80 min for the TMP increased to 3 bar. However, effective membrane fouling alleviation was achieved after ozonation treatment, and the alleviation performance was more significant as increased O_3 dosage. Compared with the original DNF effluent, the filtration time was prolonged by 1.125, 1.6, 2, 2.65 and 3.75 times with O_3 dosage of 0.1, 0.25, 1, 1.5 and 2 mg O_3/mg C, respectively. These results demonstrated the effective alleviation performance of membrane fouling by ozonation. The data of filtration tests were reported in the authors' previous study (Bai *et al.* 2020).

Modeling analysis of the membrane fouling process caused by the DNF effluent during ozonation

In order to acquire an explicit and profound understanding of the membrane fouling mechanism from the perspective of modeling analysis, four widely applied individual fouling models mentioned above (Table 1) were applied to interpret the experimental data of DNF effluent during ozonation. Figure 1 compares the fitting results of water samples treated with different ozone dosages for each model, and the model parameters (K_b, K_s, K_i and K_c), SSR and R^2 of each run are shown in Supplementary Material, Table S2. The fitting results of the cake or complete fouling model were not

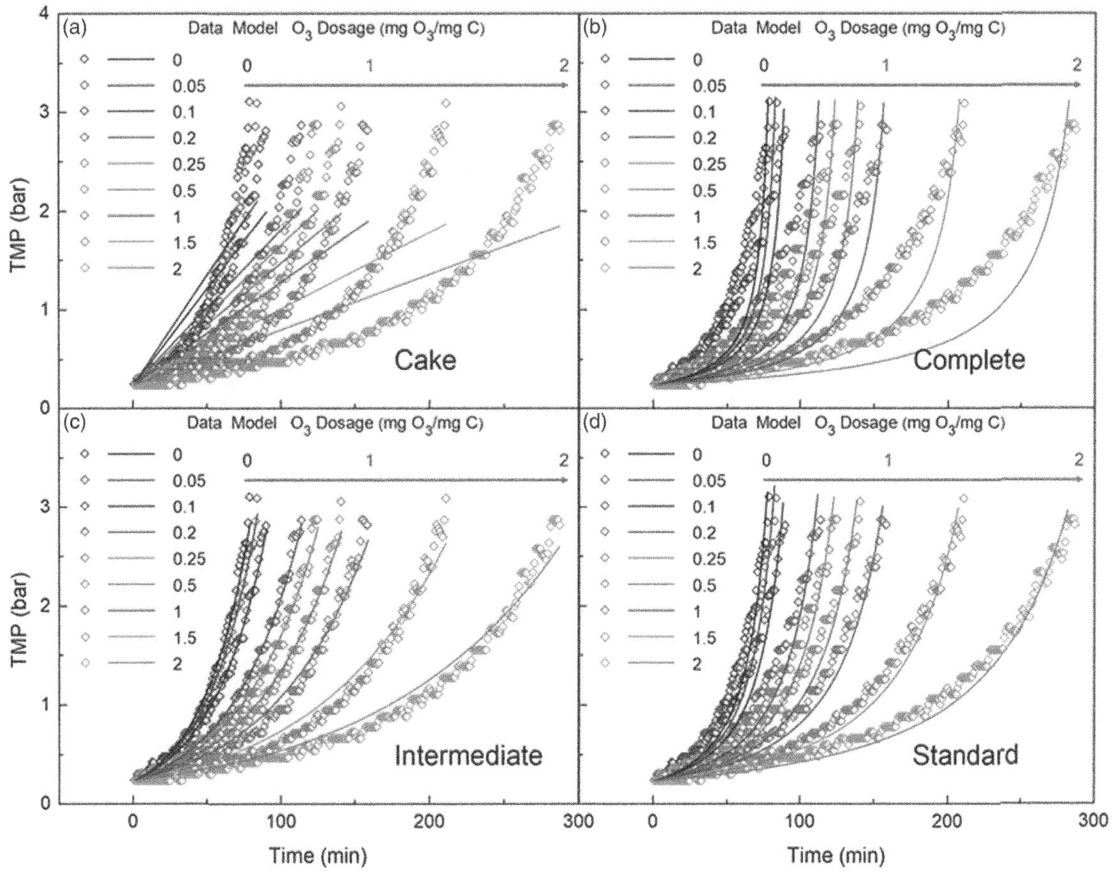

Figure 1 | Prediction of membrane fouling using the individual (a) cake, (b) complete, (c) intermediate and (d) standard model for experimental data.

promising with much lower R^2 and higher SSR values, while the standard fouling model exhibited improved fitting results especially at higher ozone dosages. Among the four applied models, the intermediate fouling model presented the best-fit performance at all ozone dosages. But the fitting performance became less satisfactory with increased ozone dosage. According to the results in Supplementary Material, Table S2, relatively low R^2 and high SSR were obtained at high ozone dosage, suggesting marked deviation between the intermediate model and experimental results after high-dosage ozonation. These results indicated that an individual fouling model was insufficient to elucidate the membrane fouling mechanism of the DNF effluent during ozonation, since several blocking laws might contribute simultaneously to the membrane fouling process (Liu & Kim 2008).

According to the fitting results of the forementioned models, we seek to analyze the data by applying the combined fouling models based on the intermediate model. Considering the improved fitting performance of the standard model at high ozone dosages, the combined intermediate-standard model was applied to describe the filtration process (Figure 2), and the specific fitting results are shown in Table 3. The combined model showed more precise fitting performance than the forementioned individual models, with all R^2 larger than 0.99 at different ozone dosages. According to these results, we speculated that the membrane fouling behavior in our study was more likely attributed to the combination of intermediate and standard blocking laws.

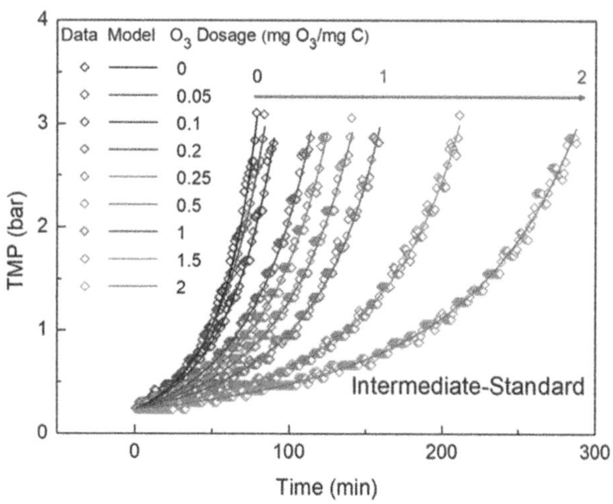

Figure 2 | Prediction of membrane fouling using the combined intermediate-standard model for experimental data.

Table 3 | Fitting parameters of each run with the combined intermediate-standard model

	Intermediate-standard				
Ozone dosage (mg O_3/mg C)	K_i (m^{-1})	K_s (m^{-1})	K_i/K_s	R^2	SSR
0	9.40	0.39	24.10	0.996	0.19
0.05	9.63	0.15	64.20	0.992	0.40
0.1	8.58	0.20	42.90	0.996	0.19
0.2	6.92	0.14	49.43	0.996	0.23
0.25	6.16	0.17	36.24	0.995	0.34
0.5	5.27	0.20	26.35	0.995	0.33
1	3.99	0.31	12.87	0.995	0.43
1.5	2.80	0.39	7.18	0.995	0.48
2	2.01	0.41	4.91	0.995	0.63

Correlations among the model parameters, fouling potential and water features of the DNF effluent

In order to further explore the influence of water features on membrane fouling and seek out the practical significance of the model parameters, correlation analysis was applied to quantify the correlations among the model parameters, fouling potential and water features of the DNF effluent (Figure 3 and Supplementary Material, Table S3). The filtration time t before the operational pressure reached 3 bar, which was the direct indicator of membrane fouling potential, had markedly negative correlation with K_i (fitting parameter of

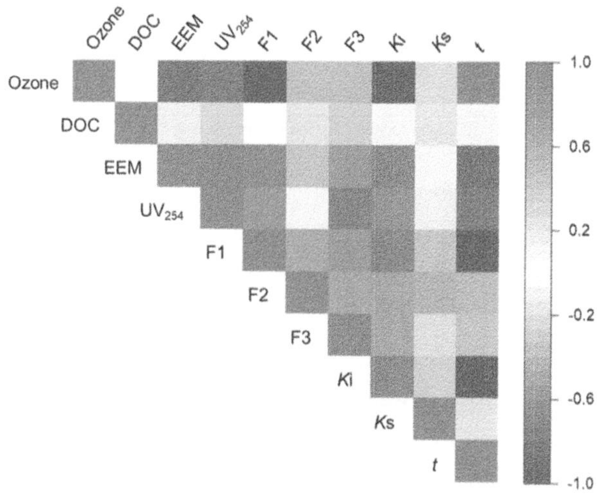

Figure 3 | Correlations among the model parameters, fouling potential and water features of the DNF effluent. EEM represented the fluorescence intensity of excitation emission matrix spectra. F1, F2 and F3 represented the proportion of high (>4,000 Da), medium (2,000–4,000 Da) and low (<2,000 Da) apparent molecular weight fractions, respectively.

intermediate blocking in intermediate-standard model), indicating that intermediate blocking played a crucial role on membrane fouling and larger K_i corresponded to higher fouling potential. By contrast, the filtration time t presented the indistinctive positive correlation with K_s (fitting parameter of standard blocking in intermediate-standard model), suggesting that standard blocking showed a minor influence on membrane fouling. Therefore, intermediate blocking had a greater predominance over standard blocking in the fouling mechanism of the DNF effluent.

Reduction of organic load and structural changes of organic matters were two common explanations for membrane fouling control by ozonation (Van Geluwe *et al.* 2011; Wei *et al.* 2016). In this study, the correlations of DOC concentration with K_i and K_s were quite low, suggesting that it was structural changes of organic matters rather than the DOC concentration that played predominant roles on membrane fouling during ozonation. As for the other water features, K_i showed the highest positive correlation (0.95) with F1 (the proportion of organic fractions >4,000 Da) but a negative correlation with F2 (the proportion of organic fractions 2,000–4,000 Da) and F3 (the proportion of organic fractions <2,000 Da). By contrast, opposite correlations of K_s with F1, F2 and F3 were obtained. EEM and UV_{254} were also positively correlated with K_i, which might be attributed to their positive correlationships with F1. These results demonstrated that K_i (corresponding to intermediate blocking) was fairly applicable to describe membrane fouling caused by high MW fractions (F1) of the DNF effluent. While K_s (corresponding to standard blocking) was more suitable to represent the membrane fouling caused by lower MW fractions (F2 and F3).

In terms of the specific hypothesis of fouling models, intermediate blocking law presumes that the mean particle size (d_p) was larger than the average pore diameter (D_p), while d_p was presumed to be smaller than D_p in standard blocking law (Sabia *et al.* 2016; Ochando-Pulido & Martínez-Ferez 2017). In other words, intermediate and standard blocking laws were widely used to describe the membrane fouling process by large and small particles, respectively. Our analytical results were in accordance with the theoretical assumptions and further established the correlation between the model parameters and water features of the DNF effluent, endowing K_i and K_s with a practical significance in our system.

RDA was carried out to comprehensively analyze the quantitative relationships among the model parameters, fouling potential and water features of the DNF effluent (Figure 4). In the RDA diagram, the angle between any two vectors represents the correlation between them (the smaller the angle, the greater the correlation). The Axis 1 and Axis 2 of RDA explained 92.7 and 3.9% of the variance, respectively. Therefore, Axis 1 could largely reflect the degree of membrane fouling. The correlation analytical results were in accordance with Figure 3. Moreover, F1 was the most important factor and could account for nearly 90% of the membrane fouling caused by the DNF effluent, which was significant at the level of 0.002. It further confirmed that the fouling of F1 corresponding to the intermediate blocking played a predominant role in membrane fouling of the DNF effluent.

Quantitative analysis of the ozonation influence on membrane fouling by model assessment

Based on RDA results, the experimental data could be classified into three groups (Figure 4). The first group was represented by the control sample without ozonation, and the data points were plotted at the top of the graph. The second group included the water samples treated by lower ozone dosages (0.05, 0.1, 0.2, 0.25 and 0.5 mg O_3/mg C) with data points plotted at the bottom left of the graph. The third group consisted of the water samples treated by

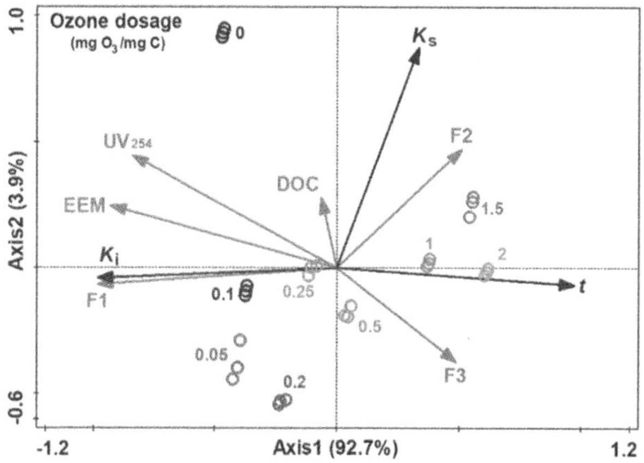

Figure 4 | Redundancy analysis (RDA) of the relationships among the model parameters, fouling potential and water features of the DNF effluent. EEM represented the fluorescence intensity of excitation emission matrix spectra. F1, F2 and F3 represented the proportion of high (>4,000 Da), medium (2,000–4,000 Da) and low (<2,000 Da) apparent molecular weight fractions, respectively.

high ozone dosages (1, 1.5 and 2 mg O_3/mg C) with data points plotted at right of the graph. The grouping results demonstrated that ozonation treatment had significant impacts on water features of the DNF effluent, resulting in the diverse fouling potential of water samples after ozonation. In addition, ozonation showed relatively higher correlations with all water features (Figure 3), suggesting the marked influence of ozonation on water features as well.

Table 3 exhibits the quantitative variations of fouling parameters in the combined intermediate-standard model during ozonation. Increased ozone dosage led to decreased K_i and increased K_s, corresponding to decreased membrane fouling caused by high MW fractions and increased fouling caused by lower MW fractions during ozonation, respectively. The respective contribution of the two individual blocking laws can be evaluated from the magnitudes of the fitting parameters (Cerón-Vivas *et al.* 2018). Decreased K_i/K_s (Table 3) during ozonation suggested the decreased contribution of intermediate blocking and increased contribution of standard blocking, which was in accordance with the fitting performances in Figure 1(c) and 1(d). The change of fouling contributions corresponded to the transformation process of high MW fractions into lower MW fractions by ozonation (Figure 5). As for the DNF effluent, high MW fractions accounted for the major part of the organic foulants and thus intermediate blocking played as a critical fouling mechanism (Figure 5(a)). After ozonation

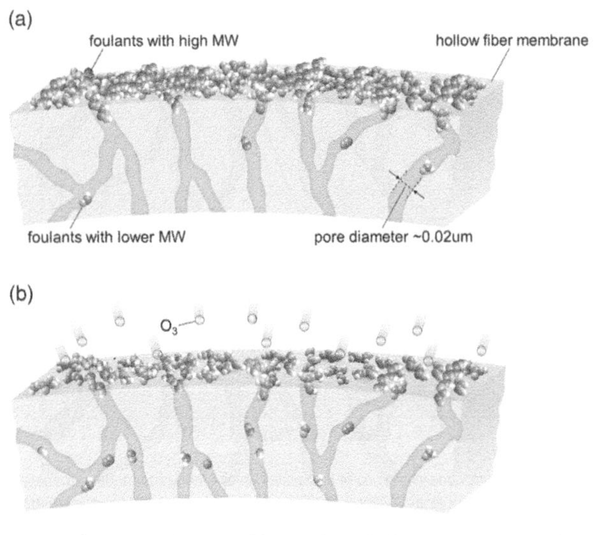

Figure 5 | Schematic diagrams of the membrane fouling process caused by the DNF effluent (a) before and (b) after ozonation treatment.

treatment, the proportion of high MW fractions decreased and that of lower MW fractions increased, and thus standard blocking need to be taken into consideration to clarify the overall fouling process (Figure 5(b)).

However, K_i was about one or two magnitudes larger than K_s at all ozone dosages, indicating greater contribution of intermediate blocking in membrane fouling. This was in accordance with the RDA results in the section 'Correlations among the model parameters, fouling potential and water features of the DNF effluent' that the intermediate blocking mechanism played a predominant role over standard blocking under our experimental conditions in spite of the inverse contribution changes of the two blocking laws after ozonation. Therefore, remarkable fouling alleviation achieved by ozonation was actually ascribed to the alleviation of intermediate blocking by the transformation of high MW fractions into lower MW fractions. Studies have also reported that ozonation could achieve significant membrane fouling alleviation by degradation of higher MW fractions (Zhu *et al.* 2010; Tang *et al.* 2017; Jin *et al.* 2019; Zhao *et al.* 2019).

In order to further describe the change of model parameters during ozonation, the Michaelis–Menten equation was modified and adopted (Equation (1)) in this study and the results are shown in Figure 6:

$$K_i = K_{i\,max} - \frac{K_{i\,max}[O_3]}{A_m + [O_3]}, \quad Ks = \frac{K_{s\,max}[O_3]}{B_m + [O_3]} \tag{1}$$

$[O_3]$ represents the concentration of ozone, $K_{i\,max}$ and $K_{s\,max}$ represent the theoretical maximal value of K_i and K_s, respectively. A_m represents the ozone concentration when K_i equals to $1/2K_{i\,max}$ and B_m represents the ozone concentration when K_s equals to $1/2K_{s\,max}$.

The fitting results of the modified Michaelis–Menten equation are shown in Equation (2):

$$K_i = 9.75 - \frac{9.75[O_3]}{0.57 + [O_3]}, \quad R^2 = 0.975$$

$$K_s = \frac{0.43[O_3]}{0.3 + [O_3]}, \quad R^2 = 0.638 \tag{2}$$

Specifically, K_i decreased significantly and showed good fit with ozone dosage ($R^2 = 0.975$). These results further implied that the change of K_i during ozonation could be

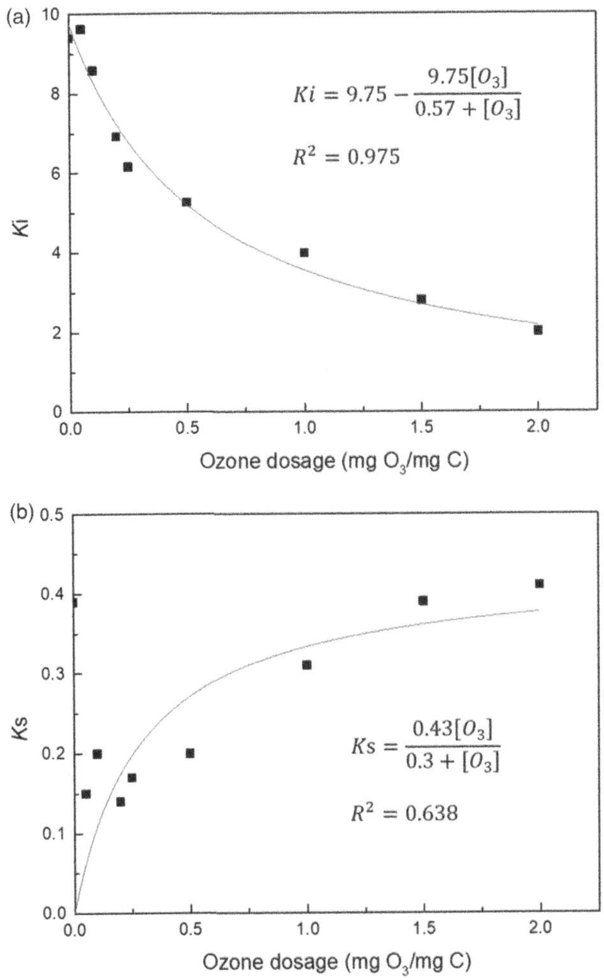

Figure 6 | Model parameters (a) K_i and (b) K_s of the combined intermediate-standard model as a function of ozone dosage.

quantitatively evaluated as a function of ozone dosage. In practical engineering, the membrane fouling control performance by ozonation, especially the degradation efficiency of high MW fractions, could be predicted by quantitative evaluation via the change of K_i by model assessment.

CONCLUSIONS

- The combined intermediate-standard model was promising to evaluate the fouling process of the DNF effluent during ozonation.
- Model parameter K_i corresponding to intermediate blocking was fairly applicable to describe the fouling caused by high MW fractions (F1, MW > 4,000 Da), while K_s corresponding to standard blocking was more suitable to describe the fouling caused by lower MW fractions (F2, MW = 2,000–4,000 Da and F3, MW < 2,000 Da) of the DNF effluent in this study.
- During ozonation, increased ozone dosage led to decreased K_i and increased K_s, indicating decreased fouling contribution of high MW fractions and increased fouling contribution of lower MW fractions, respectively.
- Intermediate blocking mechanism played a predominant role over standard blocking under the experimental conditions of this study. RDA results revealed that F1 was the most important factor to affect membrane fouling of the DNF effluent, accounting for nearly 90% of the fouling variation ($p = 0.002$).
- In practical engineering, model parameters were closely correlated with water features and the fouling control performance by ozonation could be quantitatively evaluated via the change of model parameters.

ACKNOWLEDGEMENTS

This study was supported by the Science and Technology Plan Project of Beijing (Z181100005518001) and the Key Program of the National Natural Science Foundation of China (No. 51738005).

CONFLICT OF INTEREST

The authors declare no financial or commercial conflicts of interest.

DATA AVAILABILITY STATEMENT

All relevant data are included in the paper or its Supplementary Information.

REFERENCES

Aslam, M., Lee, P.-H. & Kim, J. 2015 Analysis of membrane fouling with porous membrane filters by microbial suspensions for autotrophic nitrogen transformations. *Sep. Purif. Technol.* **146**, 284–293.

Bai, Y., Wu, Y.-H., Wang, Y.-H., Tong, X., Zhao, X.-H., Ikuno, N. & Hu, H.-Y. 2020 Membrane fouling potential of the denitrification filter effluent and the control mechanism by ozonation in the process of wastewater reclamation. *Water Res.* **173**, 115591.

Bolton, G., LaCasse, D. & Kuriyel, R. 2006 Combined models of membrane fouling: development and application to microfiltration and ultrafiltration of biological fluids. *J. Membr. Sci.* **277** (1), 75–84.

Bowen, W. R., Calvo, J. I. & Hernández, A. 1995 Steps of membrane blocking in flux decline during protein microfiltration. *J. Membr. Sci.* **101** (1), 153–165.

Cerón-Vivas, A., Kalboussi, N., Morgan-Sagastume, J. M., Harmand, J. & Noyola, A. 2018 Model assessment of the prevailing fouling mechanisms in a submerged membrane anaerobic reactor treating low-strength wastewater. *Bioresour. Technol.* **268**, 460–469.

Goldrick, S., Joseph, A., Mollet, M., Turner, R., Gruber, D., Farid, S. S. & Titchener-Hooker, N. J. 2017 Predicting performance of constant flow depth filtration using constant pressure filtration data. *J. Membr. Sci.* **531**, 138–147.

Guo, W., Ngo, H.-H. & Li, J. 2012 A mini-review on membrane fouling. *Bioresour. Technol.* **122**, 27–34.

Hermia, J. 1982 Constant pressure blocking filtration laws-application to power-law non-Newtonian fluids. *Trans. Inst. Chem. Eng.* **60** (3), 183–187.

Hlavacek, M. & Bouchet, F. 1993 Constant flowrate blocking laws and an example of their application to dead-end microfiltration of protein solutions. *J. Membr. Sci.* **82** (3), 285–295.

Ho, C. C. & Zydney, A. L. 2002 Transmembrane pressure profiles during constant flux. Microfiltration of bovine serum albumin. *J. Membr. Sci.* **209** (2), 363–377.

Iritani, E. 2013 A review on modeling of pore-blocking behaviors of membranes during pressurized membrane filtration. *Drying Technol.* **31** (2), 146–162.

Jin, X., Zhang, W., Hou, R., Jin, P., Song, J. & wang, X. C. 2019 Tracking the reactivity of ozonation towards effluent organic matters from WWTP using two-dimensional correlation spectra. *J. Environ. Sci.* **76**, 289–298.

Kang, X., Wang, X., Xia, H., Waite, T. D. & Wen, X. 2011 Combined effect of membrane and foulant hydrophobicity and surface charge on adsorptive fouling during microfiltration. *J. Membr. Sci.* **373** (1–2), 140–151.

Kim, J. & DiGiano, F. A. 2009 Fouling models for low-pressure membrane systems. *Sep. Purif. Technol.* **68** (3), 293–304.

Kirschner, A. Y., Cheng, Y.-H., Paul, D. R., Field, R. W. & Freeman, B. D. 2019 Fouling mechanisms in constant flux crossflow ultrafiltration. *J. Membr. Sci.* **574**, 65–75.

Kovalsky, P., Bushell, G. & Waite, T. D. 2009 Prediction of transmembrane pressure build-up in constant flux microfiltration of compressible materials in the absence and presence of shear. *J. Membr. Sci.* **344** (1), 204–210.

Li, R., Gao, B., Wang, W., Yue, Q. & Wang, Y. 2020 Floc properties and membrane fouling in coagulation/ultrafiltration process for the treatment of Xiaoqing River: the role of polymeric aluminum-polymer dual-coagulants. *Chemosphere* **243**, 125391.

Liu, Q.-F. & Kim, S.-H. 2008 Evaluation of membrane fouling models based on bench-scale experiments: a comparison between constant flowrate blocking laws and artificial neural network (ANNs) model. *J. Membr. Sci.* **310** (1), 393–401.

Liu, J., He, K., Zhang, J., Li, C. & Zhang, Z. 2019 Coupling ferrate pretreatment and in-situ ozonation/ceramic membrane filtration for wastewater reclamation: water quality and membrane fouling. *J. Membr. Sci.* **590**, 117310.

Ochando-Pulido, J. M. & Martínez-Ferez, A. 2017 Fouling modelling on a reverse osmosis membrane in the purification of pretreated olive mill wastewater by adapted crossflow blocking mechanisms. *J. Membr. Sci.* **544**, 108–118.

Sabia, G., Ferraris, M. & Spagni, A. 2016 Model-based analysis of the effect of different operating conditions on fouling mechanisms in a membrane bioreactor. *Environ. Sci. Pollu. Res.* **23**, 1598–1609.

Shen, Y.-x., Xiao, K., Liang, P., Sun, J.-y., Sai, S.-j. & Huang, X. 2012 Characterization of soluble microbial products in 10 large-scale membrane bioreactors for municipal wastewater treatment in China. *J. Membr. Sci.* **415–416**, 336–345.

Song, Y., Hu, Q., Sun, Y., Li, X., Wan, H., Zang, L., Jiang, K. & Gao, C. 2019 The feasibility of UF-RO integrated membrane system combined with coagulation/flocculation for hairwork dyeing effluent reclamation. *Sci. Total Environ.* **691**, 45–54.

State Environmental Protection Administration 2002 *Monitoring Method of Water and Wastewater*. 4th edn. China Environmental Science Press, Beijing, 105, pp. 246–248, 255–257.

Suarez, J. A. & Veza, J. M. 2000 Dead-end microfiltration as advanced treatment for wastewater. *Desalination* **127** (1), 47–58.

Tang, S., Zhang, Z. & Zhang, X. 2017 New insight into the effect of mixed liquor properties changed by pre-ozonation on ceramic UF membrane fouling in wastewater treatment. *Chem. Eng. J.* **314**, 670–680.

ter Braak, C. J. F. & Smilauer, P. 2012 *Canoco Reference Manual and User's Guide: Software for Ordination, Version 5.0.* Microcomputer Power, Ithaca, New York.

Tong, X., Wu, Y.-H., Wang, Y.-H., Bai, Y., Zhao, X.-H., Luo, L.-W., Mao, Y., Ikuno, N. & Hu, H.-Y. 2020 Simulating and predicting the flux change of reverse osmosis membranes over time during wastewater reclamation caused by organic fouling. *Environ. Int.* **140**, 105744.

Van Geluwe, S., Braeken, L. & Van der Bruggen, B. 2011 Ozone oxidation for the alleviation of membrane fouling by natural organic matter: a review. *Water Res.* **45** (12), 3551–3570.

Wang, W.-L., Zhang, X., Wu, Q.-Y., Du, Y. & Hu, H.-Y. 2017 Degradation of natural organic matter by UV/chlorine oxidation: molecular decomposition, formation of oxidation byproducts and cytotoxicity. *Water Res.* **124**, 251–258.

Wang, L., Liu, H., Zhang, W., Yu, T., Jin, Q., Fu, B. & Liu, H. 2018 Recovery of organic matters in wastewater by self-forming dynamic membrane bioreactor: performance and membrane fouling. *Chemosphere* **203**, 123–131.

Wei, D., Tao, Y., Zhang, Z. & Zhang, X. 2016 Effect of pre-ozonation on mitigation of ceramic UF membrane fouling caused by algal extracellular organic matters. *Chem. Eng. J.* **294**, 157–166.

Xing, J., Liang, H., Xu, S., Chuah, C. J., Luo, X., Wang, T., Wang, J., Li, G. & Snyder, S. A. 2019 Organic matter removal and membrane fouling mitigation during algae-rich surface water treatment by powdered activated carbon adsorption pretreatment: enhanced by UV and UV/chlorine oxidation. *Water Res.* **159**, 283–293.

Xiong, J., Zuo, X., Zhang, S., Liao, W. & Chen, Z. 2019 Model-based evaluation of fouling mechanisms in powdered activated carbon/membrane bioreactor system. *Water Sci. Technol.* **79** (10), 1844–1852.

Yoo, S. S. 2018 Operating cost reduction of in-line coagulation/ultrafiltration membrane process attributed to coagulation condition optimization for irreversible fouling control. *Water* **10** (8), 1076–1093.

Zhao, X., Wu, Y., Zhang, X., Tong, X., Yu, T., Wang, Y., Ikuno, N., Ishii, K. & Hu, H. 2019 Ozonation as an efficient pretreatment method to alleviate reverse osmosis membrane fouling caused by complexes of humic acid and calcium ion. *Front. Environ. Sci. Eng.* **13** (4), 55–67.

Zhu, H., Wen, X. & Xia, H. 2010 Membrane organic fouling and the effect of pre-ozonation in microfiltration of secondary effluent organic matter. *J. Membr. Sci.* **352** (1–2), 213–221.

First received 17 January 2021; accepted in revised form 19 March 2021. Available online 9 April 2021

© 2021 The Authors

Investigations on the fouling characteristic of humic acid and alginate sodium in capacitive deionization

Huizhong Zhang, Jiayu Tian, Xiujuan Hao, Dongmei Liu and Fuyi Cui

ABSTRACT

Capacitive deionization (CDI) has been investigated for brackish water desalination, selective removal of ions, and water softening. We used humic acid (HA) and alginate sodium (SA) to simulate different kinds of natural organic matter to investigate the fouling phenomena during CDI operation. Adsorption amount and energy efficiency were studied. Results showed that both SA and HA could decrease the removal of NaCl during CDI operation. There existed a slight decrease of energy consumption in SA solutions which was opposite to that in HA solutions. HA can compete with ions adsorbed by electrodes and attach to electrodes adhesively, resulting in co-ion repulsion. SA is not sensitive to electrical field and its fouling is not obvious. The amount of adsorbed Mg^{2+} would increase from 0.927 mg/g to 1.508 mg/g in ten cycles' operation and the increment of Ca^{2+} was from 1.885 mg/g to 2.878 mg/g in SA solutions. This increase of adsorption was due to the complexation between SA and cations. Simultaneously, energy consumption was decreased. In HA solutions, energy consumption of Mg^{2+} and Ca^{2+} adsorption increased. In ten cycles' operations, both HA and SA could reduce the efficiency of CDI operation. The types of organic substances are important factors in fouling of CDI electrodes.

Key words | alginate sodium, capacitive deionization, fouling, humic acid

Huizhong Zhang
Jiayu Tian (corresponding author)
Xiujuan Hao
Dongmei Liu
School of Environment, Harbin Institute of Technology,
Harbin 150090,
China
E-mail: *tjy800112@163.com*

Jiayu Tian
School of Civil Engineering and Transportation,
Hebei University of Technology,
Tianjin 300401,
China

Fuyi Cui
Chongqing University,
Chongqing 400044,
China

HIGHLIGHTS

- The mechanisms of humic acid and sodium alginate fouling were investigated in CDI operation.
- Humic acid was able to make fouling of electrodes worse during CDI operation.
- Alginate sodium acted as a medium in increasing the amount of adsorption and energy efficiency.

This is an Open Access article distributed under the terms of the Creative Commons Attribution Licence (CC BY-NC-ND 4.0), which permits copying and redistribution for non-commercial purposes with no derivatives, provided the original work is properly cited (http://creativecommons.org/licenses/by-nc-nd/4.0/).

doi: 10.2166/wrd.2021.104

GRAPHICAL ABSTRACT

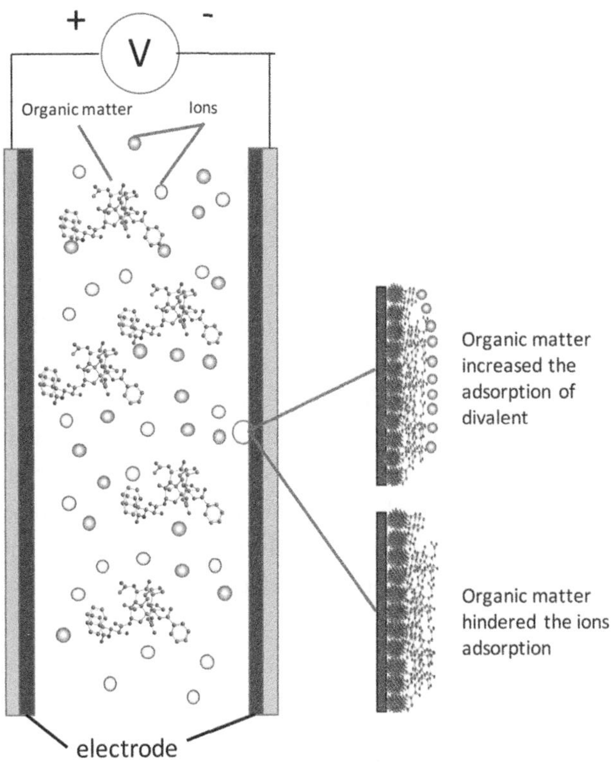

INTRODUCTION

Water scarcity posts a great challenge to the development of human society, industry, and agriculture. Desalination of brackish water and wastewater reuse are effective ways to increase the supply of clean water (Xu *et al.* 2019). Nowadays, reverse osmosis (RO), electrodialysis (ED), and vacuum membrane distillation are common methods of water treatment and desalination (Zhu *et al.* 2019). These techniques mostly have high energy requirements and may be not very efficient in desalination of medium-low salt water. In recent years, capacitive deionization (CDI) has attracted the interest of researchers for its comparatively low energy cost, easy regeneration and energy reclaim which also can be applied for selective removal of ions and water softening (Alfredy *et al.* 2019).

CDI usually works at a low voltage (lower than 2 V) in order to avoid water electrolysis. Similar to a capacitor, the voltage is applied between two parallel electrodes. In the desalination step, the movement of ions in salinity water is driven by a direct electric field. Because of electrostatic interaction, salt ions are adsorbed by electrodes with the opposite charges. They will be stored in micropores and mesopores of electrodes and form layers which are called the electrical double layers (EDLs). The adsorption amount of ions mainly depends on the thickness of EDLs. Then, the regeneration process was operated at 0 V, which is called short-circuit, or by supplying an opposite voltage (Choi *et al.* 2017). Adsorbed ions are desorbed from electrodes.

Since the deslination process mainly depends on the formation of EDLs in micropores and mesopores, the pore structure of electrodes has a great influence on salt removal efficiencies (Dykstra *et al.* 2016). Thus, the electrode materials are very important for the efficiencies of CDI. Usually, materials for electrode preparation should have plenty of mesopores and micropores, high surface areas, and high electrical conductivity. Carbon materials are ideal materials. Activated carbon, carbon fiber, mesoporous carbon, carbon nanotube, and graphene are often applied in preparing electrodes. These materials have some shortcomings in practical application. Some carbon materials are hydrophobic, thus it is very difficult for the salty water to go inside the pores. Some materials are very expensive and preparations of a lot of materials are too complicated. Some materials have low adsorption capacity of salts.

As a result, many studies have focused on developing new electrode materials. Fang *et al.*'s (2016) study used hydrophobic polyurethane elastomer as binder for preparing electrodes. This made the adsorption process more stable compared to traditional binder. Wang *et al.* (2012) prepared carbon nanofiber by electrospinning and achieved an improvement in adsorption amount. Tsai & Doong (2015) made hierarchically ordered mesoporous carbons and this material could get fast adsorption in grey and brown water treatment. These materials actually help increase the salt removal efficiencies of CDI. However, their applications in practical water treatment have not been investigated clearly.

The concentration of salt in environmental water often ranges from 1,000 mg/L to 10,000 mg/L. The concentrations of organic matter are about 1–21 mg/L as DOC (dissolved organic carbon) (Liu *et al.* 2018). During CDI operation, organic substances are inevitably in contact with carbon electrodes. Meanwhile, the carbon materials are also great adsorbents for other organic and inorganic substances. Once the electrodes adsorb the organic matter during the desalination process, properties like surface area and functional groups of carbon materials might be altered (Tang *et al.* 2014). Desalination efficiencies of some materials in CDI operation remain unknown in water with organic matter. In addition, the heterogeneous and complicated organic molecules can easily interact with cations and anions, thus affecting the transport of ions. To make the best use of some new materials in CDI operation, the impact caused by organic matter on desalination efficiencies should be found out. A large amount of new materials have been reported and there are many kinds of organic matter in practical water. Investigating this organic matter and materials at one time may be too complicated. We could first investigate the fouling phenomena and mechanisms caused by various organic matter with the help of common materials. These studies could help us predict and understand the fouling of new materials.

In this study, we used the common material activated carbon (AC) to systematically investigate the fouling mechanisms of humic acid HA and alginate sodium SA in CDI operation. HA and SA are typical organics of humic substances and polysaccharides to study the influence of organic fouling. We also studied the fouling phenomena in solutions of monovalent ions (NaCl) and divalent ions ($MgCl_2$ and $CaCl_2$). The effect of fouling was assessed by adsorption efficiency and energy consumption. SEM images and FTIR spectroscopy were applied to investigate fouling on electrodes. The interactions between organic matter and ions were compared during CDI operation. This research would help to understand the fouling phenomena on carbon materials' electrodes. It could also give some worthwhile information on CDI operation and influent water pre-treatment in practical water desalination.

MATERIALS AND METHODS

Chemical reagents and feed solution composition

NaCl, $CaCl_2$, and $MgCl_2$ are chosen to simulate the salinity of the water. The composition of feed solutions is shown in Table 1. Before the experiments, the solutions were stirred overnight to make them stable and homogeneous. Then, they were filtered through a 0.45 μm cellulose acetate membrane. The pH of the solutions was adjusted to 7. Deionized water (DI) is used throughout our experiments.

Preparation of electrodes

The commercial activated carbon (AC) (Sigma-Aldrich) powder was used to fabricate electrodes. To prepare carbon electrodes, AC, graphite (Sigma-Aldrich) as conductive material and poly (vinylidene fluoride) (PVDF) (Sigma-Aldrich) powders were mixed together at a mass ratio of 8:1:1. Then, some drops of N-methyl pyrrolidone (Aladdin) were added to make the powder slurry. The slurry was stirred for about 8 h to make it more homogeneous. The resulting slurry was coated on a graphite sheet strip with a

Table 1 | Composite of feed solutions

Solutions	Parameters		
	NaCl (mg/L)	Humic acid (mg/L)	Alginate sodium (mg/L)
S1	200	–	–
S2	200	50	–
S3	200	20	–
S4	200	–	50
S5	200	–	20
	$MgCl_2$ (mg/L)	Humic acid (mg/L)	Alginate sodium (mg/L)
S6	200	–	–
S7	200	20	–
S8	200	–	20
	$CaCl_2$ (mg/L)	Humic acid (mg/L)	Alginate sodium (mg/L)
S9	200	–	–
S10	200	20	–
S11	200	–	20
S12	–	20	–
S13	–	–	20

size of 10×4 cm. The sheets with carbon slurry were then dried in the oven at 120 °C for 2 h and then put in a vacuum oven at 80 °C for 2 h to remove the organic solvent completely. Each electrode weighed about 0.5 g and had a thickness of 200–300 μm.

Reactor configuration

CDI experiments were conducted in a flow-through system. The system consisted of a CDI unit cell, a reservoir, a peristaltic pump, and a conductivity meter. The CDI unit cell was composed of a pair of parallel activated carbon electrodes separated by a 1 mm gap. This gap could prevent a short circuit and allow liquid to flow. The electrodes were placed in plexiglass housing which had two endplates with silicone rubber sheets as gasket. Both plexiglass endplates had holes to allow water to flow. The flow rate was about 15 mL/min. The pump drove the feed solution flowing from the reservoir to the lower inlet hole of the endplate. The feed solution flew through the spacer gap between the electrodes. The total volume of solution in the reservoir was 100 mL. Then, the effluent left from the upper hole of the other endplate and flowed into the reservoir. A voltage of 1.2 V was applied between the parallel electrodes by a workstation (CHI660D, CH Instruments, Inc.) and the corresponding current was recorded. An online conductivity meter was applied to monitor the conductivity of solution in the reservoir. No pH adjustment for the solutions was carried out throughout the CDI operation. A complete CDI operation included two steps: adsorption and desorption (regeneration). In our experiments, the adsorption process took 1,300 s. Desorption, which was also called regeneration, lasted 1,100 s at a cell potential of 0 V. A cycle included one adsorption step and one desorption step. Before each cycle started, solution in the reservoir would be changed to a new one.

Experimental methods

The CDI device was carried out for ten cycles to investigate the effects of organic substances on deionization. The fouling was monitored mainly by the salt adsorption amount and energy consumption efficiency. CDI efficiencies were calculated by the changes of the ion amounts in the feed solutions. The initial and equilibrium concentrations were detected in every cycle. The salt adsorption amount per cycle (q, mg/g) was calculated according to Equation (1):

$$q = \frac{C_0 \cdot V - C_t \cdot V}{m} \tag{1}$$

where C_0 and C_t are the initial and equilibrium concentrations (mg/L), respectively. V is the volume of solution (L), and m is the weight of the electrodes (g). The total energy consumption (W) of an adsorption step was calculated by Equation (2) and energy consumption efficiency (Wq) of the CDI unit cells was calculated using Equation (3) (Chen *et al.* 2018):

$$W = \frac{U \int_0^t I dt}{3600m} \tag{2}$$

where W is the energy consumption (Wh), U is the applied voltage (V), I is the correspondent current (A) and t is adsorption time (s):

$$Wq = \frac{1000W}{q} \tag{3}$$

Also, in some experiments, the carbon electrodes were used to contact with organic matter before conducting the CDI process. The adsorption of SA and HA was performed individually at a voltage of 1.2 V for 2 h, 10 h, and 4 days. Then pre-fouled electrodes were used in CDI operation.

Analysis

Conductivity and concentration analysis

The changes in the conductivity of solutions were analyzed by a conductivity meter (DDSJ-308A, Leici company). Cation concentrations were determined by inductive couple plasma optical emission spectrometry (ICP-OES, Perkin Elmer Optima 5300DV). In NaCl solutions, the concentrations of NaCl were detected via the concentrations of chloride ions. Anion concentrations were determined by ion chromatography (883-basic IC plus, Metrohm). In addition, the variation of the organic matter was also detected. The concentrations of HA were measured by a UV-Vis spectrophotometer (Beijing Puxi, T6) at a wavelength of 254 nm. Alginate sodium was determined using the phenol-sulfuric acid colorimetric method at 490 nm (Xie *et al.* 2013).

Characterization of electrodes and particles in solutions

Zeta potentials and nanoparticle size were detected with Zetasizer nano ZS (Malvern, Nano-Z) at 25 °C. Solutions with organic matter and different cations were investigated. Solutions for ζ-potential and nanoparticle detecting were adjusted to different pH values by NaOH and HCl. All the samples were shaken for 24 h. pH would be determined again before measuring the zeta potential and nanoparticle size. We studied the fresh and fouled electrodes using Fourier transform infrared spectroscopy (Spotlight 400, Pekin-Elmer, USA). The morphology of fresh and fouled electrodes was studied by SEM (SUPRA55, ZEISS, Germany).

Electrochemical analysis

Electrochemical properties of electrodes were evaluated using a CHI660D electrochemical workstation (Shanghai Chenhua Instruments Co., China) in a three-electrode system at 25 °C. The Ag/AgCl electrode was used as the reference electrode and a piece of platinum foil was used as the counter electrode. The working electrode was activated carbon electrode. Prior to the tests, adsorption experiments were conducted at a voltage of 1.2 V for 10 h in $CaCl_2$ or $MgCl_2$ solutions with organic substances. The cyclic voltammetry (CV) of electrodes was investigated at the scan rate of 50 mV/s in solutions containing 5 mM $CaCl_2$ or $MgCl_2$ and 200 mg/L of SA or HA. The potential was set between −0.4 and 0.6 V. The specific capacitance was calculated from Equation (4) (Hou & Huang 2013):

$$C = \frac{\int I dV}{2vm\Delta V} \tag{4}$$

where C (F) is the specific capacitance, I (A) is the response current density, ΔV (V) is the potential window, v (V/s) is the scan rate, and m (g) is the mass of the electrode material.

RESULTS AND DISCUSSION

ζ-potential and nanoparticle size of organic substances molecules

Foulants in environmental water can be categorized into different kinds, such as humic substances, polysaccharides, and protein. The typical humic substance is humic acid. HA is a kind of hydrophobic organic acid (Chen *et al.* 2018). It contains abundant functional groups, for example, aromatic nucleus. Alginate sodium (SA), which consists of β-D-mannuronic and α-L-guluronic acids, is a kind of hydrophilic organic matter and is a typical polysaccharide substance (Lee & Mooney 2012). In this study, these two substances act as the natural organic substances in the water. The measured ζ-potentials of HA and SA are shown in Figure 1(a). It was found that both HA and SA had negative ζ-potentials at all pH ranging from 3 to 10. HA and SA are macromolecule organic substances with numerous functional groups like carboxylic and oxhydryl groups which make them soluble in water and have negative charge (Lee & Mooney 2012; Tang *et al.* 2014). The potential of HA was about −15 mV to −25 mV and that of SA was about −5 mV to −15 mV. This demonstrates that HA produced more charges and could be easily driven by electrostatic force. After the addition of

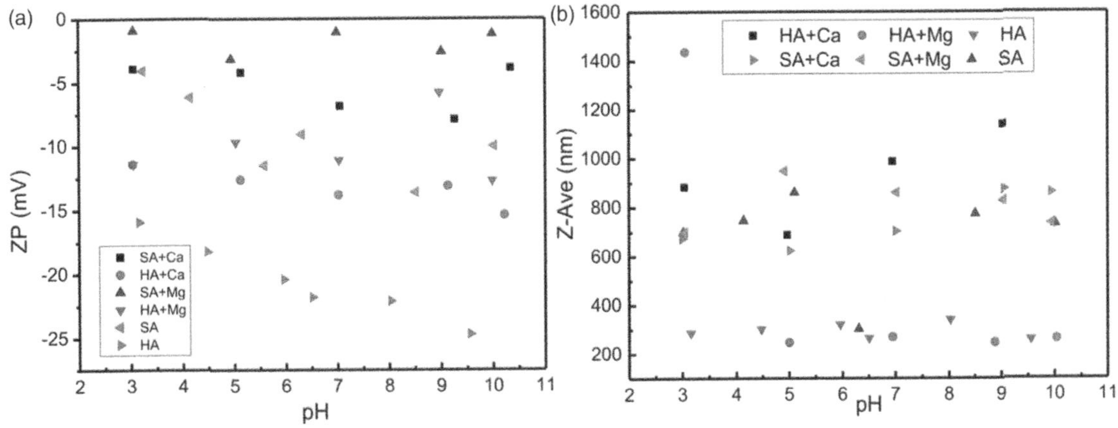

Figure 1 | ζ-potential and nanoparticle size of HA and SA at various solution pH and in different solutions.

divalent cations, ζ-potential turned to the positive direction gradually. Both Ca^{2+} and Mg^{2+} could interact with HA and SA forming complexes (Alberts & Filip 1998; Topuz *et al.* 2012). The cations complexed on organic molecules can neutralize the negative charges.

Simultaneously, the nanoparticle size was also analyzed and shown in Figure 1(b). SA had larger particle size than HA. Alginate is a kind of anioinic polymer and has a liner structure (Lee & Mooney 2012). HA has a curled molecular structure resulting from the inner electrostatic interaction. The divalent cations complexed with HA molecules and made the curled molecular stretched (Yang *et al.* 2011). The complexation might hinder the mobility of ions. In SA solutions, the urionic acid on SA made it interact easily with divalent cations and gelation. The gelation captures cations in water leading to reduction of free ions in solutions which can easily be mobilized in the electrical field and be adsorbed by electrodes. The gel can attach to electrodes, slowing down the transmission of ions.

NaCl adsorption efficiency with and without organic substances

NaCl adsorption efficiency

We investigated the adsorption of NaCl with and without organic substances. The conductivity during the experiments was recorded (Figure 2). The concentration of NaCl was directly proportional to the conductivity. The changes of conductivity could represent the variation of the concentration of NaCl in the solutions. When the voltage of 1.2 V was applied, electrodes adsorbed ions and the conductivity of the NaCl solution dropped. When the conductivity was stable, the adsorption step ended. Then, the desorption step started and adsorbed ions were desorbed from electrodes and conductivity began to decrease. As the number of cycles increased, the variation of conductivity was quite different from that in the first cycle. Especially in the ninth and tenth cycles, when the voltage was applied, conductivity was increased instead of dropping. This was due to the fact that ions absorbed by electrodes could not be removed totally at each regeneration step and some ions might be absorbed by electrodes freely no matter what kind of charges they brought at the regeneration step. When the next adsorption step started, ions, bringing the same charges with the electrodes would go back to water driven by repulsive force. This phenomenon was called co-ion repulsion and it often happened in CDI operation. Co-ion repulsive force could consume the energy provided by the voltage of 1.2 V which should have been used for deionization, leading to the decrease of the adsorption amount in the next cycle. From the experiments in S2 and S3, it can be seen that HA in water could enhance the co-ion repulsion. In solution S2, co-ion repulsion could be observed in five cycles. In S3, only in three cycles did the co-ion repulsion take place. Due to the higher concentration of HA, co-ion repulsion became obvious. In S4 and S5, a similar phenomenon was observed. SA also enhanced the co-ion repulsion

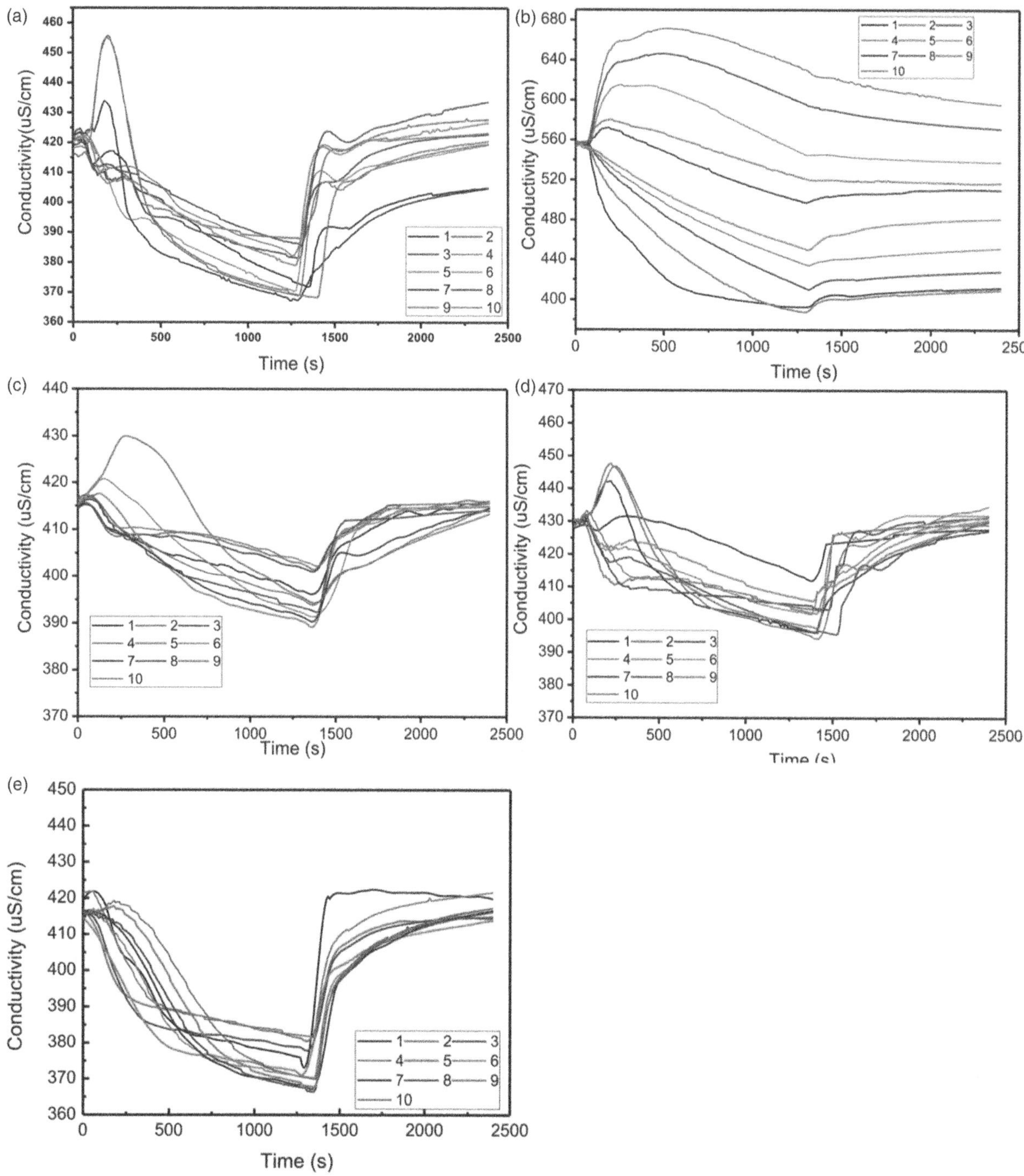

Figure 2 | The variation of conductivity and concentration in NaCl solutions with different constituents: (a) S1, (b) S2, (c) S3, (d) S4, (e) S5.

during the adsorption step. These results showed that with the concentration of organic substances increasing, co-ion repulsion was enhanced. However, in SA fouling, co-ion repulsion was not as severe as that in HA fouling.

Figure 3 exhibits the salt adsorption amount per cycle in different kinds of NaCl solutions. Salt adsorption in pure NaCl solution is from 6.87 mg/g to 2.62 mg/g. There was a reduction of the adsorption amount in pure solution because

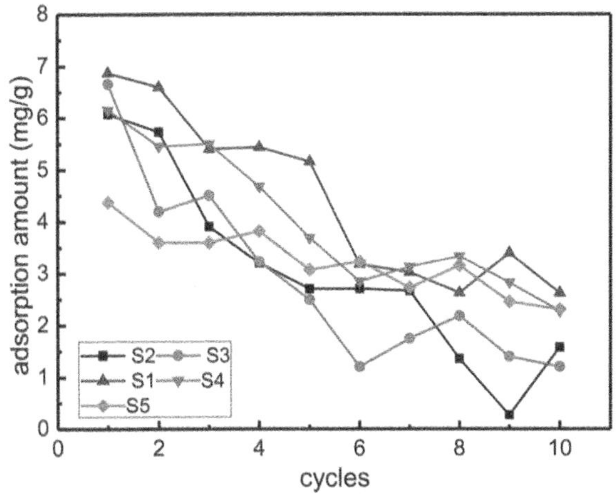

Figure 3 | The salt adsorption amount in different solutions during ten cycles.

of the inevitable redox reaction on electrodes (Lu *et al.* 2017), which would change functional groups on electrodes. These functional groups could adsorb more ions during the generation step and lead to co-ion repulsion. In pure NaCl solution, the adsorption amount decreased about 62%. In solutions with 50 mg/L HA, the adsorption amount decreased about 96%. In solutions with 20 mg/L HA, the adsorption amount decreased about 82%. This proved that with the increasing concentration of HA, the adsorption amount of salts was reduced faster than that in pure NaCl solution. HA had an adverse impact on salts' removal efficiencies. In solutions of 20 mg/L SA, the adsorption amount decreased 47%. In solutions of 50 mg/L SA, the adsorption amount decreased about 63%. Different from HA, the existence of SA in water did not reduce the adsorption amount of salts too much. From these results, it can be concluded that the existence of some organic substances clearly reduces the deionization of the CDI process. With the concentration of HA increasing, the adverse impact became obvious. HA would lead to more severe fouling than SA. From Figure 1, both HA and SA have negative ζ-potential in water. The organic matter competed with NaCl ions in water adsorbed by electrodes. A part of the energy was consumed by the adsorption of organic substances. Because SA produced fewer charges, the competition between organic matter and ions was weakened. SA was a kind of hydrophilic organic matter. Adsorption of SA could make the electrodes hydrophilic. Also, salty water could easily get inside the pores of electrodes.

Effect of voltage on adsorption of organic substances

HA has lower ζ-potential than SA dosage, meaning that HA has stronger electrostatic attraction. Figure 4 shows the effect of voltage on adsorption of organic matter. In HA solution, the adsorption amount was from 0.412 mg/g to 0.161 mg/g at voltage of 1.4 V. At 1.2 V, the adsorption amount of HA decreased to 0.173–0 mg/g and at 1.0 V, the adsorption amount was stable with 0.112 mg/g to 0 mg/g. There are many interactions between the organic substances and adsorbents, such as π-π bonding, hydrophobic interactions, and electrostatic interactions (Yang & Xing 2009). The adsorption of organic matter on carbon-based adsorbents was affected by

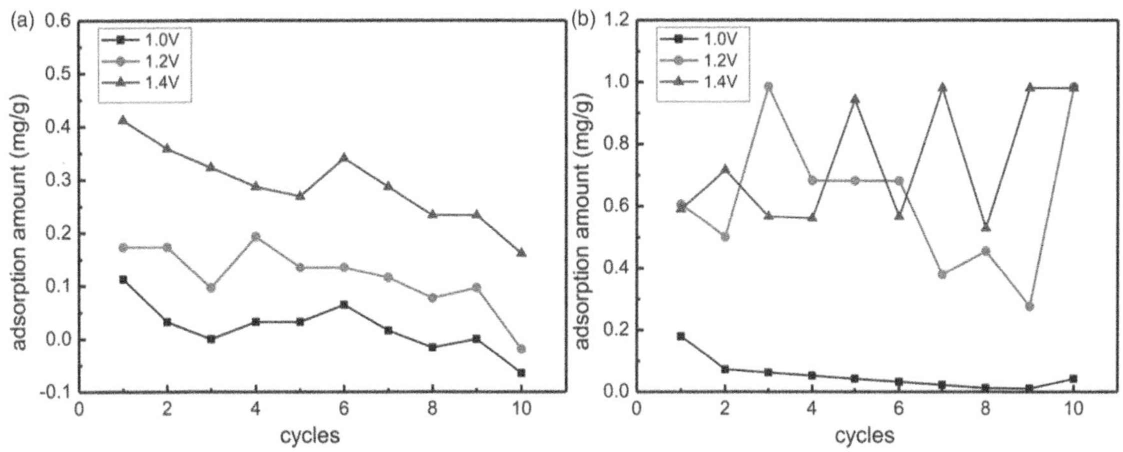

Figure 4 | Adsorption amount of HA and SA in electric field: (a) S12, (b) S13.

the water quality, the contaminants, and properties of adsorbents. The results from our work demonstrated that in the electrical field the dominant interaction was electrostatic force. In SA solution, at 1.0 V, the adsorption amount was nearly 0 mg/g. This proved that SA was not as sensitive as HA to voltage. The adsorption amount of SA could probably be attributed to other interactions with electrodes.

In Figure 5(a) and 5(b), it can be seen that the concentration at adsorption equilibrium was higher than that at desorption equilibrium. This meant that adsorbed organic substances hardly desorbed from the electrodes during the regeneration step. In HA solutions, the electrodes kept adsorbing organic substances during the desorption steps. This could prove that there were not only electrostatic interactions but also other interactions between organic substances and electrodes during CDI operation. Thus, HA and SA could accumulate on carbon electrodes without being desorbed from electrodes when the voltages were 0 V. The adsorbed organic substances can interact with ions in water through electrostatic force in the regeneration step (Li *et al.* 2002). This interaction makes the electrodes adsorb more ions than that in pure solution, leading to more obvious co-ion repulsion.

The adsorption amount of organic matters in solutions with salts is shown in Figure 6. Electrodes adsorbed organic matters and ions at the same time during adsorption steps. The adsorption amount of HA in S2 was about 0.58–0.22 mg/g, which was higher than that in S12. The higher concentration of HA promoted the adsorption of organic matter. Similarly, the adsorption amount of SA in S4 was higher than that in S13. The adsorption amount of HA in S3 was about 0.2 mg/g–0 mg/g and the adsorption amount of SA in S5 was about 0.6 mg/g–0.25 mg/g. There was no obvious difference between adsorption in solutions with and without salts. Although there was competition in adsorption between organic matters and salts, the other interaction forces like π-π bonding and hydrophobic interactions could help electrodes adsorb some HA and SA.

Characterization of organic fouled electrodes

To explore the foulants, we studied the fouling electrodes. The electrodes were fouled in organic solutions at 1.2 V for about 4 days to make the fouling obvious. In Figure 7(a) and 7(c), HA and SA both had stretching at about 3,350 cm^{-1} representing alcohol/phenol O-H, at about 1,560 cm^{-1} representing carboxylate C=O and at 1,250–1,500 cm^{-1} representing in-plane O-H. The bending at 1,090 cm^{-1} and 1,030 cm^{-1} represented aliphatic C-O stretching in HA and polysaccharide C-O in SA (Liu *et al.* 2018). In Figure 7(b), fouled electrodes had an obvious peak intensity at about 1,560 cm^{-1}. In Figure 7(d), a similar

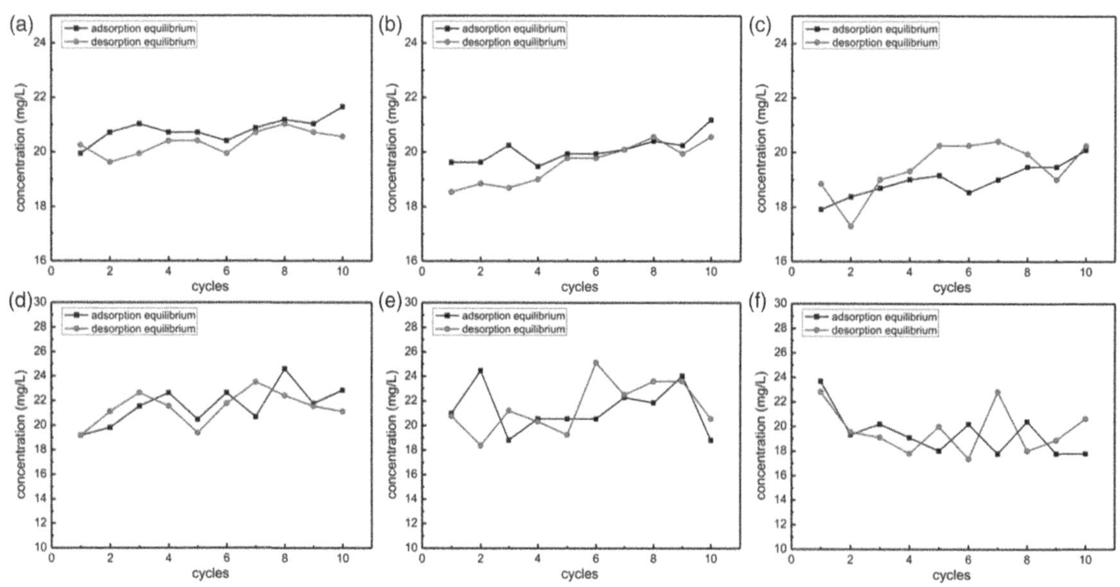

Figure 5 | The variation of concentration in HA and SA solution at voltage of 1.0 V, 1.2 V, and 1.4 V: HA solution (a) 1.0 V, (b) 1.2 V, (c) 1.4 V and SA solution (d) 1.0 V, (e) 1.2 V, (f) 1.4 V.

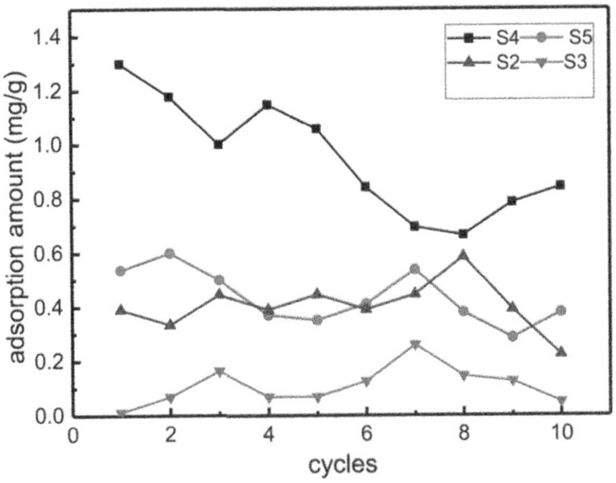

Figure 6 | The variation of HA and SA concentration in solutions with salts.

stretching is also shown. It proved that carboxylate functional groups appeared on electrodes and made them produce more negative charges in solutions. Then, the electrodes could adsorb ions during generation steps resulting in severe co-ion repulsion.

The SEM images of fresh and fouled electrodes are shown in Figure 8. With the fouling time increasing, organic matter could aggregate on electrodes. AC particles on electrodes were coated by SA and HA. The fouling layer formed on electrodes. Due to the deposition of organic matter, some micropores and mesopores on electrodes were covered.

Effect of organic fouling on energy consumption

The energy consumption efficiencies in NaCl solutions are shown in Figure 9. It can be seen that as cycles increased energy consumption efficiencies kept decreasing. In pure solution, with CDI devices running, the performance of

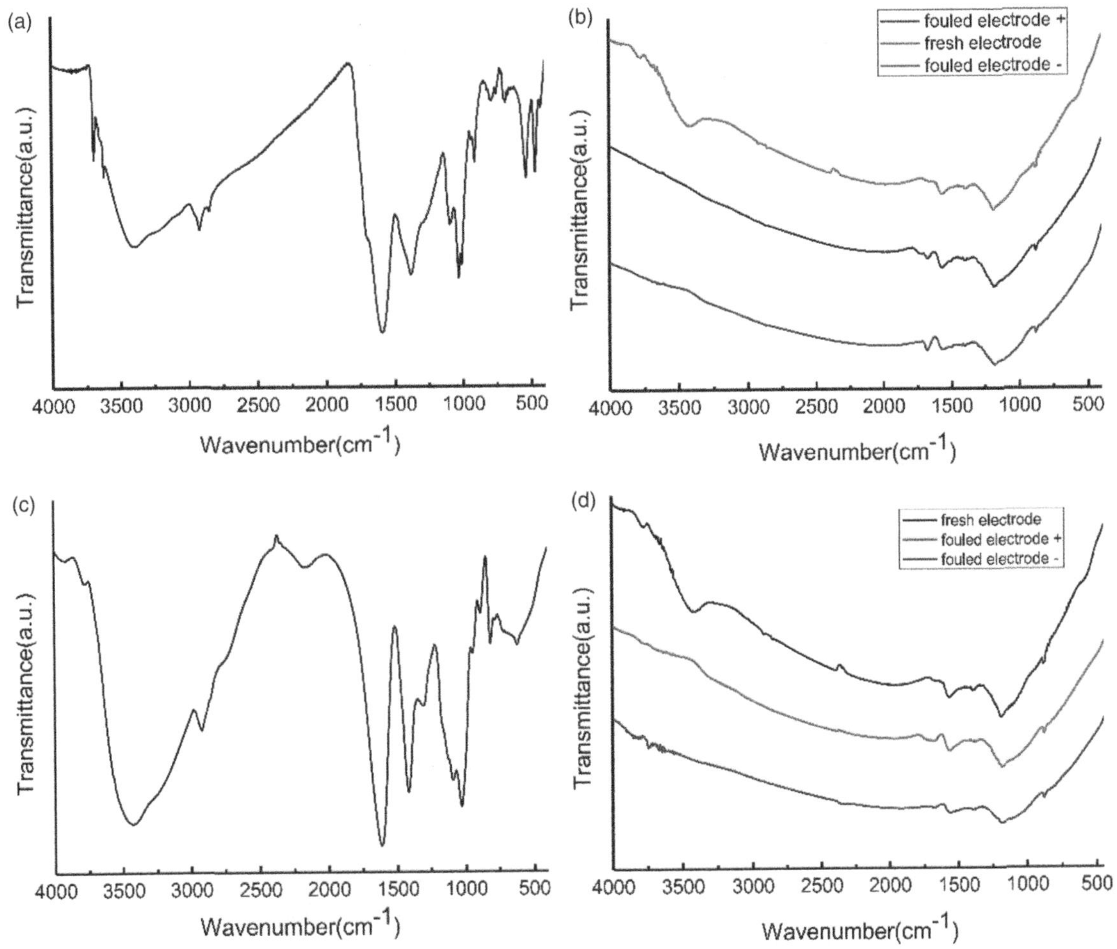

Figure 7 | FTIR spectrum of organic substances and fouled electrodes: (a) HA, (b) fouled electrodes by HA, (c) SA, (d) fouled electrodes by SA.

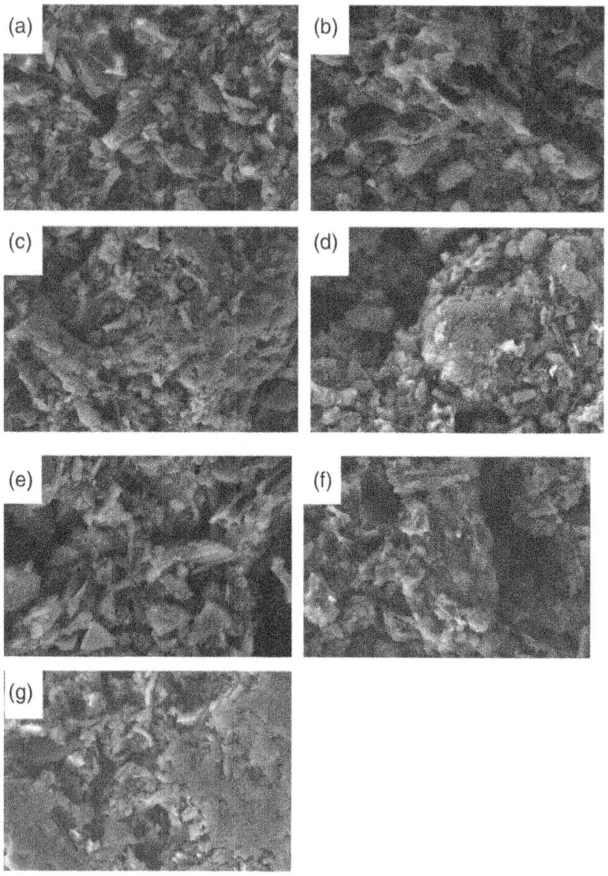

Figure 8 | The SEM images of fresh and fouled electrodes: (a) fresh electrodes, (b) 2 h HA fouling, (c) 10 h HA fouling, (d) 4 d HA fouling, (e) 2 h SA fouling, (f) 10 h SA fouling, (g) 4 d SA fouling.

electrodes was decreasing because of the inevitable redox reaction on them (Shapira *et al.* 2016). At the beginning of CDI operation, the energy efficiencies in S1 solution was 1.4637 Wh/g higher than that in organic matter solutions of around 0.82 Wh/g. In HA solution, the energy consumption efficiency rose sharply to nearly 5 Wh/g, which was much higher than efficiency in pure NaCl and SA solutions. In SA solution, the efficiency substantially increased from 0.8279 Wh/g to 2.1175 Wh/g. At the beginning, conductivities of S2 and S3 were a little bit higher than that of pure solution, possibly due to dissolution of organic substances. Higher conductivity helped reduce the resistance of solution leading to lower energy consumption on the solution. At the end of CDI operation, a certain amount of HA in water accumulated on electrodes resulting in obvious co-ion repulsion which wastes too much energy.

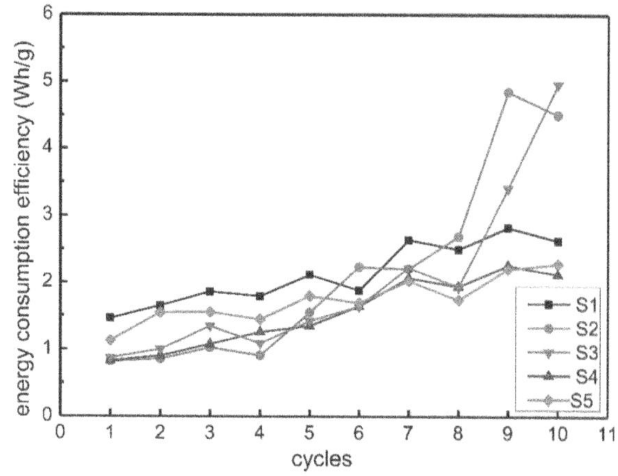

Figure 9 | Energy consumption efficiency in NaCl solutions.

In the electrical field, the dominant interaction between organic substances and electrodes was electrostatic force. Organic substances in water competed with NaCl ions during the deionization process and decreased the adsorption amount of NaCl ions. The adsorbed organic substances, like HA, enhanced co-ion repulsion. These two interactions raised the energy consumption during the deionization process. Different to HA, SA molecules have higher ζ-potential than HA molecules. Alginate did not attach to electrodes as adhesively as HA did. SA in solutions helped increase the conductivity in water and decrease the resistance of solutions leading to lower energy consumption than that in pure NaCl. Also, SA attached to the electrodes and made it become hydrophilic.

$Mgcl_2$ and $CaCl_2$ adsorption efficiency with and without organic substances

$Mgcl_2$ and $CaCl_2$ adsorption efficiency

The variations of $MgCl_2$ and $CaCl_2$ concentrations in pure solutions are demonstrated in Figure 10(b) and 10(d). There was no significant difference in the adsorption of Mg^{2+} and Ca^{2+} between the ten cycles. This could be explained by the fact that Mg^{2+} and Ca^{2+} both had a larger hydrated radius than did Na^{+} (Hou & Huang 2013). Ca^{2+} and Mg^{2+} could not enter into micropores of electrodes. They could be desorbed easily in regeneration steps. In $CaCl_2$ solutions, apparent co-ion repulsion was also observed. The calcium hydroxide

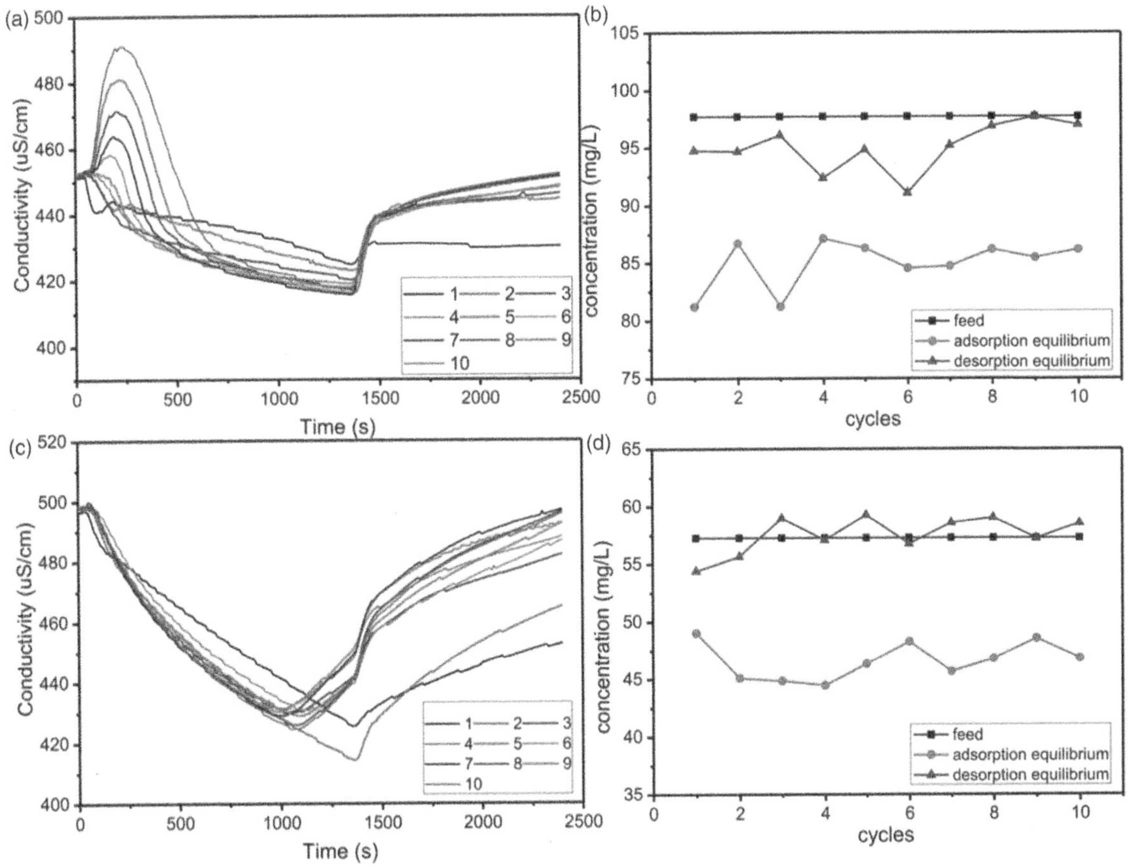

Figure 10 | The variation of conductivity and concentration in $CaCl_2$ (a) and (b) and $MgCl_2$ (c) and (d) solutions.

sensitive to the change of pH could form easily on electrodes, which would encourage Ca^{2+} to accumulate on electrodes leading to co-ion repulsion (Mossad & Zou 2013).

Variations of $MgCl_2$ concentration without organic substances are shown in Figure 11. Compared to S6, in S7 co-ion repulsion happened (Figure 11(a)). However, the co-ion repulsion was not obvious in SA solution (Figure 11(b)). These results suggest that in $MgCl_2$ solutions HA could make co-ion repulsion serious, which was similar to that in NaCl solutions. Salt adsorption efficiencies are shown in Figure 11(c), and it can be seen that Mg^{2+} adsorption in S6 was higher than that in S7. Different from Na^+, divalent ions Mg^{2+} would interact with HA through complex formation caused by acidic functional groups on HA,

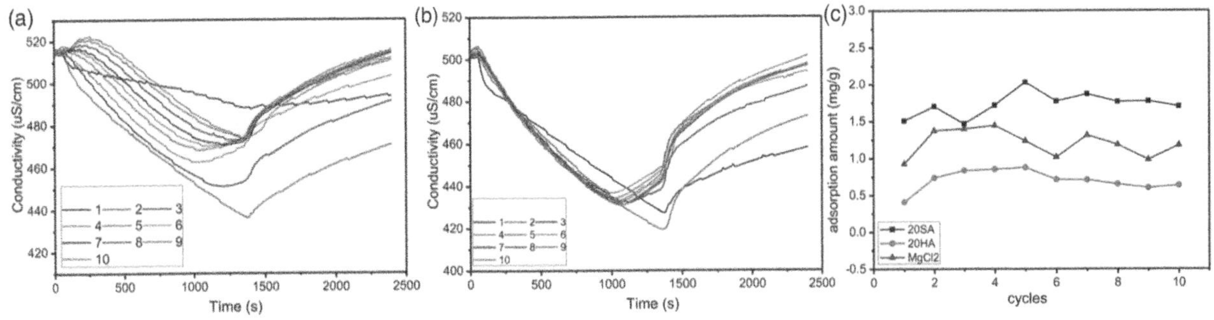

Figure 11 | The variation of conductivity and the adsorption amount of Mg^{2+} in different solutions: conductivity in (a) S7, (b) S8 and (c) adsorption amount.

which is called metal–humic complexes' effect (Alberts & Filip 1998). Mg^{2+} was able to form 'inner sphere' complexes. Owing to complexation, free magnesium ions available to be absorbed in water became less, leading to the decrease of adsorption amount (Topuz *et al.* 2012). In addition, from Figure 1, Mg^{2+} ions could neutralize the charge of HA molecules. The adsorbed molecules produced fewer charges. In the regeneration step, electrodes would adsorb fewer ions than in NaCl solution. Co-ion repulsion is also weakened. Some humic molecules will gather on electrodes. These macromolecules could lower the mobility of ions and block some pores on the electrodes. As a result, the salt adsorption amount in S7 is lower than that in S6. In pure solution, the Mg^{2+} adsorption amount was from 1.443 mg/g to 0.927 mg/g. In SA solution the adsorption amount was from 2.029 mg/g to 1.508 mg/g. It can be seen that in ten cycles, SA in water could help the adsorption of Mg^{2+}. Magnesium ions could interact with alginate resulting in ionic bridges (Topuz *et al.* 2012). Experimental results in this work showed that alginate absorbed by electrodes would bind magnesium ions and SA leading to increasing adsorption amount. In addition, SA is a kind of hydrophilic substance (Yan & Bai 2005). Mg-alginate gel attached on electrodes could decrease the hydrophobicity of electrodes to help salt removal (Fang *et al.* 2016).

In $CaCl_2$ solutions (Figure 12), adsorption in S9 was from 1.885 mg/g to 1.207 mg/g. In S10, the adsorption decreased to 1.010 mg/g to 0.483 mg/g. In S11, the adsorption amount was from 3.084 mg/g to 1.162 mg/g. Ca^{2+} and Mg^{2+} were similar ions, and both of them could complex with the negatively charged organic substances. But Ca^{2+} and alginate could interact with each other quickly (in a few minutes) and formed a stable gel with 'egg-box' structure (Lee & Mooney 2012). The bounds between calcium and alginate were firmer than those between magnesium and alginate (Topuz *et al.* 2012). The Ca-alginate could form a gel-like layer, which could not only lower the mobility of Ca^{2+} but also affect the structure of pores on electrodes. Thus, in the tenth cycle, the adsorption amount sharply decreased. ζ-potential of SA-Ca was lower than that of HA-Mg. Ca-SA gel could quickly accumulate on electrodes. Although SA could help the CDI adsorption amount, after long-time operation, gel would still accumulate on electrodes leading to fouling and reduce the adsorption amount.

CDI efficiency with organic fouling electrodes

From Figures 11 and 12, some organic substances in water might help increase the adsorption of divalent ions. To investigate the mechanism of the interaction between Mg^{2+} and Ca^{2+} and organic substances during the CDI process, the adsorption of Mg^{2+} and Ca^{2+} by the electrodes prefouled by organic substances is shown in Figure 13. In HA solutions and HA-fouled electrodes, the adsorption of Mg^{2+} is 1.036 mg/g–0.682 mg/g by 2 h HA-fouling electrodes and 0.437 mg/g–0.224 mg/g by 10 h-fouling electrodes. The adsorption amount of Ca^{2+} is nearly 0 mg/g. These results showed that HA adsorbed on electrodes only had an adverse influence on the adsorption amount of electrodes. When the fouling time increased the adsorption amount decreased. Apart from interaction and competition adsorption between ions and HA, HA would form a fouling layer on the electrodes. Adsorbed HA hindered transport of ions into pores of electrodes leading to the decrease of ions adsorption amount. It could be observed that the adsorption

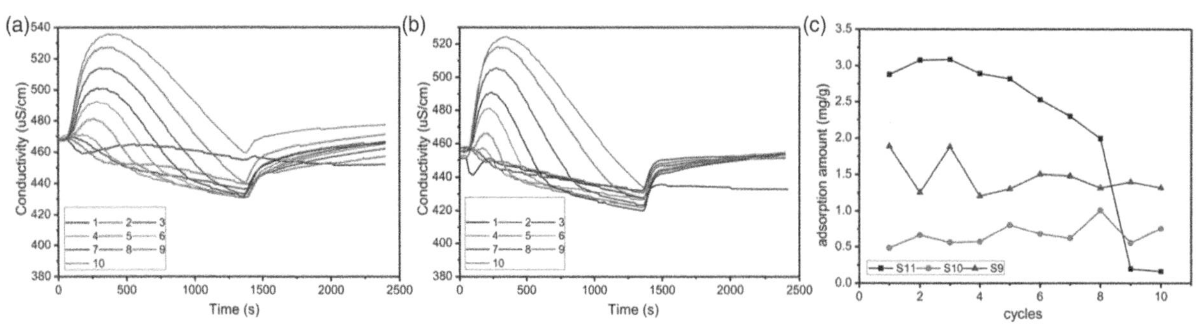

Figure 12 | The variation of conductivity and the adsorption amount of Ca^{2+} in different solutions: conductivity in (a) S10, (b) S11 and (c) adsorption amount.

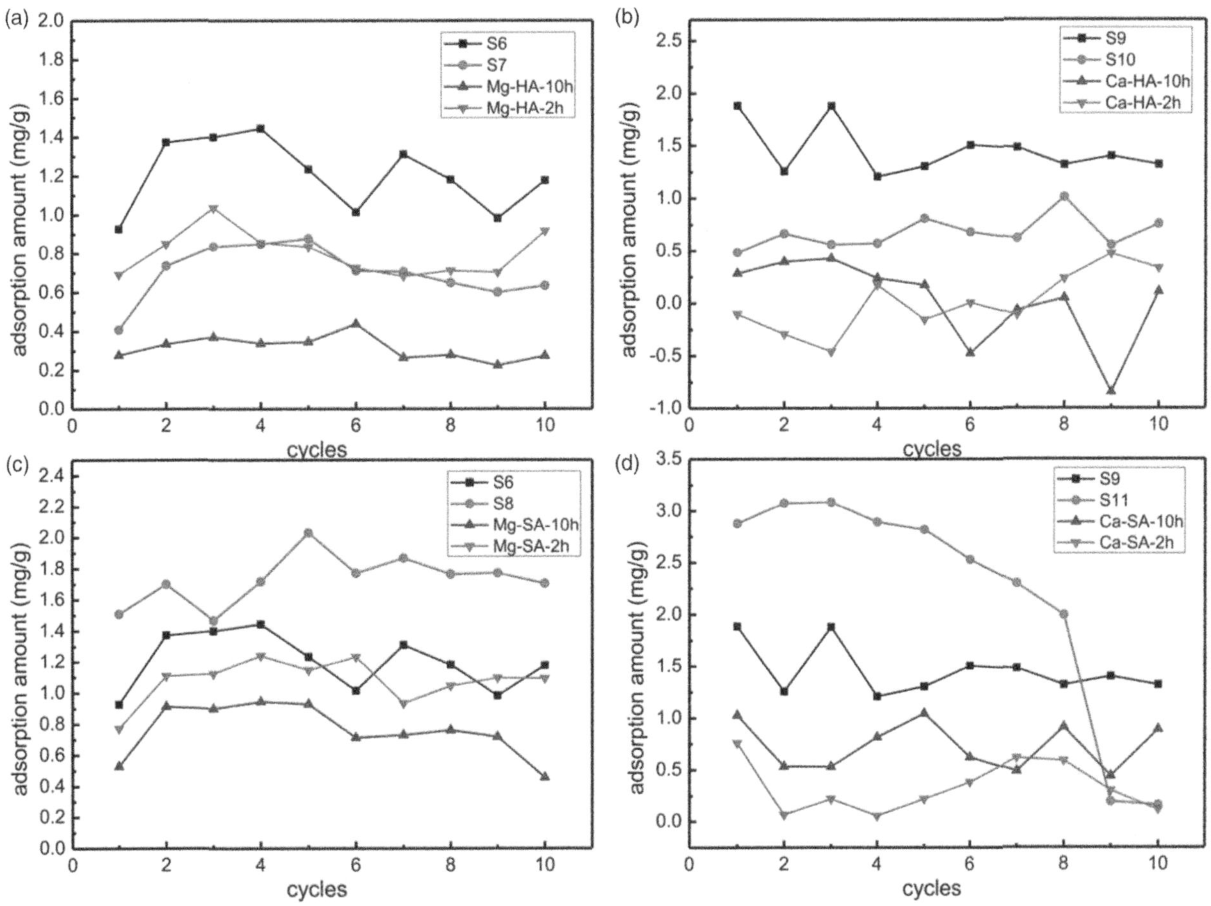

Figure 13 | Adsorption amount of Mg^{2+} and Ca^{2+} by organic fouling electrodes.

amount by 10 h-fouling electrodes is lower than that by 2 h-fouling electrodes.

In terms of SA fouling electrodes, the adsorption of Mg^{2+} was 1.241 mg/g–0.771 mg/g by 2 h-fouling electrodes, and 0.944 mg/g–0.457 mg/g by 10 h-fouling electrodes. The adsorption amount of Ca^{2+} was 0.759 mg/g–0.054 mg/g by 2 h-fouling electrodes and 1.044 mg/g–0.440 mg/g by 10 h-fouling electrodes. In $CaCl_2$ solution, 10 h-fouling electrodes adsorbed more Ca^{2+} ions than did 2 h-fouling electrodes. The increase of adsorption amount of ions was actually owing to SA on electrodes (Yoon *et al.* 2016). 10 h-fouling electrodes attracted more alginate. Bounds between calcium and alginate were more stable than between magnesium and alginate. These results proved that even alginate sodium could help increase adsorption by complexing divalent ions, the promotion depended on cations in water. If the interaction between the cations and organic substances was not strong enough to overcome the adverse impact of foulants on electrodes, efficiency of CDI would still decrease.

Electrochemical characteristics of electrodes

The CDI performance of the electrodes in different solutions was evaluated by cyclic voltammetry, as shown in Figure 14. Clearly, no evident oxidation/reduction peak could be seen in CV curves. This revealed that ions adsorption on electrodes was mainly caused by electrostatic interaction instead of Faradic reaction in solutions with organic substances. The specific capacitance could represent the double-layer capacitance of electrodes. According to previous reports, an ideal double-layer capacitor could exhibit a rectangular shape. The curves in Figure 14 were distorted. To exhibit a perfect CV curve, materials should have structure with abundant mesopores. Ions could enter into mesopores and form

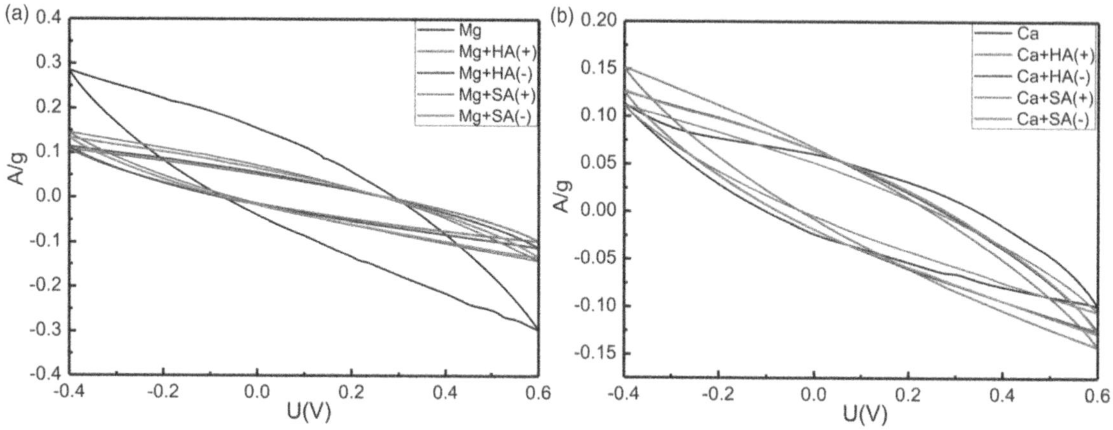

Figure 14 | CV curves of fresh and organic fouled electrodes in 5 mM $MgCl_2$ or $CaCl_2$ and 20 mg/L organic matter solutions. Positive and negative signs represent the polarities of electrodes fouled in solutions before electrochemical tests.

EDLs and EDLs would not overlap in mesopores (Hou & Huang 2013). Electrodes here belonged to microporous materials. Moreover, organic substances in water affected ion transport near the surface of electrodes. Symmetry in profiles meant that adsorption and desorption of salts were efficient without any chemical reactions.

The specific capacitance of those electrodes was calculated. In $MgCl_2$ solutions, electrodes had the specific capacitance of 1.39 F/g. In solutions with HA, the specific capacitance of positive and negative electrodes for both was 0.60 F/g. The positive and negative electrodes had the specific electrodes of 0.47 F/g and 0.52 F/g with SA, respectively. In $CaCl_2$ solutions, the specific capacitance of clean electrodes was 0.59 F/g. Positive and negative electrodes in HA solution had the capacitance of 0.54 F/g and 0.58 F/g, respectively. In SA solutions, the capacitance of positive and negative electrodes was 0.4 F/g and 0.59 F/g. The specific capacitance of fouled electrodes is lower than clean ones. The reduction of specific capacitance was due to a decreased amount of micropores and mesopores of electrodes. It proved that both SA and HA affected the formation of EDLs. They decreased the pores on electrodes and hindered the ion transport.

Effect of organic fouling on energy consumption

Energy consumption efficiency in $CaCl_2$ and $MgCl_2$ is shown in Figure 15. It can be seen that due to an increase of adsorption amount, energy consumption in SA was lower than that in pure and HA solution. In SA solution, the adsorption amount increased because of the complexation between alginate and divalent ions. However, complexation led to the formation of alginate gel; in particular, Ca-alginate gel was very stable. After ten cycles of the CDI process, the gel accumulated on electrodes. Then, some Ca-alginate complexes could accumulate on electrodes hindering ions entering the pores of activated carbon and increasing the resistance of electrodes. This interaction made energy consumption efficiency rise sharply at the tenth cycle.

In view of the energy consumption analysis, it is necessary to take action to remove organic substances. Analysis of influent water must be carried out to figure out the kind of organic substances and ions.

CONCLUSION

In this study, we systematically investigated the role of organic substances on CDI fouling by varying the ion compositions of the feed solution. The adsorption efficiency and energy consumption were studied in diverse feed solutions. We obtained the following conclusions:

(1) During CDI operation, in NaCl solutions, both SA and HA could influence deionization. With the concentration of organics increasing, the adsorption amount of ions decreased. Organic substances were adsorbed on electrodes and hardly desorbed. There existed adsorption competition between organics and ions.

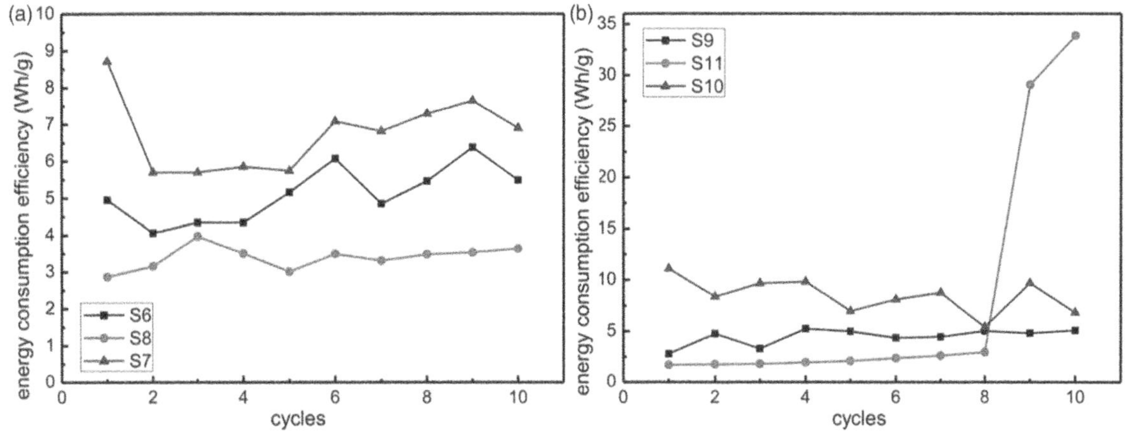

Figure 15 | Energy consumption efficiency in $CaCl_2$ and $MgCl_2$ solutions.

(2) In addition, adsorbed organic molecules caused serious co-ion repulsion during the CDI process, decreasing the efficiency of the adsorption amount and energy consumption.
(3) In $MgCl_2$ and $CaCl_2$ solutions, the adsorption amount was also affected by organic substances. Adsorption of Ca^{2+} and Mg^{2+} with the absence of organic matter was at a stable stage.
(4) In SA fouling, the adsorption amount was increased because of the complexation between SA and divalent ions. Alginate and Ca^{2+} and Mg^{2+} can interact and formed Ca-alginate and Mg-alginate gel. Alginate acted as binding bridges connecting electrodes and ions. However, after long-time operation, complexation gel would accumulate on electrodes, reducing the efficiency of the CDI device.

ACKNOWLEDGEMENTS

This work was supported by the National Natural Science Foundation of China (No. 51908181&No.51678187), the Natural Science Foundation of Hebei Province (No. E2019202011 and No. E2019202012), and the Natural Foundation of Tianjin (No. 19JCJQJC63000).

DATA AVAILABILITY STATEMENT

All relevant data are included in the paper or its Supplementary Information.

REFERENCES

Alberts, J. J. & Filip, Z. 1998 Metal binding in estuarine humic and fulvic acids: FTIR analysis of humic acid-metal complexes. *Environmental Technology* **19**, 923–931.

Alfredy, T., Jande, Y. A. C. & Pogrebnaya, T. 2019 Removal of lead ions from water by capacitive deionization electrode materials derived from chicken feathers. *Journal of Water Reuse and Desalination* **9**, 282–291.

Chen, L., Wang, C., Liu, S., Hu, Q., Zhu, L. & Cao, C. 2018 Investigation of the long-term desalination performance of membrane capacitive deionization at the presence of organic foulants. *Chemosphere* **193**, 989–997.

Choi, S., Chang, B., Kang, J. H., Diallo, M. S. & Choi, J. W. 2017 Energy-efficient hybrid FCDI-NF desalination process with tunable salt rejection and high water recovery. *Journal of Membrane Science* **541**, 580–586.

Dykstra, J. E., Zhao, R., Biesheuvel, P. M. & van der Wal, A. 2016 Resistance identification and rational process design in capacitive deionization. *Water Research* **88**, 358–370.

Fang, C., Liu, P., Chung, L., Shao, H., Ho, C., Chen, R., Fan, H., Liang, T., Chang, M. & Horng, R. 2016 A flexible and hydrophobic polyurethane elastomer used as binder for the activated carbon electrode in capacitive deionization. *Desalination* **399**, 34–39.

Hou, C. & Huang, C. 2013 A comparative study of electrosorption selectivity of ions by activated carbon electrodes in capacitive deionization. *Desalination* **314**, 124–129.

Lee, K. Y. & Mooney, D. J. 2012 Alginate: properties and biomedical applications. *Progress in Polymer Science* **37**, 106–126.

Li, F., Yuasa, A., Ebie, K., Azuma, Y., Hagishita, T. & Matsui, Y. 2002 Factors affecting the adsorption capacity of dissolved organic matter onto activated carbon: modified isotherm analysis. *Water Research* **36**, 4592–4604.

Liu, X., Whitacre, J. F. & Mauter, M. S. 2018 Mechanisms of humic acid fouling on capacitive and insertion electrodes for electrochemical desalination. *Environmental Science and Technology* **52**, 12633–12641.

Lu, D., Cai, W. & Wang, Y. 2017 Optimization of the voltage window for long-term capacitive deionization stability. *Desalination* **424**, 53–61.

Mossad, M. & Zou, L. 2013 Study of fouling and scaling in capacitive deionisation by using dissolved organic and inorganic salts. *Journal of Hazardous Materials* **244–245**, 387–393.

Shapira, B., Avraham, E. & Aurbach, D. 2016 Side reactions in capacitive deionization (CDI) processes: the role of oxygen reduction. *Electrochimica Acta* **220**, 285–295.

Tang, W., Zeng, G., Gong, J., Liang, J., Xu, P., Zhang, C. & Huang, B. 2014 Impact of humic/fulvic acid on the removal of heavy metals from aqueous solutions using nanomaterials: a review. *Science of the Total Environment* **468–469**, 1014–1027.

Topuz, F., Henke, A., Richtering, W. & Groll, J. 2012 Magnesium ions and alginate do form hydrogels: a rheological study. *Soft Matter* **8**, 4877.

Tsai, Y. & Doong, R. 2015 Activation of hierarchically ordered mesoporous carbons for enhanced capacitive deionization application. *Synthetic Metals* **205**, 48–57.

Wang, G., Pan, C., Wang, L., Dong, Q., Yu, C., Zhao, Z. & Qiu, J. 2012 Activated carbon nanofiber webs made by electrospinning for capacitive deionization. *Electrochimica Acta* **69**, 65–70.

Xie, J., Liu, X., Shen, M., Nie, S., Zhang, H., Li, C., Gong, D. & Xie, M. 2013 Purification, physicochemical characterisation and anticancer activity of a polysaccharide from Cyclocarya paliurus leaves. *Food Chemistry* **136**, 1453–1460.

Xu, H., de Koning, J. & Geng, Y. 2019 Reliability and efficiency of an advanced tertiary treatment process for wastewater reclamation. *Journal of Water Reuse and Desalination* **9**, 385–395.

Yan, W. L. & Bai, R. 2005 Adsorption of lead and humic acid on chitosan hydrogel beads. *Water Research* **39**, 688–698.

Yang, K. & Xing, B. 2009 Adsorption of fulvic acid by carbon nanotubes from water. *Environmental Pollution* **157**, 1095–1100.

Yang, S., Hu, J., Chen, C., Shao, D. & Wang, X. 2011 Mutual effects of Pb(II) and humic acid adsorption on multiwalled carbon nanotubes/polyacrylamide composites from aqueous solutions. *Environmental Science and Technology* **45**, 3621–3627.

Yoon, H., Lee, J., Kim, S., Kang, J., Kim, S., Kim, C. & Yoon, J. 2016 Capacitive deionization with Ca-alginate coated-carbon electrode for hardness control. *Desalination* **392**, 46–53.

Zhu, Z., Peng, D. & Wang, H. 2019 Seawater desalination in China: an overview. *Journal of Water Reuse and Desalination* **9**, 115–132.

First received 31 October 2020; accepted in revised form 31 December 2020. Available online 4 February 2021

 © 2021 The Authors

Microcoagulation improved the performance of the UF–RO system treating the effluent from a coastal municipal wastewater treatment plant: a pilot-scale study

Tong Yu, Chenlu Xu, Feng Chen, Haoshuai Yin, Hao Sun, Lihua Cheng IWA and Xuejun Bi IWA

ABSTRACT

Microcoagulation has recently been considered as a promising pretreatment for an ultrafiltration (UF) process from numerous studies. To investigate the effects of microcoagulation on the performance of the UF–reverse osmosis (RO) system treating wastewater with high and fluctuant salinity, different dosages of coagulant (poly-aluminum chloride) were added prior to the UF unit in a pilot-scale UF–RO system for a 10-week period operation. Microcoagulation obviously improved the contaminant removal and cleaning efficiencies, including water backwash, chemical enhanced backwash and cleaning in place processes. Organic fouling was dominated during the initial stage of the RO membrane fouling. The microbial communities of water samples and foulant on the RO membrane were similar to those of seawater and foulant on the RO membranes from seawater RO plants. The microbial community of the foulant on the membrane was similar to that of UF permeate and RO concentrate. These results demonstrated that microcoagulation could improve the performance of the UF–RO system treating the effluent with high and fluctuant salinity from a coastal municipal wastewater treatment plant.

Key words | cmWWTP, membrane cleaning, membrane fouling, microcoagulation, pilot UF–RO system

Tong Yu
Chenlu Xu
Feng Chen
Haoshuai Yin
Hao Sun
Lihua Cheng IWA
Xuejun Bi IWA (corresponding author)
School of Environmental and Municipal Engineering,
Qingdao University of Technology,
Qingdao 266000,
China
E-mail: *xuejunb@126.com*

HIGHLIGHTS

- A pilot-scale ultrafiltration (UF)–reverse osmosis (RO) system with microcoagulation was applied.
- Microcoagulation-UF had a better efficiency of contaminant removal than direct UF.
- Microcoagulation improved the efficiency of water backwash, chemical enhanced backwash and cleaning in place.
- Organic fouling was dominated in the initial stage of the RO membrane fouling.
- The microbial community of the foulant was similar to that of UF permeate and RO concentrate.

This is an Open Access article distributed under the terms of the Creative Commons Attribution Licence (CC BY-NC-ND 4.0), which permits copying and redistribution for non-commercial purposes with no derivatives, provided the original work is properly cited (http://creativecommons.org/licenses/by-nc-nd/4.0/).

doi: 10.2166/wrd.2021.099

GRAPHICAL ABSTRACT

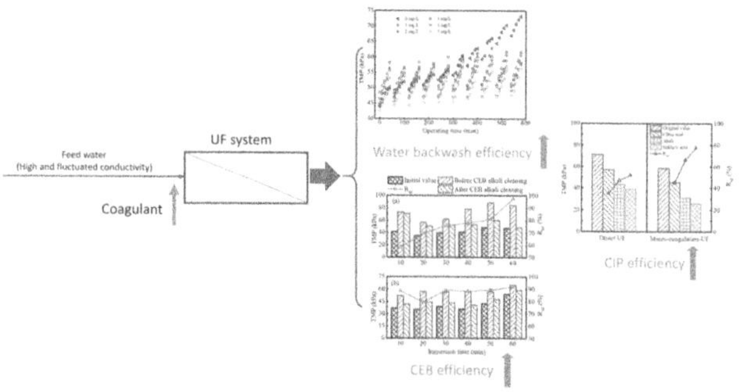

INTRODUCTION

Wastewater reclamation is one of the important means to overcome the water crisis. In recent years, to obtain high-quality reclaimed water, an ultrafiltration (UF)–reverse osmosis (RO) process has been developed as an effective approach due to its stable permeate quality and automatic operation (Tang *et al.* 2016; Anis *et al.* 2019; Ezugbe & Rathilal 2020). Despite the wide applicability of the UF–RO process, the fouling of UF and RO membranes adversely impacts the performance of the UF–RO system, which severely hinders the development of the UF–RO process (Flemming 1997; Schneider *et al.* 2005; Gao *et al.* 2011; Xu *et al.* 2019).

The feed water (FW) composition is one of the major factors governing the system performance and membrane fouling. Therefore, the pretreatment of the UF–RO process should also consider the difference in FW quality (Khan *et al.* 2013b, 2014). In a coastal municipal wastewater treatment plant (cmWWTP), the salinity of the effluent is usually higher than that of other plants and fluctuates periodically due to seawater intrusion. Therefore, when using the UF–RO process to treat the effluent from cmWWTP, additional considerations should be taken to the performance and pretreatment of the UF–RO system.

To optimize the performance of the UF–RO system and alleviate membrane fouling, several pretreatment methods were investigated and applied. Coagulation, which could remove colloid and high-molecular-weight organics, was recognized as an effective pretreatment method for alleviating the membrane fouling, and continuous attention was paid by recent studies (Humbert *et al.* 2007; Huang *et al.* 2017; Li *et al.* 2019). Comparing with the traditional coagulation process, microcoagulation has advantages of less coagulant dosage, shorter coagulation time and less footprints (Guigui *et al.* 2002; Yu *et al.* 2015; Ferrer *et al.* 2016; Aly *et al.* 2018; Xu *et al.* 2020). It seems that the microcoagulation is more suitable for the pretreatment of UF. However, in the UF–RO system treating cmWWTP effluent with high and fluctuant salinity, there were few reports on the effect of microcoagulation on pollutant removal and membrane fouling alleviation, especially in pilot-scale studies.

In this study, a pilot-scale UF–RO system was applied to investigate the influences of microcoagulation-UF (mcUF) on contaminant removal in the cmWWTP effluent with high and fluctuant salinity. Furthermore, the backwash and chemical cleaning efficiencies of UF and mcUF were compared. After 10 weeks of operation of the UF–RO system, the UF and RO membranes were autopsied to evaluate initial-stage membrane foulants (e.g. inorganics, organics and bacteria) under this water quality condition.

METHODS

Description of cmWWTP and the pilot-scale UF–RO system

A pilot-scale UF–RO system was conducted at a cmWWTP, which was located in Qingdao, Shandong Province, China.

The cmWWTP adopted anaerobic–anoxic–oxic with moving-bed biofilm reactor as a secondary treatment process for the biological removal of nitrogen and phosphorus (Supplementary Material, Figure S1). The FW for the pilot-scale UF–RO system was collected from cmWWTP effluent with additional coagulation, filtration and UV disinfection treatments. The pilot-scale UF–RO system, as shown in Figure 1, has a capacity of 48 m^3/day and a water recovery rate of 70%. The characteristics of the UF (inge dizzer® 5000 plus 0.9MB50) and RO membranes (DOW FILMTEC™ LC HR-4040) used in the pilot system are shown in Supplementary Material, Table S1. The RO array is two stages with two pressure vessels in parallel followed by one single pressure vessel. One pressure vessel contains six 4-inch spiral-wound modules housed in series (Supplementary Material, Figure S2).

Experimental setup

The experimental setup of microcoagulation, water backwash, chemical enhanced backwash (CEB) and cleaning in place (CIP) processes is shown in Figure 1. Polyaluminum chloride (PAC) was used as a coagulant for microcoagulation. During the experimental period, different doses of PAC (1, 2, 3, 4 and 5 mg/L, respectively) were added prior to the UF unit, with a retention time of 60 s. The water backwash was performed every 45 min with a flux of 230 L/(m^2·h) and a duration of 40 s. The CEB process was performed on a daily basis. The CEB process sequence was as follows: alkaline cleaning (NaOH, pH 12, 40 s), alkaline immersion (10–60 min), water backwash (40 s), acid cleaning (H_2SO_4, pH 2, 40 s), acid immersion (40 min) and water backwash (40 s). UF permeate (UFP) was used as cleaning water for the water backwash and CEB process. RO permeate was used as cleaning water for the CIP process. The CIP process sequence was as follows: acid treatment (citric acid, pH 2, cycle 3 h – immersion 4 h), followed by water treatment (rinse to neutral), then alkaline treatment (NaOH, pH 12, cycle 3 h – immersion 14 h), and a second water treatment (rinse to neutral), then a second acid treatment (H_2SO_4, pH 2, cycle 3 h – immersion 12 h) and finally a third water treatment (rinse to neutral).

Sample preparation

Water samples

Three different water samples, including FW, UFP and RO permeate and RO concentrate (ROC), were collected every 1 or 2 days from the pilot-scale UF–RO system. Water samples were filtered using a 0.2-μm nylon membrane (Whatman, UK) to eliminate suspended solids and stored at 4 °C prior to analyses.

Membrane samples

A laboratory-scale UF membrane filament, which was in parallel with the pilot UF module, was collected (Supplementary Material, Figure S3). Three different kinds of solutions (1 M HCl, 1 M NaOH and ultrapure water (UPW)) were used to dissolve the deposits on the fouled UF membranes. Three 20-cm UF membrane filaments were, respectively, soaked in 300 mL each of 1 M HCl, 1 M NaOH and UPW for 2 h with moderate stirring. The pH of the three soaking solutions (with desorbed UF membrane foulants) was adjusted to 7.0 using NaOH and HCl and then was filtered immediately by 0.2-μm nylon membranes (Whatman, UK). The three obtained permeate samples were referred to as

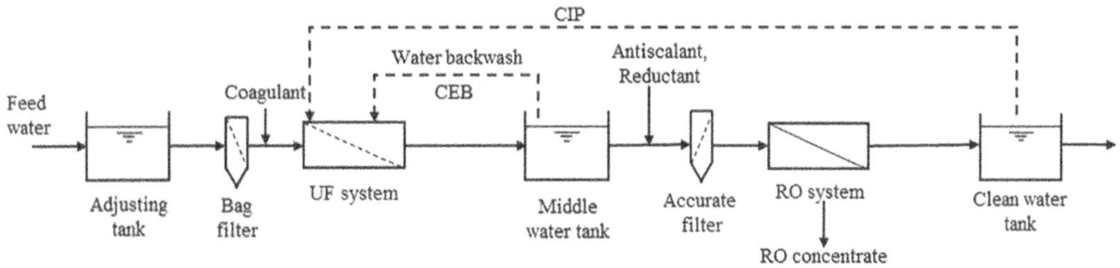

Figure 1 | Schematic diagram of the major process in the pilot UF–RO system.

S_{HCl}, S_{NaOH} and S_{UPW} (Tang *et al.* 2014). Six RO membrane modules in the first stage were collected, drained and weighed. The first RO membrane module in the first stage was collected and stored at 4 °C to minimize the changes of constituents on the RO membrane surface (Supplementary Material, Figure S2). The foulants at the inlet and outlet parts of the fouled RO membrane were collected by physical scrapping. The inlet part of the RO membrane (M1) was nearest to RO influent and the outlet part of the RO membrane (M2) were nearest to ROC.

Analytical methods

Basic index

Conductivity (Cond.), pH, transmembrane pressure (TMP) and membrane flux were measured by an online monitor of the UF–RO system. Salt rejection was defined by Equation (1). The recovery of membrane specific flux (R_{SF}) was defined by Equation (2).

$$\text{Salt rejection} = (\text{Cond}_{\cdot\text{influent}} - \text{Cond}_{\cdot\text{permeate}})/\text{Cond}_{\cdot\text{influent}} \times 100\% \quad (1)$$

$$R_{SF} = \text{SF}/\text{SF}_0 \times 100\% \quad (2)$$

where SF is the specific flux after cleaning, L/(m^2 h kPa), and SF_0 is the initial specific flux, L/(m^2 h kPa).

Turbidity was measured by an SGZ-20B portable turbidimeter (Mingbolm, China). Dissolved organic carbon (DOC) and total nitrogen (TN) were measured by a multi N/C 2100 analyzer (Analytik Jena, Germany). UV absorbance at 254 nm (UV_{254}) was measured by a Uvmini-1240 spectrophotometer (Shimadzu, Japan).

Morphological analysis

Photographs of the UF and RO membranes were taken by a camera. The surface morphology and interface element analysis of fouled RO membranes were conducted by quanta 250 FEG scanning electron microscopy (SEM; FEI, USA) equipped with an energy dispersive spectroscopy (EDS).

Loss on ignition

Loss on ignition (LOI) analysis was conducted for the fouled RO membrane. The foulants on the membranes were dried at 110 °C and heated to 550 °C for 12 h. The mass loss and the remaining mass on ignition at 550 °C represented the content of organics and inorganics, respectively (Tan *et al.* 2017).

Elemental analysis

An inductively coupled plasma optical emission spectroscopy (ICP-OES; Varian VISTA-MPX, USA), and a 761 Compact Ion Chromatography (IC, Metrohm, Switzerland) were used for inorganic element analysis. The element deposition ratio (EDR) as described by Tang *et al.* (2014)is defined as the percentage of the amount of an inorganic element deposited on the membrane over the total amount of the inorganic element in the RO influent for 70-day operation (Equation (3)).

$$\text{EDR} = (C_{im} \times P_i)/(C_{iw} \times \text{Flux} \times 24\,\text{h} \times 70\text{day}) \times 100\% \quad (3)$$

where C_{im} is the content of inorganic foulants on the membrane measured by LOI analysis, mg/m^2; P_i is the proportion of an inorganic element in the total inorganic matter measured by SEM-EDS, %; C_{iw} is the concentration of an inorganic element in the RO influent measured by ICP-OES, mg/L; and Flux, L/(m^2 h).

Microbial analyses

Foulants from 5 × 5 cm^2 membrane samples were suspended in 25 mL of phosphate-buffered saline. Heterotrophic plate counts (HPCs) of membrane foulant samples and water samples were performed on R2A medium (Sigma-Aldrich, USA) according to Reasoner & Geldreich (1985).

The genomic DNA was extracted from the samples of 500 mL of FW, UFP, ROC and 50 mg of foulants on RO membranes using the OMEGA D5525-01 E.Z.N.A.™ Water DNA kit (USA) according to the manufacturer's protocols. The V4 region of 16S rRNA gene was PCR amplified using the primer pairs (515F/806R) (Caporaso *et al.* 2012). Pyrosequencing of 16S rRNA genes was performed on the

Illumina MiSeq platform (USA). Sequences were processed by Mothur v.1.32.0 in order to reduce the noise and sequencing artifacts, as described by Pang *et al.* (2016). Sequences were clustered into operational taxonomic units based on 97% sequence similarity by comparison to the Silva reference database (db119) (July 2014).

RESULTS AND DISCUSSION

Water characteristics

Water characteristics through the pilot UF–RO system, in terms of pH, conductivity, DOC, UV_{254}, TN, turbidity, metal ions and anions, were monitored in order to investigate the relationship between the FW characteristics and foulants on membranes (Supplementary Material, Table S2). The conductivity, sodium, chloride and sulfate in the FW of the UF–RO system (cmWWTP effluent) were 1,811–5,420 μs/cm, 414.1–856.5 mg/L, 402.9–1,579.8 mg/L and 198.8–349.2 mg/L, respectively, which were significantly higher than those of other common mWWTPs (Xu *et al.* 2010; Chon *et al.* 2012; Ayache *et al.* 2013; Wang *et al.* 2019; Tong *et al.* 2020), which could be resulted by the periodical seawater intrusion of this cmWWTP. The UF membrane removed turbidity to a certain extent but had no ability to remove organic matter and ions. The RO membrane removed all kinds of pollutants effectively, indicating that the UF–RO system could operate normally under this water quality condition (Supplementary Material, Table S2).

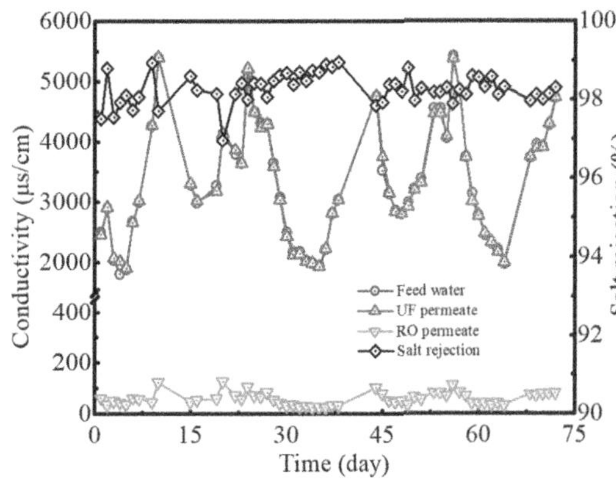

Figure 2 | The conductivity and salt rejection of the UF–RO system during the operation period with the salinity fluctuation of water quality.

Effect of salinity fluctuation on the UF–RO system

With FW conductivity fluctuating periodically, the RO permeate conductivity and salt rejection fluctuated slightly (Figure 2). The data showed that the flux decreased approximately from 14 to 10 L/(m^2·h) and the salt rejection decreased approximately from 99 to 98%, as the conductivity of FW increased from 1,811 to 5,420 μs/cm, respectively (Figure 3). The results indicated that the UF–RO system could guarantee the permeate quality (salt rejection above 98%) under such a conductivity fluctuation condition. Therefore, the wastewater reclamation UF–RO

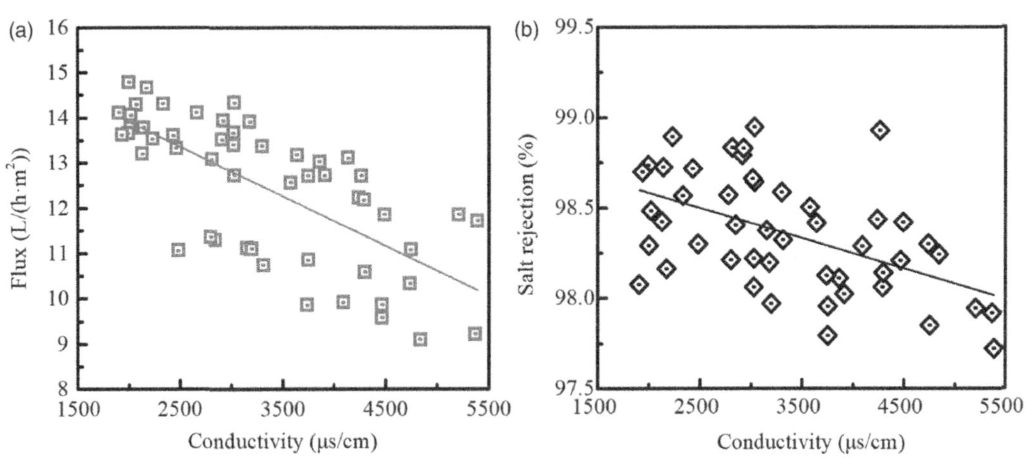

Figure 3 | Effect of FW conductivity on flux (a) and salt rejection (b).

process could be applied for the production of high-quality reclaimed water. However, if the conductivity of the FW increases further, it may affect the performance stability of the UF–RO system and the water quality of the RO permeate (Figure 3).

Effect of microcoagulation on the removal of contaminants

Coagulation is a widely used conventional technique in reclaimed water treatment processes for removing colloid and high-molecular-weight organics (Zhao *et al.* 2014; Yu *et al.* 2016). Comparing with direct UF, except for TN, mcUF had better removal efficiencies of turbidity, DOC and UV_{254}, which was more favorable for the subsequent RO operation (Figure 4). These results were consistent with another study on mcUF for treating activated carbon filter backwash water (Zhang *et al.* 2017). The removal rate of turbidity by mcUF with all the given PAC dosages was consistently higher than that of direct UF (66.1%), where at the dosage of 4 mg/L PAC, the removal rate of turbidity by mcUF reached the highest of 91.7%. Similarly, the removal rates of DOC and UV_{254} by direct UF were 6.2 and 4.8% respectively, while the removal rates of DOC and UV_{254} by mcUF with different PAC dosages were consistently higher.

It should be noted that the conductivity had an effect on the removal efficiency of organic matter by coagulation. In this study, the effect of conductivity on the removal of organic matters by microcoagulation using the actual cmWWTP effluent with two kinds of conductivities (4,840 and 2,220 μs/cm) was conducted via the laboratory-scale experiment. It was found that higher conductivity could enhance the removal efficiency of organic matter by coagulation in certain degree (data not shown). In the process of pilot-scale test, the conductivities corresponding to different coagulant concentrations are shown in Supplementary Material, Table S3. It could be noted that with the increase

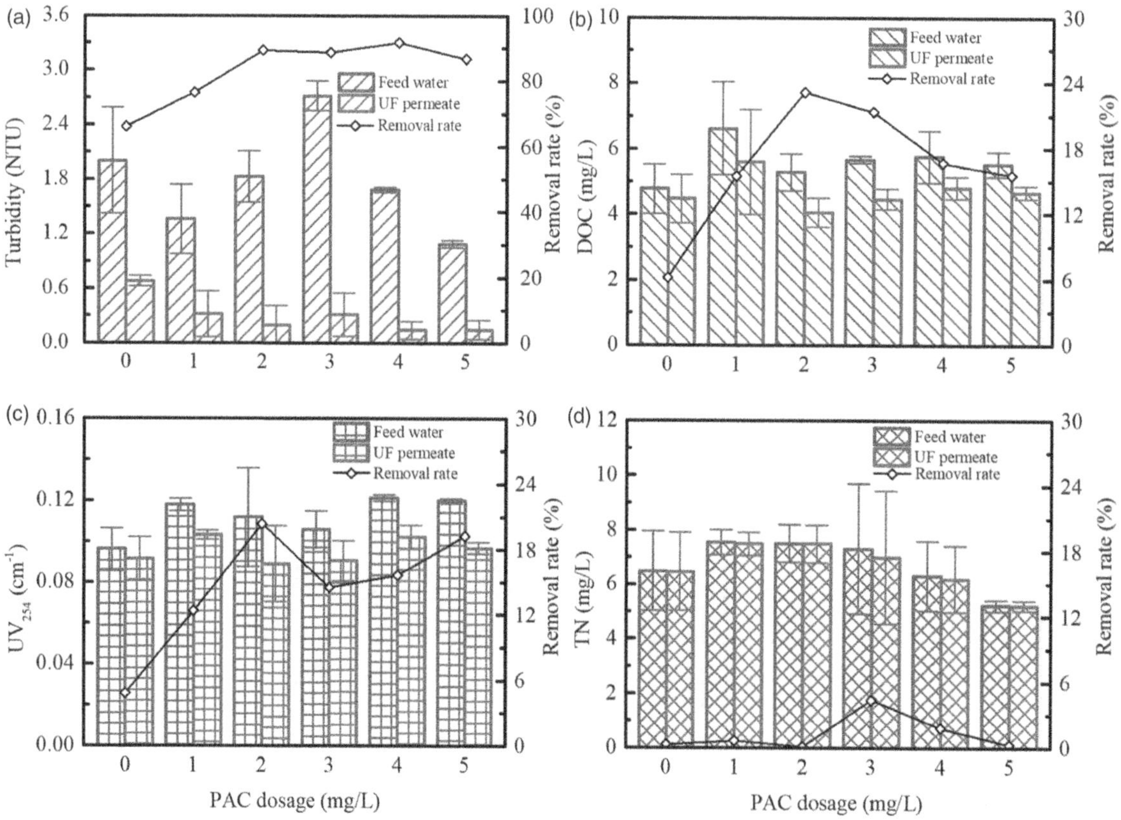

Figure 4 | Removal of turbidity (a), DOC (b), UV_{254} (c), and TN (d) by direct UF (0 mg/L) and mcUF with different dosages of PAC (1–5 mg/L).

of coagulant concentration after 2 mg/L, the removal efficiency of organic matter decreased, which may be caused by the decrease of conductivity.

Effect of microcoagulation on UF cleaning

To investigate the effect of microcoagulation on UF performance and the efficiency of cleaning, mcUF with different dosages of PAC was run for 10 cycles (filtration for 45 min and backwash with UFP for 40 s), respectively. As shown in Figure 5, the TMP of direct UF remarkably increased after each cycle, while the TMP of mcUF with different dosages of PAC showed a slighter increase. After 10 cycles, the TMP of direct UF was significantly higher than that of mcUF.

The increase of TMP for mcUF with different PAC dosages was higher than direct UF within one cycle (Figure 5). However, the water backwash efficiency of mcUF was significantly better than that of direct UF. Conductivity had effect on the turbidity and organic matters removal by microcoagulation. The water backwash of UF membrane was also affected by conductivity fluctuation. Therefore, there was no obvious rule for the efficiency of water backwash with different coagulant concentrations. However, it was clear that microcoagulation could improve the efficiency of water backwash. Taking dosage of 4 mg/L PAC as an instance, although the TMP of mcUF increased approximately 10 kPa within one cycle, which was much higher than that of direct UF of approximately 5 kPa, most of them could be removed by the water backwash more easily. The R_{SF} maintained above 99% during most cycles, which were higher than those of direct UF of 95% (Supplementary Material, Figure S4). Similarly, researchers reported that the floc cake layer on the membrane surface formed by coagulation could be easily removed by backwashing and flushing (Dong *et al.* 2007; Xu *et al.* 2020). This result indicated that the performance of mcUF was more stable than that of direct UF.

The effects of CEB alkali cleaning on mcUF (PAC 4 mg/L) and direct UF with different immersion times were compared. With the alkali immersion time increased, the R_{SF} of direct UF increased from 60% to nearly 100%, which indicated that the foulant on the UF membrane surface could not be easily removed by alkali cleaning with short-time immersion. While, the R_{SF} of mcUF remained stable at approximately 90% even with immersion time of 10 min (Figure 6). These results showed that the CEB alkali cleaning efficiency of mcUF was more stable and had a better cleaning effect with shorter immersion time.

CIP is a more complex cleaning process, which was applied when the TMP could not be recovered by the water backwash and CEB process. The effects of the CIP process on mcUF (PAC 4 mg/L) and direct UF were

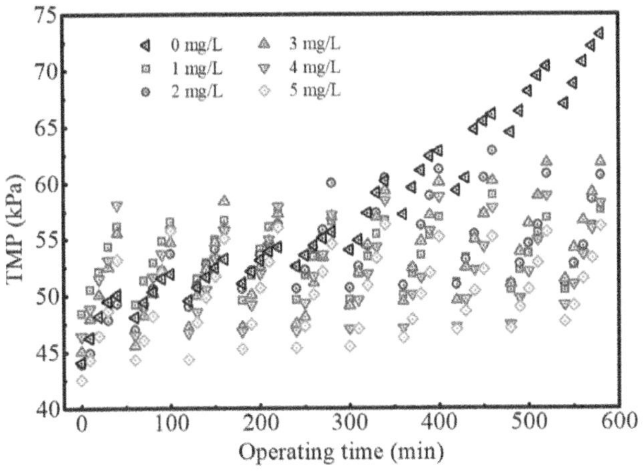

Figure 5 | TMP variation of direct UF and mcUF with different PAC dosages in 10 cycles.

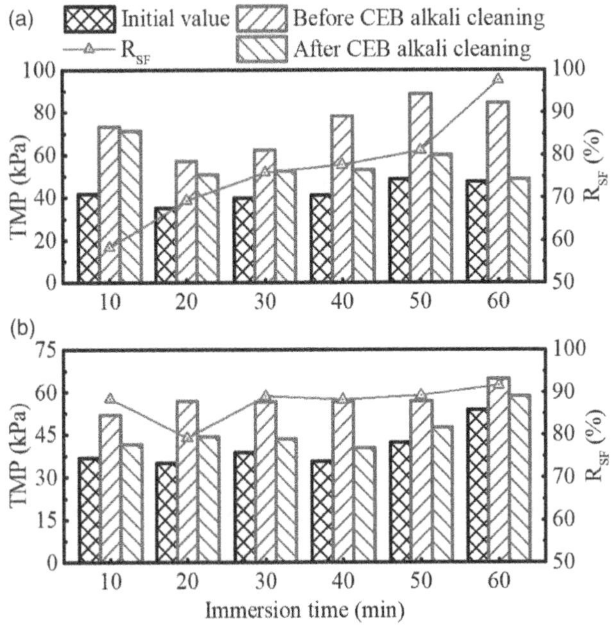

Figure 6 | The TMP decrease and R_{SF} of direct UF (a) and mcUF (b) by CEB alkali cleaning.

compared. The R_{SF} of each step in CIP of mcUF was much better than that of direct UF. The R_{SF} of the whole CIP process of mcUF was nearly 30% higher than that of direct UF (Figure 7). This result showed that microcoagulation could improve the cleaning of CIP effectiveness of UF significantly.

Membrane fouling analyses

UF membrane fouling analysis

Three different kinds of solutions (1 M HCl, 1 M NaOH and UPW) were used to dissolve the deposits on fouled UF membranes. The organic contaminants on the UF membrane obtained by immersion in pure water are significantly lower than those of the S_{HCl} and S_{NaOH}. The absorbance of UV_{254} and the concentration of TN in the S_{HCl} are significantly lower than those of the S_{NaOH} (Table 1). The result of simple immersion experiment also showed that alkali cleaning is more effective than acid cleaning for the organic contaminant removal. In terms of inorganic contaminants, acid cleaning could remove Ca^{2+} and Mg^{2+} effectively (Supplementary Material, Table S4).

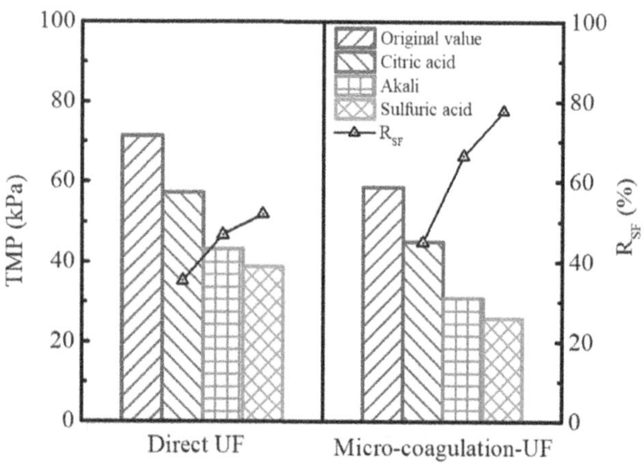

Figure 7 | The TMP decrease and R_{SF} of direct UF (a) and mcUF (b) by the CIP process.

Table 1 | Organic foulants on the UF membrane

Filtrate type	UV_{254} (cm^{-1})	DOC (mg/L)	TN (mg/L)
S_{UPW}	0.006	1.68	0.14
S_{NaOH}	0.113	8.44	1.29
S_{HCl}	0.013	8.81	0.24

RO membrane fouling analysis

Microcoagulation could improve the removal efficiency of organic matters. Theoretically, it could improve the performance of the RO system. However, due to the short experimental period, the RO system had been running well, so the data could not reflect the effect of microcoagulation. To share more information on the RO membrane fouling in the initial stage of operation under the water quality with high and fluctuated conductivity, the RO membrane autopsy and analyses were conducted in this study. After 10 weeks of operation of the UF–RO system, to investigate RO fouling in the initial stage, the six RO membrane modules in the first stage were collected, drained and weighed. The first membrane module was the heaviest among all the six membrane modules (data not shown). Therefore, the first membrane module was selected for anatomy to investigate the most severe fouling membrane. Macroscopic observation showed that there were significantly more foulants in the inlet part (M1) of RO membrane than those in the outlet part (M2) (Supplementary Material, Figure S5). Similarly, microscopy observation also showed more foulants at the inlet part (Supplementary Material, Figure S5).

The contents of the organic and inorganic components of the foulants were measured by LOI analysis with data summarized in Table 2. The contents of the organic components of M1 and M2 were 1.55 ± 0.33 and 0.81 ± 0.31 g/m^2, respectively, accounting for approximately 98.1 ± 0.2 and 96.3 ± 0.7% of the foulant dry weights, whereas the contents of the inorganic component were 0.03 ± 0.01 and 0.03 ± 0.02 g/m^2. The LOI analysis showed that the organic content in the inlet part

Table 2 | The content of organic and inorganic components on the RO membrane

	M1	M2
Dry weight (DW, g/m^2)	1.58 ± 0.34[a]	0.81 ± 0.33
Organic content (g/m^2)	1.55 ± 0.33	0.78 ± 0.31
Organic content (% of DW)	98.1 ± 0.2	96.3 ± 0.7
Inorganic content (g/m^2)	0.03 ± 0.01	0.03 ± 0.02
Inorganic content (% of DW)	1.9 ± 0.2	3.7 ± 0.7

[a]Errors indicate the standard deviation of three replicates of the deposit tested.

(M1) was twice as high as that in the outlet part (M2). In terms of the total amount, organic fouling was the major problem for the RO membrane. These results were accorded with the results of studies on RO membranes in full-scale plants (Tang *et al.* 2014, 2016).

In terms of the deposition ratio of the elements, consistent with other anatomical analyses, Fe could deposit on the RO membrane much more easily than other elements (Supplementary Material, Table S5) (Tang *et al.* 2014; Tan *et al.* 2017). Baker & Dudley (1998) recognized that the presence of Fe bonding with organic matter and extracellular polymeric substances (EPS) was a major contributor for membrane fouling. For this reason, the removal of Fe in the RO influent by pretreatment would be crucial for the performance of the UF–RO system.

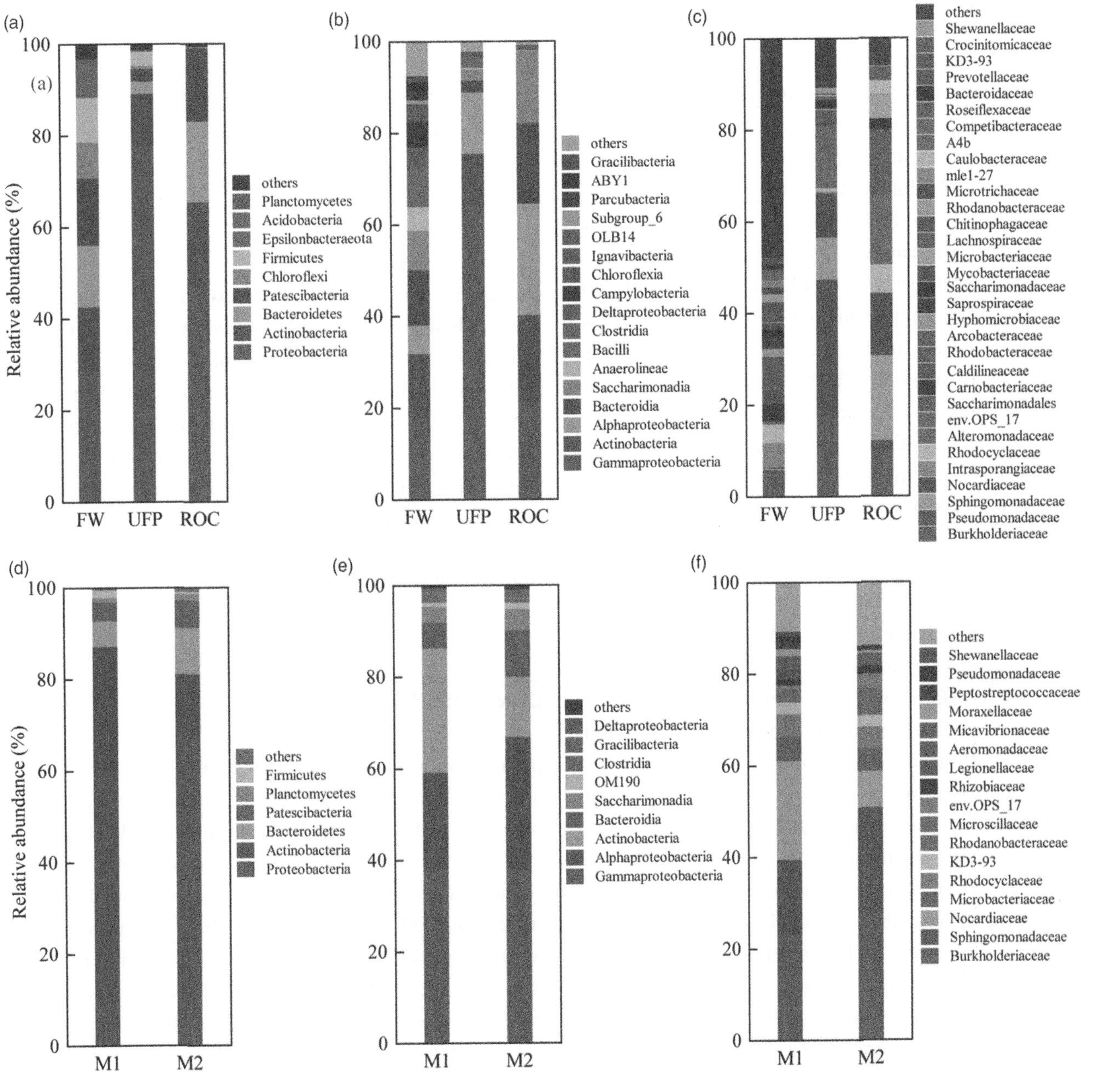

Figure 8 | Microbial communities of the water samples (FW, UFP and ROC) and the membrane samples (M1 and M2) at the phylum level (a and d), class level (b and e) and family level (c and f). 'Others' in the legend indicates the sum of unidentified bacteria and other minor bacteria populations.

HPCs of the water samples and RO membrane samples were performed. The HPC values of FW, UFP and ROC were $1.7\times10^2\pm7.2\times10^1$, $5.2\times10^1\pm6.0\times10^0$ and $2.5\times10^2\pm4.5\times10^1$ CFU/mL, respectively. Similar to the organic contaminant on the RO membrane, the HPC of M1 ($2.9\times10^6\pm4.6\times10^5$ CFU/cm^2) is twice as high as that of M2 ($1.3\times10^6\pm2.9\times10^5$ CFU/cm^2) (Supplementary Material, Table S6).

Regarding the structure of microbial communities in the water samples (FW, UFP and ROC) and the RO membrane samples (M1 and M2), triplicate sequencing results are averaged and shown in Figure 8. Proteobacteria was predominant in all water and membrane samples (Figure 8(a) and 8(d)). In terms of the water samples, Gammaproteobacteria, Actinobacteria and Alphaproteobacteria were three dominant classes (Figure 8(b)). The microbial community structure of the water samples was closer to that of seawater, rather than that of the secondary effluent of mWWTP (Khan *et al.* 2013a, 2013b; Pang *et al.* 2016). After UF treatment, the microbial diversity of the water decreased significantly (Supplementary Material, Table S7). Meanwhile, the microbial community structure also changed significantly (Figure 8(a)–8(c)). At the family level, Burkholderiaceae, Pseudomonadaceae, Sphingomonadaceae and Nocardiaceae were the four dominant families in the UFP and ROC (Figure 8(c)). Similar to the water samples, Gammaproteobacteria, Actinobacteria and Alphaproteobacteria were the three dominant classes, which are obviously different from the microbial community of the foulant on the RO membrane from wastewater treatment RO plant, but similar to that of seawater RO plant (Figure 8(e)) (Khan *et al.* 2014). At the family level, similar to the samples of the UFP and ROC, Burkholderiaceae, Pseudomonadaceae, Sphingomonadaceae and Nocardiaceae were the four dominant families on the RO membrane (Figure 8(f)).

The Bray–Curtis dissimilar heatmap showed that the dissimilarities of two membrane parts (M1 and M2) were very low (Figure 9). The microbial communities of UFP and ROC were clustered with that of RO membrane foulant rather than with the FW. However, many researchers reported that the microbial communities significantly differed from the UFP to the RO membranes (Ayache *et al.* 2013; Khan *et al.* 2015; Tan *et al.* 2017). Yu *et al.* (2017) reported that the microbial community of RO membrane foulant after

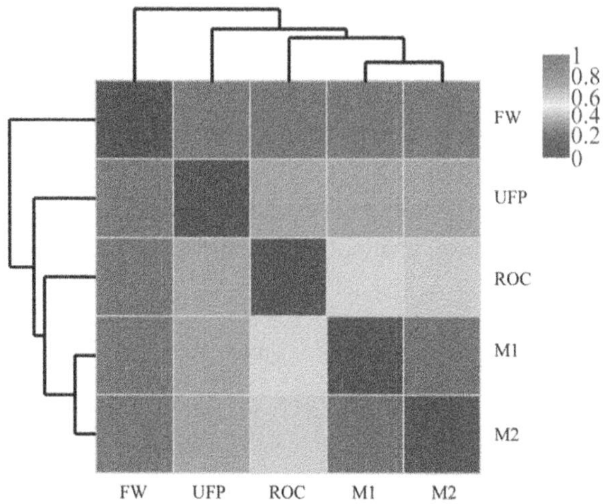

Figure 9 | Bray–Curtis dissimilar heatmap of the water samples and the membrane samples. The Bray–Curtis index ranges from 0 to 1.0, indicating that two samples are composed of the same species. 1 indicates that two samples do not have any of the same species.

membrane chemical cleaning was closer to the water sample than that of the non-cleaned membrane surface. It was speculated that the bacteria deposited on the membrane surface at the initial stage was closer to that of the water sample. This study proved that in the initial stage of membrane fouling, the bacteria on the membrane surface mainly deposited and enriched from the RO FW (UFP). With the increase of operation time, the microbial community on the membrane surface may be gradually differed from that of water.

CONCLUSIONS

In this work, we evaluated the effect of microcoagulation on the performance of the UF–RO system treating the effluent with high and fluctuant salinity from a cmWWTP using a pilot-scale system. mcUF had better removal efficiencies of contaminants than direct UF, which was favorable for the subsequent RO operation. Meanwhile, microcoagulation remarkably improved the cleaning efficiencies of water backwash, CEB and CIP processes. Thus, microcoagulation could improve the performance of the UF–RO system treating the effluent with high and fluctuant salinity from a cmWWTP. The RO membrane anatomical analysis showed that organic fouling was dominated in the initial stage of the RO membrane fouling. The microbial communities of water samples and foulant on the RO membrane were similar to those of seawater and foulant on the RO membranes from seawater RO plants. The microbial community of the foulant on the membrane was similar to that of UFP and ROC.

ACKNOWLEDGEMENTS

This study was supported by the China Postdoctoral Science Foundation (no. 2019M652347) and the Major Science and Technology Program for Water Pollution Control and Treatment (no. 2017ZX07101002-06).

DATA AVAILABILITY STATEMENT

All relevant data are included in the paper or its Supplementary Information.

REFERENCES

Aly, S. A., Anderson, W. B. & Huck, P. M. 2018 Biological filtration with and without prior in-line coagulation to reduce UF fouling by secondary effluent. *Journal of Water Reuse and Desalination* **8** (2), 176–191.

Anis, S. F., Hashaikeh, R. & Hilal, N. 2019 Reverse osmosis pretreatment technologies and future trends: a comprehensive review. *Desalination* **452**, 159–195.

Ayache, C., Manes, C., Pidou, M., Croue, J.-P. & Gernjak, W. 2013 Microbial community analysis of fouled reverse osmosis membranes used in water recycling. *Water Research* **47** (10), 3291–3299.

Baker, J. S. & Dudley, L. Y. 1998 Biofouling in membrane systems – a review. *Desalination* **118** (1–3), 81–89.

Caporaso, J. G., Lauber, C. L., Walters, W. A., Berg-Lyons, D., Huntley, J., Fierer, N., Owens, S. M., Betley, J., Fraser, L. & Bauer, M. 2012 Ultra-high-throughput microbial community analysis on the Illumina HiSeq and MiSeq platforms. *The ISME Journal* **6** (8), 1621–1624.

Chon, K., Cho, J., Shon, H. K. & Chon, K. 2012 Advanced characterization of organic foulants of ultrafiltration and reverse osmosis from water reclamation. *Desalination* **301**, 59–66.

Dong, B.-Z., Chen, Y., Gao, N.-Y. & Fan, J.-C. 2007 Effect of coagulation pretreatment on the fouling of ultrafiltration membrane. *Journal of Environmental Sciences* **19**, 278–283.

Ezugbe, E. O. & Rathilal, S. 2020 Membrane technologies in wastewater treatment: a review. *Membrane* **10** (5), 89.

Ferrer, O., Dekker, R., Mesa, J., Martín-Alonso, J., Cortina, J. L. & Gibert, O. 2016 Micro-coagulation effects on direct ultrafiltration of challenging raw river water. *Journal of Chemical Technology Biotechnology* **91** (12), 2959–2973.

Flemming, H.-C. 1997 Reverse osmosis membrane biofouling. *Experimental Thermal and Fluid Science* **14** (4), 382–391.

Gao, W., Liang, H., Ma, J., Han, M., Chen, Z.-L., Han, Z.-S. & Li, G.-B. 2011 Membrane fouling control in ultrafiltration technology for drinking water production: a review. *Desalination* **272** (1–3), 1–8.

Guigui, C., Rouch, J. C., Durand-Bourlier, L., Bonnelye, V. & Aptel, P. 2002 Impact of coagulation conditions on the in-line coagulation/UF process for drinking water production. *Desalination* **147** (1–3), 95–100.

Huang, B.-C., Guan, Y.-F., Chen, W. & Yu, H.-Q. 2017 Membrane fouling characteristics and mitigation in a coagulation-assisted microfiltration process for municipal wastewater pretreatment. *Water Research* **123**, 216–223.

Humbert, H., Gallard, H., Jacquemet, V. & Croue, J.-P. 2007 Combination of coagulation and ion exchange for the reduction of UF fouling properties of a high DOC content surface water. *Water Research* **41** (17), 3803–3811.

Khan, M. T., de O. Manes, C.-L., Aubry, C., Gutierrez, L. & Croue, J. P. 2013a Kinetic study of seawater reverse osmosis membrane fouling. *Environmental Science & Technology* **47** (19), 10884–10894.

Khan, M. T., Manes, C.-L. d. O., Aubry, C. & Croué, J.-P. 2013b Source water quality shaping different fouling scenarios in a full-scale desalination plant at the Red Sea. *Water Research* **47** (2), 558–568.

Khan, M. T., Busch, M., Molina, V. G., Emwas, A.-H., Aubry, C. & Croue, J.-P. 2014 How different is the composition of the fouling layer of wastewater reuse and seawater desalination RO membranes?. *Water Research* **59**, 271–282.

Khan, M. T., Hong, P.-Y., Nada, N. & Croue, J. P. 2015 Does chlorination of seawater reverse osmosis membranes control biofouling? *Water Research* **78**, 84–97.

Li, Z.-H., Li, Y., Gao, S.-X., Liang, W. & Guo, P.-S. 2019 Mitigated membrane fouling and enhanced removal of extracellular antibiotic resistance genes from wastewater effluent via an integrated pre-coagulation and microfiltration process. *Water Research* **159**, 145–152.

Pang, Y.-C., Xi, J.-Y., Xu, Y., Huo, Z.-Y. & Hu, H.-Y. 2016 Shifts of live bacterial community in secondary effluent by chlorine disinfection revealed by Miseq high-throughput sequencing combined with propidium monoazide treatment. *Applied Microbiology and Biotechnology* **100**, 6435–6446.

Reasoner, D. & Geldreich, E. 1985 A new medium for the enumeration and subculture of bacteria from potable water. *Applied and Environmental Microbiology* **49** (1), 1–7.

Schneider, R. P., Ferreira, L. M., Binder, P. & Ramos, J. R. 2005 Analysis of foulant layer in all elements of an RO train. *Journal of Membrane Science* **261** (1–2), 152–162.

Tan, Y.-J., Sun, L.-J., Li, B.-T., Zhao, X.-H., Yu, T., Ikuno, N., Ishii, K. & Hu, H.-Y. 2017 Fouling characteristics and fouling control of reverse osmosis membranes for desalination of dyeing wastewater with high chemical oxygen demand. *Desalination* **419**, 1–7.

Tang, F., Hu, H.-Y., Sun, L.-J., Wu, Q.-Y., Jiang, Y.-M., Guan, Y.-T. & Huang, J.-J. 2014 Fouling of reverse osmosis membrane for municipal wastewater reclamation: autopsy results from a full-scale plant. *Desalination* **349**, 73–79.

Tang, F., Hu, H.-Y., Sun, L.-J., Sun, Y.-X., Shi, N. & Crittenden, J. C. 2016 Fouling characteristics of reverse osmosis membranes at different positions of a full-scale plant for municipal wastewater reclamation. *Water Research* **90**, 329–336.

Tong, X., Cui, Y., Wang, Y.-H., Bai, Y., Yu, T., Zhao, X.-H., Ikuno, N., Luo, H.-J., Hu, H.-Y. & Wu, Y.-H. 2020 Fouling properties of reverse osmosis membranes along the feed channel in an industrial-scale system for wastewater reclamation. *Science of the Total Environment* **713**, 136673.

Wang, Y.-H., Wu, Y.-H., Tong, X., Yu, T., Peng, L., Bai, Y., Zhao, X.-H., Huo, Z.-Y., Ikuno, N. & Hu, H.-Y. 2019 Chlorine disinfection significantly aggravated the biofouling of reverse osmosis membrane used for municipal wastewater reclamation. *Water Research* **154**, 246–257.

Xu, P., Bellona, C. & Drewes, J. E. 2010 Fouling of nanofiltration and reverse osmosis membranes during municipal wastewater reclamation: membrane autopsy results from pilot-scale investigations. *Journal of Membrane Science* **353** (1–2), 111–121.

Xu, H.-B., de Koning, J. & Geng, Y. 2019 Reliability and efficiency of an advanced tertiary treatment process for wastewater reclamation. *Journal of Water Reuse and Desalination* **9** (4), 385–395.

Xu, L., Wei, C., Siddique, M. S. & Yu, W. 2020 Insight into the effect of in-situ galvanic micro-coagulation on membrane fouling mitigation treating surface water. *Journal of Membrane Science* **610**, 118234.

Yu, W.-Z., Qu, J.-H. & Gregory, J. 2015 Pre-coagulation on the submerged membrane fouling in nano-scale: effect of sedimentation process. *Chemical Engineering Journal* **262**, 676–682.

Yu, T., Li, G., Lin, W., Hu, H.-Y. & Lu, Y. 2016 Coagulation increased the growth potential of various species bacteria of the effluent of a MBR for the treatment of domestic wastewater. *Environmental Science and Pollution Research* **24** (6), 5126–5133.

Yu, T., Meng, L., Zhao, Q.-B., Shi, Y., Hu, H.-Y. & Lu, Y. 2017 Effects of chemical cleaning on RO membrane inorganic, organic and microbial foulant removal in a full-scale plant for municipal wastewater reclamation. *Water Research* **113**, 1–10.

Zhang, J.-N., Lin, T. & Chen, W. 2017 Micro-flocculation/sedimentation and ozonation for controlling ultrafiltration membrane fouling in recycling of activated carbon filter backwash water. *Chemical Engineering Journal* **325**, 160–168.

Zhao, X., Huang, H., Hu, H.-Y., Su, C., Zhao, J. & Liu, S.-M. 2014 Increase of microbial growth potential in municipal secondary effluent by coagulation. *Chemosphere* **109**, 14–19.

First received 31 October 2020; accepted in revised form 28 December 2020. Available online 25 January 2021

 © 2021 The Authors

Advances in efficient desalination technology of capacitive deionization for water recycling

Yuhao Tong, Shuang Zhou, Jun Zhou, Guanteng Zhang, Xiaolin Li, Chunxia Zhao and Pengyan Liu

ABSTRACT

Available freshwater resources are becoming harder to obtain due to climate change, population growth, industrial development, and water pollution. The main technologies in the field of wastewater desalination include reverse osmosis, electrodialysis, thermal distillation, and adsorption etc. Capacitive deionization technology (CDI) belongs to a novel electrochemical desalination technology with low energy consumption and low environmental impact, simple equipment structure and convenient operation. With the importance of wastewater desalination highlighted, some great technological progress of CDI has been made in electrode materials, reactor structure and the hybrid process. In this paper, the development of CDI technology was expounded from three aspects to achieve the goal of strong adaptability, low cost and strong adsorption capacity by analysis of the latest research papers. Corresponding improved methods of CDI are summarized to solve the main technology bottlenecks such as the inefficient and vulnerable electrode materials, low selectivity and unreasonable unit structure, and limitations of single CDI unit for promoting the continuous development of CDI technology.

Key words | capacitive deionization, desalination, electrode materials, reactor structure, water recycling

Yuhao Tong
Shuang Zhou
Jun Zhou
Guanteng Zhang
Xiaolin Li
Chunxia Zhao (corresponding author)
Pengyan Liu
College of Chemistry and Environmental Science, Hebei University, Institute of Life Science and Green Development, Hebei University, Key Laboratory of Analytical Science and Technology of Hebei Province,
Baoding 071002,
China
E-mail: *zhaocxhbu@edu.cn*

HIGHLIGHTS

- The main technology bottlenecks of CDI were analyzed from three aspects.
- Review the development of CDI based on electrode material, novel process and hybrid process.
- New electrode materials are introduced and analyzed for the future development of CDI.
- The energy consumption of novel FCDI-NF hybrid process is significantly reduced.

INTRODUCTION

Freshwater resources are essential for the sustainable development of human society. Rapid population growth, urbanization, and climate change are putting tremendous stress on freshwater resources (Fan *et al.* 2020); approximately one-fifth and one-third of the world's population live in areas of water scarcity and in countries with moderate to high water stress, respectively, according to the report of UN Water (Chen *et al.* 2020). In the past decades, the usage and treatment of water in China has grown tremendously due to the large population and rapid economic development (Xu *et al.* 2020). Water recycling can be an effective option for saving water resources, reducing the environmental impacts from the discharge of treated wastewater, and reducing the cost and energy involved in water resource management (Lyu *et al.* 2016; Takeuchi & Tanaka 2020). The development of novel

This is an Open Access article distributed under the terms of the Creative Commons Attribution Licence (CC BY 4.0), which permits copying, adaptation and redistribution, provided the original work is properly cited (http://creativecommons.org/licenses/by/4.0/).

doi: 10.2166/wrd.2021.101

water regeneration technology is of great significance to alleviate the crisis of freshwater resources.

Traditional nonthermal distillation methodologies, such as reverse osmosis (RO), adsorption/ion exchange, and membrane desalination (MD) (Burakov *et al.* 2018; Folaranmi *et al.* 2020; Shi *et al.* 2020), have wide applications in desalination and water purification, while membrane fouling, adsorbent regeneration and high energy consumption are always problems to be solved (Naushad *et al.* 2015). Capacitive deionization (CDI) is a promising new technology for water desalination and ions removal compared to other traditional desalination methodologies, due to its low energy consumption (i.e. room temperatures, low pressures and low voltages), low capital and processing costs, easy operation and maintenance processes, and environmentally friendly characteristics (i.e. without chemicals or hazardous substances) (Khalil *et al.* 2020). CDI has emerged as a promising alternative desalination technology of reverse osmosis, electrodialysis and thermal distillation (Zhang *et al.* 2018; Tang *et al.* 2019). The electrosorption/desorption processes of CDI for deionization is based on electric double layer theory. Ions in the feed water will be attracted on the electrode with opposite charge by an external electrical potential (very low voltage, 1.2 V), and desorbed into the concentrate from the electrode by a reversing potential for a continuous desalination process (Faisal *et al.* 2020).

The technical principle of CDI is shown in Figure 1. Ions are removed by transferring from aqueous solution to

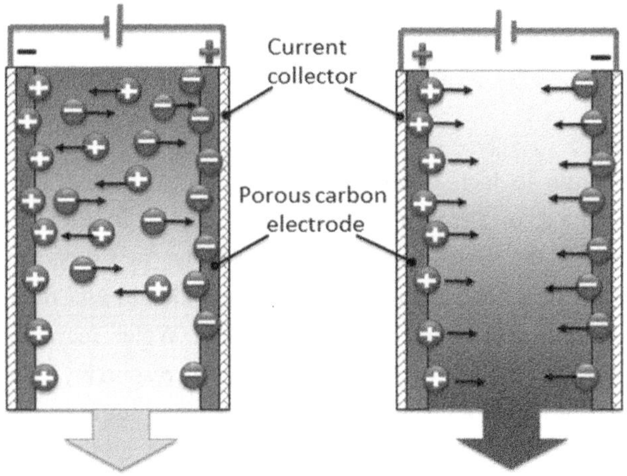

Figure 1 | Schematic diagram of the working and regeneration principle of CDI.

the electrode surface, the charge balance is achieved on the electric double layer, and the effluent is purified (Figure 1(a)). The ions or charged particles adsorbed on the surface of the electrode material are released into solution to realize the electrode regeneration through reverse connection or short-circuit voltage (Figure 1(b)). During the charging and discharging process of the CDI electrodes, the absorption and desorption of ions are completed, respectively. Therefore, the electrode materials with huge surface, ultra-high capacitance and high desalination efficiency are essential to CDI technology for water resources recycling.

Porous carbon electrodes materials are widely used in electrostatic adsorption of ions in the CDI process. Significant efforts have been made in recent decades in the development of customized porous carbon materials for CDI including carbon nanotubes, graphene, MXene, aerogels, xerogels, mesoporous carbon, and activated carbons. Although posing unique benefits, so far these materials have generally failed to provide a significant competitive advantage to traditional adsorption and RO in either capital or operating cost due to low salt adsorption capacity and carbon electrode oxidation (Landon *et al.* 2019). Nowadays, how to promote the desalination efficiency of CDI device is still a difficult problem for many researchers to develop materials with ultra-high capacitance or new processes with more efficient desalination performance (Zhao *et al.* 2020).

The desalting performance of CDI also depends on the operating parameters including external voltage, inflow velocity, thickness of interlayer between two electrodes and design of flow channel, and quality, species and concentration of solution ions etc. (Chen *et al.* 2018). Commonly used CDI desalination performance evaluation indicators mainly include electroadsorption capacity, charge efficiency, average desalination rate, and long-term stable desalting belonging to reactor structure optimization (Yasin *et al.* 2018; Maheshwari *et al.* 2020; Wang *et al.* 2020).

In this paper, the three main technology bottlenecks have been analyzed for the desalination performance of CDI device such as the inefficient and vulnerable electrode materials with low selectivity, unreasonable unit structure and limitations of single CDI unit (Figure 2). Furthermore, the corresponding improved methods of CDI were

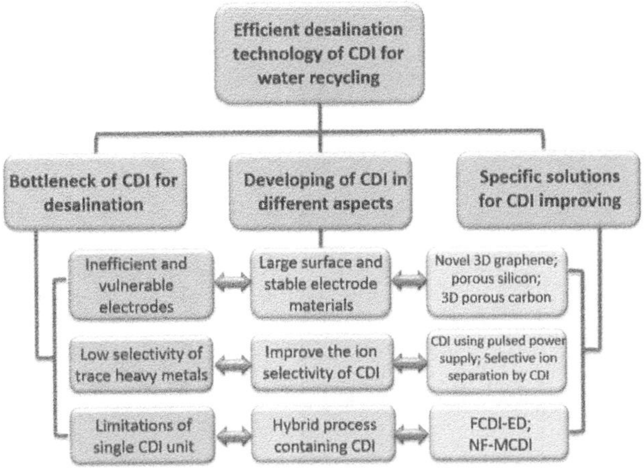

Figure 2 | Overview of the bottlenecks, developments and solutions of CDI for water recycling.

summarized for promoting the continuous development of CDI technology based on the electrode material, novel process and hybrid process.

NEW CDI ELECTRODE MATERIAL

Among the various factors that affect the desalination performance of CDI, the most important one is the selection of electrode materials. The pore size distribution, specific surface area, hydrophilicity, stability, specific capacitance and resistance of the electrode material all have the most direct impact on the capacitance adsorption effect (Shi *et al.* 2020; Zhao *et al.* 2020). This section focuses on solving the problems of instability and low performance of traditional CDI electrodes. Based on the research of the past five years, researchers have developed new electrode materials with a large specific surface area and stable electrochemical performance. Next, three representative new electrode materials are introduced and analyzed, the advantages and disadvantages of each new electrode material are explored, and the future development of CDI electrode materials is expected.

Electrospun network electrode

Conventional CDI electrode materials include activated carbon, graphite and carbon nanofibers (CNFs), but these materials have the problems of high manufacturing cost and low adsorption capacity, which limits the large-scale application of CDI. Researchers find that the high surface area and porosity of the electrode material are very important to enhance the specific capacitance and cycling stability of the capacitor to achieve high performance (Jiang & School of Petrochemical Engineering 2020). Electrospinning, as a simple yet effective technique to fabricate continuous CNFs, has attracted much attention in the CDI field (Feng *et al.* 2019). The combination of free-standing e-CNFs and PCP with hierarchical pore structure should be a promising electrode material for CDI application (Liu *et al.* 2015). A new CNFs reinforced 3D PCP network (e-CNF-PCP) was fabricated through electrospinning followed by thermal treatment and used as electrode materials for CDI with an EC of 16.98 mg/g in 500 mg/L NaCl solution (Figure 3), which shows great improvement compared to those of PCP and e-CNFs (Liu *et al.* 2016). The CDI performances and manufacturing difficulties of e-CNF-PCP were compared with those of other carbon electrode materials reported in the literature, as shown in Table 1.

However, due to the complex processing technology and high cost of CNF materials, it is difficult to realize large-scale industrial application. Therefore, in the future development of this technology, researchers should focus on the realization of low-cost electrospinning and thermal treatment process to reduce the manufacturing cost of e-CNF-PCP, so that e-CNF-PCP can be widely used as a CDI electrode (Wang *et al.* 2020).

Novel 3D graphene electrode

In order to achieve optimal desalination during CDI, CDI electrodes should possess high electrical conductivity, large surface area, good wettability to water, narrow pore size distribution and efficient pathways for ion and electron transportation (Shi *et al.* 2016). Graphene is one of the thinnest materials in the world, and the theoretical thickness value, band length, and bond angle are 0.335 nm, 0.142 nm, and 120°, respectively. Compared with 2D graphene, 3D graphene is an ideal electrode material for CDI because of its large surface area, good mechanical strength and flexibility (Yu *et al.* 2016). The researchers fabricated a novel CDI electrode based on a three-dimensional graphene (3DG) architecture by

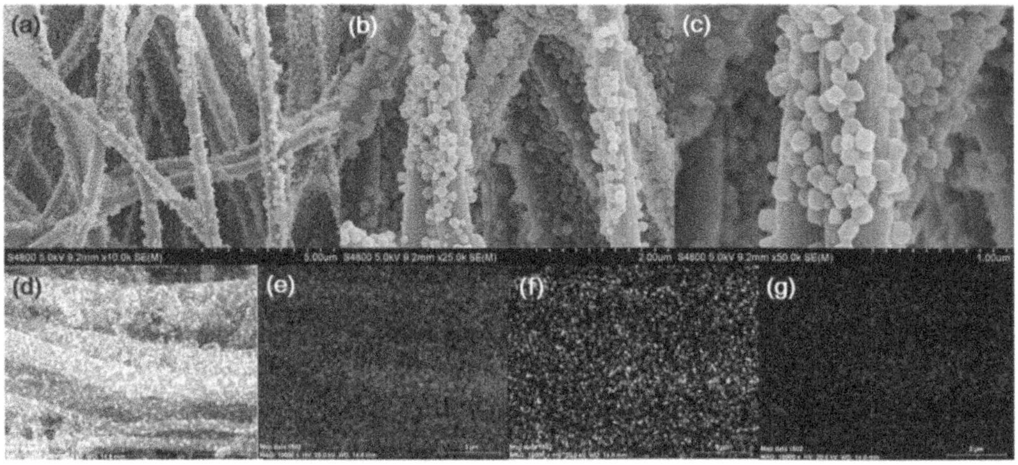

Figure 3 | (a–c) FESEM images of e-CNF-PCP at different magnifications and elemental mapping images of (d) e-CNF-PCP, (e) C element, (f) N element and (g) O element (Liu *et al.* 2016).

Table 1 | Comparison of CDI performances and manufacturing difficulties among various carbon electrode materials from the literature (Liu *et al.* 2016)

Electrode material	CDI performances NaCl concentration (mg/L)	Electrosorption capacity (mg/g)	Manufacturing difficulties
CAs	~500	2.9	Supercritical drying, binder
AC	500 1,000 1,500 2,000	9.72 10.80 11.00 11.76	Coating process, binder
CNTs	500 1,000 1,500 2,000	2.57 3.71 4.76 5.24	Chemical vapor deposition, binder
e-CNF-PCP	100 300 500	10.03 12.56 16.98	–

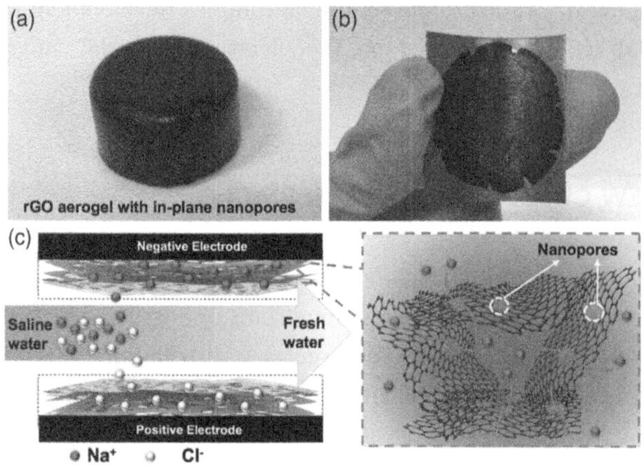

Figure 4 | Photographs of (a) as-prepared rGO hydrogel with in-plane nanopores (NP-3DG) and (b) CDI electrode, (c) Schematic diagram of the CDI process (Shi *et al.* 2016).

constructing interconnected graphene sheets with in-plane nanopores (NP-3DG). As compared to 3DG, NP-3DG features a larger specific surface area and therefore higher specific capacitance (Figure 4). The NP-3DG electrode can be used as a promising electrode material for high-performance CDI applications. The results of NP-3DG exhibit ultrahigh CDI performance with a superior electrosorption capacity of 22.09 mg/g, high adsorption rate, good deionization cyclic stability and high salt removal percentage. Furthermore, due to its excellent electrosorption performance, it is expected that NP-3DG will also play an important role in the removal of charged hazardous species and other heavy metal ions (Khan *et al.* 2018).

As a new type of 3D carbon nanomaterial, NP-3DG has incomparable advantages for use in CDI electrodes. Because it is still in the initial stage of development, there are some problems such as high cost, imperfect theoretical system and unsatisfactory treatment effect (Zhang *et al.* 2018). In recent years, although some progress has been made in the design and preparation of graphene CDI electrode materials, further research on the construction of graphene pore structure and the comprehensive performance of CDI graphene electrode materials is needed to promote the large-scale practical development and application of graphene electrode materials in CDI.

High efficiency particle free porous silicon network electrode

An energy saving water desalination process using low-cost materials is a key step for the sustainable demand of clean water resources in the future. Here, porous silicon has been proven to be a kind of material with rich content, low cost and strong biocompatibility (Gu *et al.* 2016). With appropriate surface passivation of the porous silicon material to prevent surface corrosion in aqueous environments, the results showed that porous silicon templates can enable salt removal of 0.36 and 0.52% in brackish water (10 mM, or 0.06% salinity) and ocean water (500 mM, or ~2.9% salinity) by CDI, respectively. The use of porous silicon for CDI enables new routes to directly couple water desalination technology with microfluidic systems or photovoltaics cells (Figure 5), while mitigating adverse effects of water contamination occurring from nanoparticulate-based CDI electrodes (Metke *et al.* 2016).

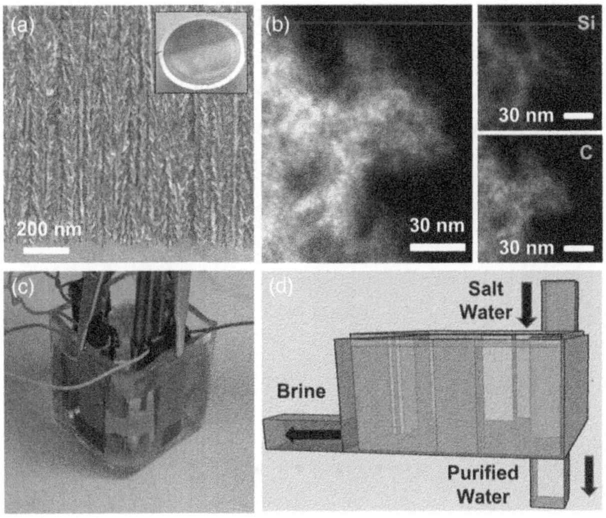

Figure 5 | Porous silicon-based CDI electrodes: (a) Cross-sectional SEM image of a carbon passivated porous silicon electrode for water desalination; (b) Analytical elemental TEM analysis a carbon passivated porous silicon material showing uniform carbon passivation on the material; (c) Photograph of the CDI testing cell combining four pairs of silicon-based electrodes, and (d) illustration of a CDI module in the context of the testing apparatus for continuous flow-through CDI water desalination using porous silicon (Metke *et al.* 2016).

According to Figure 6, the porous silicon network electrode described in this section has a specific salt removal rate of 3.8–5.2 mg/g, ranging from fresh water to seawater salt concentration, and only 1.45 Wh/L of energy input is needed to convert seawater into drinking water (Yu *et al.* 2017; Zhao *et al.* 2019a, 2019b).

In summary, compared with the traditional carbon-based electrode, the main advantage of a porous silicon network is that the silicon content is huge, resulting in lower cost, and it has a tunable pore morphology and structure that can optimize CDI performance in different flow architectures and mitigate the effect of fouling. Therefore, a porous silicon network can be used in developing countries on a large scale, thus becoming a low-cost and easy to operate basic seawater purification equipment.

NEW CDI PROCESS

Although CDI has made great breakthroughs in electrode materials, architecture and system integration, it has no advantages over traditional desalination technologies in terms of energy efficiency, water recovery rate, desalination concentration range and capital cost. However, CDI has illustrated great potential for selective removal of target ions because of its tailorable electrode materials and ability to couple with selective ion exchange membranes (Zhang *et al.* 2020; Park *et al.* 2021). In this section, two novel CDI processes were introduced for their higher ability of selective ions removal, and the application feasibility was explored.

Novel CDI using a pulsed power supply

In recent years, capacitive deionization technology (CDI) has been widely used in many fields due to its environmental protection, low energy consumption, simple operation, and safety (Cho *et al.* 2019). For the desalination of low-concentration brine (50 mg/L), electrodialysis, reverse osmosis and ion exchange are usually used. However, the above technologies have high energy consumption and high additional costs. Therefore, the CDI desalination was proposed to provide new ideas for the preparation of ultrapure water and the advanced treatment of wastewater (Yang *et al.* 2017).

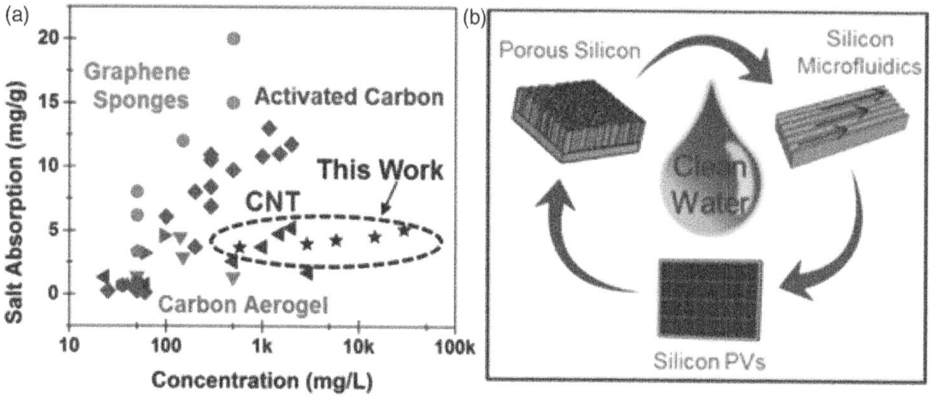

Figure 6 | High performance of porous silicon CDI electrodes and integration vision: (a) Comparison of average specific salt removal capacity for porous silicon electrodes compared to other notable forms of carbon; (b) Scheme emphasizing the vision of porous silicon integration with microfluidics and/or solar cells for technological water desalination platforms (Metke *et al.* 2016).

Pulsed electric field is a new technology to treat liquid and semi-solid objects with high electric field intensity (10–50 kV/cm), short pulse width (0–100 μs) and high pulse frequency (0–2000 Hz) (Sistat *et al.* 2015) Under the pulsed electric field, ions transfer from solution to the electrode or the electro-dialysis membranes (Kim *et al.* 2019). The CDI cell, including the reactor and the activated carbon coated electrode, was designed. The pulsed power supply was applied in CDI for the electric field. In Figure 7, the removal rate of direct current-CDI (DC-CDI) and pulsed-capacitive deionization (pulsed-CDI) are compared (Jin *et al.* 2019).

In all, compared with the traditional direct current-CDI, novel pulsed-CDI has better ion removal rate and lower energy consumption, which proves that pulsed CDI can be an excellent alternative to traditional DC-CDI. However, due to the high cost and complex operation of a pulsed electric field, and the diversity of capacitive deionization electrode materials, repeated testing is needed to find the best voltage condition. Therefore, in the future, researchers should focus on enhancing the compatibility of the pulsed electric field so that it can better adapt to various CDI electrode materials, so as to realize large-scale industrial application.

Selective ion separation by CDI based technologies

Selective ion extraction from aqueous solution is of great significance for water purification as well as resource recovery (Kavitha *et al.* 2019). However, despite the great

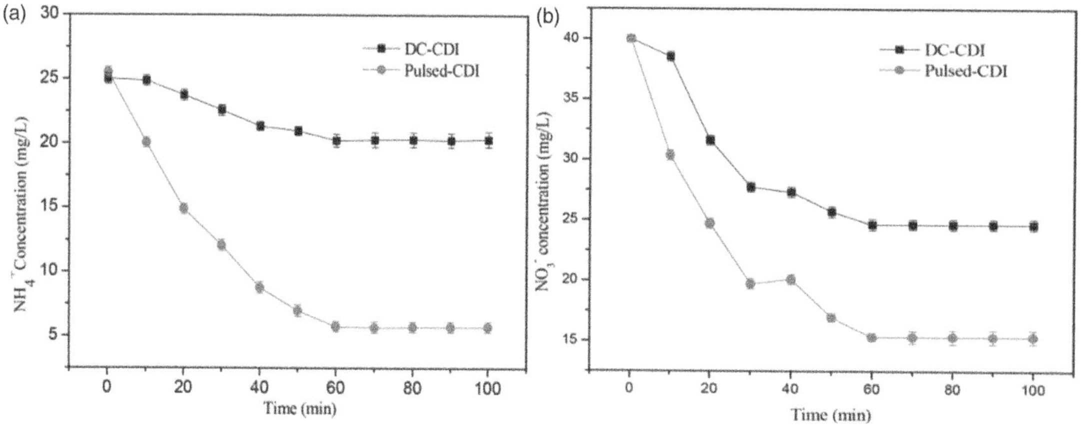

Figure 7 | Performance comparison of DC-CDI and pulsed-CDI (a frequency of 100 Hz and a duty ratio of 50%) in desalination. (a) NH_4^+, (b) NO_3^- (Jin *et al.* 2019).

development in the material, architecture, and system integration in the past 15 years, CDI was reported to have no advantages on energy efficiency, water recovery, salinity range, salt rejection, and capital cost compared with reverse osmosis (RO) (Jain *et al.* 2018). Nevertheless, due to its tailorable electrode materials and ability to couple selective ion exchange membranes/coatings on the electrode interface, the great potential of CDI has been shown for selective ions removal, which cannot be effectively solved by other desalination technologies including RO, electrodialysis or thermal evaporation (Zhang *et al.* 2020). Therefore, how to improve the ion selectivity of CDI has become an important development goal of CDI.

Hence, asymmetric activated carbon (AC) electrodes have been designed and grafted by sulfonic groups and amine groups on the surface of AC through covalent bonds as a cation-selective electrode and an anion-selective electrode respectively for enhanced capacitive deionization (Choi *et al.* 2019; Cen *et al.* 2020). Ten pairs of asymmetric electrodes are assembled with sulfonic AC as positive electrodes and aminated AC as negative electrodes. The scheme of co-ion minimization is illustrated in Figure 8. The electrosorption capacity and charge efficiency of asymmetric AC electrodes have been significantly improved because of not only preventing co-ion expulsion effect but also promoting the wettability and accelerating salt solution infiltration. These results will be beneficial to solve the technical bottlenecks and accelerate the practical engineering application of CDI. These electrodes have higher charge efficiency (0.84) and higher electrosorption capacity (23.2 $mg{\cdot}g^{-1}$) compared to pristine AC electrodes (0.54 of charge efficiency and 11.0 $mg{\cdot}g^{-1}$ of electrosorption capacity) due to the wettability and hydrophilicity effectively enhanced by grafting ion-selective functional groups (Yan *et al.* 2018).

Asymmetric AC electrodes show better cycling stability (Zuo *et al.* 2018). The modification of the ion-selective functional groups on the surfaces of AC could reduce the co-ion effects effectively and improve the salt removal efficiency from the solution. This will provide a new opportunity for the practical development of capacitive deionization technology.

In all, compared with the traditional symmetrical electrode CDI, CDI using the asymmetric activated carbon electrode has higher charge efficiency and higher electrosorption capacity. However, many factors, including the oxidation of carbon-based materials, decomposition of intercalation structures, blockage of electrode pores caused by fouling and scaling, and membrane fouling resulting from heavy metals, would lead to the decline of desalination performance and also affect the stability of selective separation. Therefore, long-term operation stability should be evaluated, and protection measurements (such as water pretreatment, developing new materials, etc.) can also be adopted to enhance the system stability in the future (Zhang *et al.* 2020).

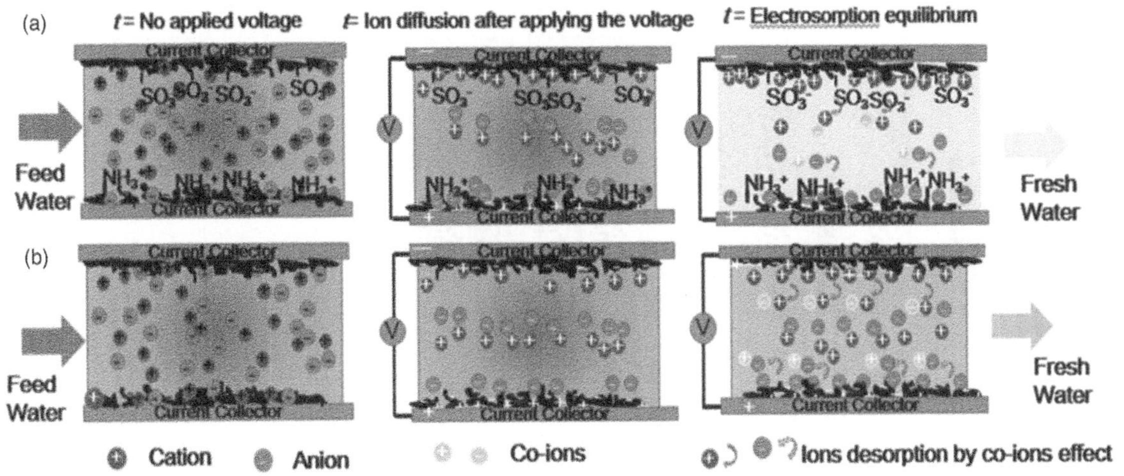

Figure 8 | Scheme illustration of the time-dependent CDI model of (a) asymmetric AC electrodes and (b) symmetric activated carbon electrodes (Yan *et al.* 2018).

HYBRID CDI PROCESS

Although CDI has been developed for many years, there are still some problems that are difficult to solve, such as difficult regeneration and low treatment efficiency. These problems are usually attributed to the limitations of the device itself, which cannot be fundamentally solved by conventional device adjustment (Campione *et al.* 2020; Wu *et al.* 2020). The purpose of this section is to introduce the coupling application of CDI with other water treatment technologies. Through the advantages of other devices, the shortcomings of traditional CDI unit are supplemented. This method effectively frees CDI from its own limitations, solves the problems in traditional CDI and makes the industrial development of CDI have practical significance.

Electrodialysis and FCDI coupling technology in continuous desalination

Membrane capacitive deionization (MCDI) and capacitive deionization (CDI) both adopt the fixed electrode mode, so the adsorption capacity is low and needs frequent regeneration, which does not fundamentally solve the problems of continuous desalination and low water efficiency caused by the small adsorption capacity of the fixed electrode. Researchers put forward the concept of flow capacitive deionization. Activated carbon (AC) was used as the mobile electrode. The scale of the project was realized by increasing the amount of the flow electrode (Jeon *et al.* 2019). By using electrodialysis (ED) as the regeneration unit of mobile electrode, and using flow capacitance deionization (FCDI) for desalination, the continuous and stable desalination process of brine can be realized (Cho *et al.* 2019). FCDI plays a role in the salt concentration of medium and low concentration brine (to obtain concentrated solution and desalination water), while ED carries out intermittent high-efficiency and rapid desalination of concentrated solution with high salt concentration. The FCDI-ED coupling process can realize industrial application of low energy consumption, high water efficiency and continuous treatment of brine (Hassanvand *et al.* 2017; Chen & Hou 2019; Gao *et al.* 2020). The processing route of FCDI-ED is shown in Figure 9.

However, the disadvantages of FCDI-ED technology are complex process, high cost and easy failure, so it is difficult to carry out large-scale industrial application at this stage. Therefore, researchers should focus on improving the stability of the process and using renewable materials, so as to make the FCDI-ED coupling process better developed.

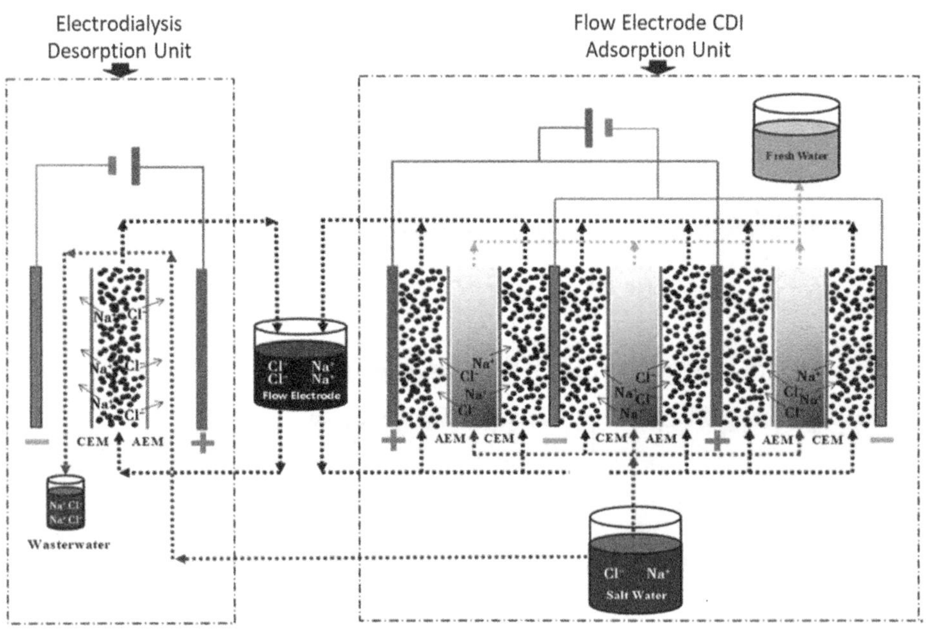

Figure 9 | Schematic diagram of deionization principle of FCDI-ED process (Mo *et al.* 2017).

Novel nanofiltration and membrane capacitive deionization (NF-MCDI) hybrid system

Among various desalination processes, CDI has advantages in terms of cost and energy efficiencies and ease of operation and maintenance, as it can remove salts by adsorbing the ions on the surface of an electrode with a relatively low cell voltage of 1–1.4 V and without high pressure pumps or heat sources (Suss *et al.* 2015). However, single CDI has demonstrated low removal efficiency with a high-concentration (TDS > 3,000 mg/L) because of the ion saturation on the electrode (Choi *et al.* 2019). Using the combined RO-MCDI system, TDS of seawater was decreased from 35,000 to 200 mg/L (Abdelkader *et al.* 2018). However, the operation cost of this process is high, so it is difficult to use in a large-scale seawater desalination plant.

In recent years, a novel hybrid process that combines CDI with nanofiltration (NF) has received growing attention for implementation in energy-efficient desalination systems. Researchers proposed an MCDI unit with flow-electrodes followed by NF, abbreviated as FCDI-NF (Choi *et al.* 2017; Oatley-Radcliffe *et al.* 2017; Sami *et al.* 2018). The energy consumption of FCDI-NF (70% water recovery with 0.460 kWh m^{-3}) was lower by 16% when treating water in comparison with the practical minimum energy consumption of the RO (0.550 kWh m^{-3}) unit for 10,000 ppm NaCl solution desalination to drinking water range. Moreover, the FCDI-NF system attained the drinking water TDS standard (≤500 mg/L) in the final effluent. The schematic of nanofiltration and membrane capacitive deionization (NF-MCDI) hybrid system configuration is shown in Figure 10. The location and sequence of NF unit and CDI unit should be comprehensively considered by the influent quality, requirements of effluent and energy consumption.

After study, researchers found that the standalone NF system was insufficient for salty water desalination due to its comparatively low removal rate of <60% (NaCl). However, it was improved through combination with the MCDI unit to adsorb the remaining salts in the NF-treated water, thereby achieving a total salt removal rate of up to 95% (Jeon *et al.* 2019). Therefore, compared with the single NF or FCDI process, the desalination rate of FCDI-NF is significantly improved, and the energy consumption of the process is also significantly reduced.

However, this hybrid system may require a higher capital cost than the typical RO membrane system because of the price of the electrodes and ion-exchange membranes (Marchetti *et al.* 2017). Also, the NF-MCDI system performance is influenced by the feed concentration, MCDI flow rate and the NF recovery rate. Therefore, the complexity of the system will result in a significant increase in the capital cost and maintenance complexity of the specific equipment used to couple the MCDI system with the NF system. In future work, how to overcome the limitations of industrial applications brought about by the high cost and high complexity of FCDI-NF devices is very important.

CONCLUSION AND PROSPECTS

In less than 20 years CDI technology has developed into the most attractive research field and it is expected that in the

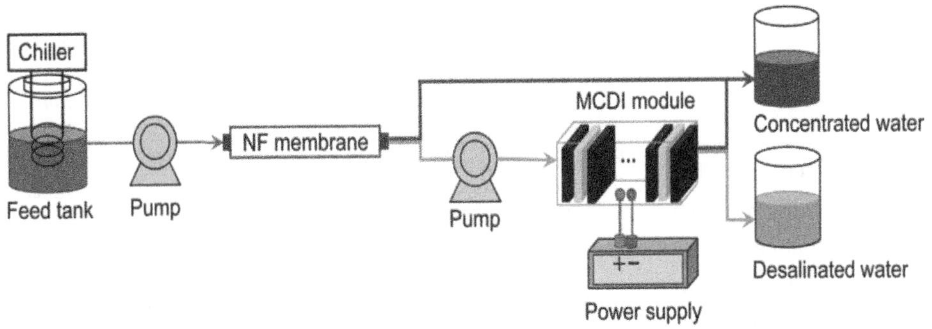

Figure 10 | Schematic of nanofiltration and membrane capacitive deionization (NF-MCDI) hybrid system configuration (Jeon *et al.* 2019).

next few years CDI will further explosive growth in a more mature and advanced direction. Nowadays, CDI has begun to adopt a variety of new technologies and electrode materials, which have unique functions and characteristics, and have made significant progress.

Through the systematic review of the current research it is found that improvement of the structure and composition of electrode materials can improve the desalination effect of CDI technology. However, further research is needed to realize the large-scale industrialization of CDI technology in the following aspects. (1) It is necessary to expand the research on the mechanism of influencing factors of adsorption performance and the kinetics of electro adsorption reaction, so as to provide theoretical basis for the study of electrode materials. (2) Improving the ability of CDI to remove ions selectively can develop CDI into a desalination device that can specifically remove specific ions, thus making up for the deficiency that traditional desalination technology can not specifically remove ions. (3) The development of electrode materials with strong adaptability, low cost and strong adsorption capacity can provide convenience for future research and development, and further promote the application of CDI technology.

The exciting development in the field of CDI is the emergence of FCDI batteries. The selection and synthesis of carbon electrode materials in FCDI have revolutionary changes because there is no need for polymer binders which have side effects on ion adsorption. The FCDI system not only has strong salt adsorption capacity to treat high salinity water, but also can continuously produce fresh water and regenerate electrodes by using non-charged carbon-based materials to continuously supplement the electroactive region. However, FCDI is still in the initial development stage, and there are many possibilities for its future development. One of the main current challenges for FCDI is the low efficiency of charge transfer between collector and flow electrode, and much work needs to be done to solve this problem.

In conclusion, despite the challenges, CDI is still a promising water treatment technology and exciting research field. With the continuous solution to these challenges and the emergence of new CDI units, CDI may be regarded as the key to solve the global water shortage and water pollution in the future.

ACKNOWLEDGEMENTS

This work was supported by financial support from the National Natural Science Foundation of China (51308179); the Natural Science Foundation of Hebei Province (E2020201036, B2018201224); the research fund of Hebei Education Department (KCJSZ2020008); and the Laboratory Opening Project of Hebei University (sy202054).

DATA AVAILABILITY STATEMENT

All relevant data are included in the paper or its Supplementary Information.

REFERENCES

Abdelkader, B. A., Antar, M. A. & Khan, Z. 2018 Nanofiltration as a pretreatment step in seawater desalination: a review. *Arabian Journal for Science and Engineering* **43** (9), 4413–4432.

Burakov, A. E., Galunin, E. V., Burakova, I. V., Kucherova, A. E., Agarwal, S., Tkachev, A. G. & Gupta, V. K. 2018 Adsorption of heavy metals on conventional and nanostructured materials for wastewater treatment purposes: a review. *Ecotoxicology and Environmental Safety* **148**, 702–712.

Campione, A., Cipollina, A., Toet, E., Gurreri, L., Bogle, I. D. L. & Micale, G. 2020 Water desalination by capacitive electrodialysis: experiments and modelling. *Desalination* **473**, 114150.

Cen, B., Li, K., Lv, C. & Yang, R. 2020 A novel asymmetric activated carbon electrode doped with metal-organic frameworks for high desalination performance. *Journal of Solid State Electrochemistry* **24** (3), 687–697.

Chen, T. & Hou, C. 2019 Low energy desalination using a novel capacitive deionization-electrodialysis process. *Abstracts of Papers of the American Chemical Society* **257**, 313.

Chen, Z., Zhang, H., Wu, C., Luo, L., Wang, C., Huang, S. & Xu, H. 2018 A study of the effect of carbon characteristics on capacitive deionization (CDI) performance. *Desalination* **433**, 68–74.

Chen, Z., Wu, G., Wu, Y., Wu, Q., Shi, Q., Ngo, H. H., Vargas Saucedo, O. A. & Hu, H. 2020 Water Eco-Nexus cycle system (WaterEcoNet) as a key solution for water shortage and water environment problems in urban areas. *Water Cycle* **1**, 71–77.

Cho, Y., Yoo, C. Y., Lee, S. W., Yoon, H., Lee, K. S., Yang, S. & Kim, D. K. 2019 Flow-electrode capacitive deionization with highly enhanced salt removal performance utilizing high-aspect ratio functionalized carbon nanotubes. *Water Research* **151**, 252–259.

Choi, S., Chang, B., Kang, J. H., Diallo, M. S. & Choi, J. W. 2017 Energy-efficient hybrid FCDI-NF desalination process with tunable salt rejection and high water recovery. *Journal of Membrane Science* **541**, 580–586.

Choi, J., Dorji, P., Shon, H. K. & Hong, S. 2019 Applications of capacitive deionization: desalination, softening, selective removal, and energy efficiency. *Desalination* **449**, 118–130.

Faisal, A. A. H., Al-Wakel, S. F. A., Assi, H. A., Naji, L. A. & Naushad, M. 2020 Waterworks sludge-filter sand permeable reactive barrier for removal of toxic lead ions from contaminated groundwater. *Journal of Water Process Engineering* **33**, 101112.

Fan, W., Yang, X., Wang, Y. & Huo, M. 2020 Loopholes in the current reclaimed water quality standards for clogging control during aquifer storage and recovery in China. *Water Cycle* **1**, 13–18.

Feng, J., Xiong, S. & Wang, Y. 2019 Atomic layer deposition of TiO_2 on carbon-nanotube membranes for enhanced capacitive deionization. *Separation and Purification Technology* **213**, 70–77.

Folaranmi, G., Bechelany, M., Sistat, P., Cretin, M. & Zaviska, F. 2020 Towards electrochemical water desalination techniques: a review on capacitive deionization, membrane capacitive deionization and flow capacitive deionization. *Membranes (Basel)* **10** (5), 96.

Gao, F., Wang, L., Wang, J., Zhang, H. & Lin, S. 2020 Nutrient recovery from treated wastewater by a hybrid electrochemical sequence integrating bipolar membrane electrodialysis and membrane capacitive deionization. *Environmental Science – Water Research & Technology* **6** (2), 383–391.

Gu, H., Rahardianto, A., Gao, L. X., Christofides, P. D. & Cohen, Y. 2016 Ultrafiltration with self-generated RO concentrate pulse backwash in a novel integrated seawater desalination UF-RO system. *Journal of Membrane Science* **520**, 111–119.

Hassanvand, A., Wei, K., Talebi, S., Chen, G. Q. & Kentish, S. E. 2017 The role of ion exchange membranes in membrane capacitive deionization. *Membranes* **7** (3), 54.

Jain, A., Kim, J., Zuo, K., Li, Q. & Verduzco, R. 2018 Ion-selective and high capacity electrodes for membrane capacitive deionization. *Abstracts of Papers of the American Chemical Society* **255**, 469.

Jeon, S., Lee, J., Jo, K., Kim, C., Lee, C. & Yoon, J. 2019 Novel reuse strategy in flow-electrode capacitive deionization with switch cycle operation to enhance desalination performance. *Environmental Science & Technology Letters* **6** (12), 739–744.

Jiang, X. & School of Petrochemical Engineering L. U. O. T. 2020 Electrospinning preparation of Fe_3O_4/porous carbon nanofibres for use as supercapacitor electrode materials. *International Journal of Electrochemical Science* **15** (5), 4602–4618.

Jin, Z., Hu, X., Sun, T. & Zhang, W. 2019 Capacitive deionization using a pulsed power supply for water treatment. *Desalination and Water Treatment* **160**, 23–31.

Kavitha, J., Rajalakshmi, M., Phani, A. R. & Padaki, M. 2019 Pretreatment processes for seawater reverse osmosis desalination systems – a review. *Journal of Water Process Engineering* **32**, 100926.

Khalil, K. A., Barakat, N. A. M., Motlak, M. & Al-Mubaddel, F. S. 2020 A novel graphene oxide-based ceramic composite as an efficient electrode for capacitive deionization. *Scientific Reports* **10** (1), 9676.

Khan, Z. U., Yan, T., Shi, L. & Zhang, D. 2018 Improved capacitive deionization by using 3D intercalated graphene sheet–sphere nanocomposite architectures. *Environmental Science Nano* **5** (4), 980–991.

Kim, Y. H., Tang, K., Chang, J., Sharma, K., Yiacoumi, S., Mayes, R. T., Bilheux, H. Z., Santodonato, L. J. & Tsouris, C. 2019 Potential limits of capacitive deionization and membrane capacitive deionization for water electrolysis. *Separation Science and Technology* **54** (13), 2112–2125.

Landon, J., Gao, X., Omosebi, A. & Liu, K. 2019 Progress and outlook for capacitive deionization technology. *Current Opinion in Chemical Engineering* **25**, 1–8.

Liu, Y., Xu, X., Wang, M., Lu, T., Sun, Z. & Pan, L. 2015 Metal-organic framework-derived porous carbon polyhedra for highly efficient capacitive deionization. *Chemical Communications* **51** (60), 12020–12023.

Liu, Y., Ma, J., Lu, T. & Pan, L. 2016 Electrospun carbon nanofibers reinforced 3D porous carbon polyhedra network derived from metal-organic frameworks for capacitive deionization. *Scientific Reports* **6**, 32784.

Lyu, S., Chen, W., Zhang, W., Fan, Y. & Jiao, W. 2016 Wastewater reclamation and reuse in China: opportunities and challenges. *Journal of Environmental Science (China)* **39**, 86–96.

Maheshwari, K. & Agrawal, M. 2020 Advances in capacitive deionization as an effective technique for reverse osmosis reject stream treatment. *Journal of Environmental Chemical Engineering* **8** (6), 104413.

Marchetti, P., Peeva, L. & Livingston, A. 2017 The selectivity challenge in organic solvent nanofiltration: membrane and process solutions. *Annual Review of Chemical and Biomolecular Engineering* **8**, 473–497.

Metke, T., Westover, A. S., Carter, R., Oakes, L., Douglas, A. & Pint, C. L. 2016 Particulate-free porous silicon networks for efficient capacitive deionization water desalination. *Scientific Reports* **6**, 24680.

Mo, H., Tang, Y., Chen, Y. & Wan, P. 2017 Technical research in continuous desalting by coupling flow capacitive deionization and electrodialysis. *Modem Chemical Industry* **39** (5), 91–95. (in Chinese).

Naushad, M., Mittal, A., Rathore, M. & Gupta, V. 2015 Ion-exchange kinetic studies for Cd(II), Co(II), Cu(II), and Pb(II) metal ions over a composite cation exchanger. *Desalination and Water Treatment* **54** (10), 2883–2890.

Oatley-Radcliffe, D. L., Walters, M., Ainscough, T. J., Williams, P. M., Mohammad, A. W. & Hilal, N. 2017 Nanofiltration membranes and processes: a review of research trends over the past decade. *Journal of Water Process Engineering* **19**, 164–171.

Park, G., Hong, S. P., Lee, C., Lee, J. & Yoon, J. 2021 Selective fluoride removal in capacitive deionization by reduced graphene oxide/hydroxyapatite composite electrode. *Journal of Colloid and Interface Science* **581** (A), 396–402.

Sami, S. K., Seo, J. Y., Hyeon, S., Shershah, M. S. A., Yoo, P. & Chung, C. 2018 Enhanced capacitive deionization performance by an rGO–SnO2 nanocomposite modified carbon felt electrode. *RSC Advances* **8** (8), 4182–4190.

Shi, W., Li, H., Cao, X., Leong, Z. Y., Zhang, J., Chen, T., Zhang, H. & Yang, H. Y. 2016 Ultrahigh performance of novel capacitive deionization electrodes based on a three-dimensional graphene architecture with nanopores. *Scientific Reports* **6**, 18966.

Shi, L., Xiao, L., Hu, Z. & Zhan, X. 2020 Nutrient recovery from animal manure using bipolar membrane electrodialysis: study on product purity and energy efficiency. *Water Cycle* **1**, 54–62.

Sistat, P., Huguet, P., Ruiz, B., Pourcelly, G., Mareev, S. A. & Nikonenko, V. V. 2015 Effect of pulsed electric field on electrodialysis of a NaCl solution in sub-limiting current regime. *Electrochimica Acta* **164**, 267–280.

Suss, M. E., Porada, S., Sun, X., Biesheuvel, P. M., Yoon, J. & Presser, V. 2015 Water desalination via capacitive deionization: what is it and what can we expect from it? *Energy & Environmental Science* **8** (8), 2296–2319.

Takeuchi, H. & Tanaka, H. 2020 Water reuse and recycling in Japan – history, current situation, and future perspectives. *Water Cycle* **1**, 1–12.

Tang, W., Liang, J., He, D., Gong, J., Tang, L., Liu, Z., Wang, D. & Zeng, G. 2019 Various cell architectures of capacitive deionization: recent advances and future trends. *Water Research* **150**, 225–251.

Wang, S., Wang, S., Wang, G., Che, X., Li, D., Li, C. & Qiu, J. 2020 Ion removal performance and enhanced cyclic stability of sno2/CNT composite electrode in hybrid capacitive deionization. *Materials Today Communications* **23**, 100904.

Wu, G., Yin, Q., Guo, F. & Hu, Z. 2020 Slow growers possess a high pollutant removal potential through granule formation for wastewater treatment. *Water Cycle* **1**, 63–69.

Xu, A., Wu, Y., Chen, Z., Wu, G., Wu, Q., Ling, F., Huang, W. E. & Hu, H. 2020 Towards the new era of wastewater treatment of China: development history, current status, and future directions. *Water Cycle* **1**, 80–87.

Yan, T., Xu, B., Zhang, J., Shi, L. & Zhang, D. 2018 Ion-selective asymmetric carbon electrodes for enhanced capacitive deionization. *RSC Advances* **8** (5), 2490–2497.

Yang, S., Kim, H., Jeon, S., Choi, J., Yeo, J., Park, H., Jin, J. & Kim, D. K. 2017 Analysis of the desalting performance of flow-electrode capacitive deionization under short-circuited closed cycle operation. *Desalination* **424**, 110–121.

Yasin, A. S., Mohamed, I. M. A., Mousa, H. M., Park, C. H. & Kim, C. S. 2018 Facile synthesis of TiO_2/ZrO_2 nanofibers/nitrogen co-doped activated carbon to enhance the desalination and bacterial inactivation via capacitive deionization. *Scientific Reports* **8** (1), 541.

Yu, F., Wang, C. & Ma, J. 2016 Applications of graphene-modified electrodes in microbial fuel cells. *Materials (Basel)* **9** (10), 807.

Yu, Z., Yu, Z., Zheng, D., Zheng, D., Zhang, K., Zhang, K., Yang, T., Yang, T., Chen, Y., Chen, Y., Li, X. & Li, X. 2017 Optimally catalyzed porous-silicon electrode of self-breathing micro fuel cells. *Microsystem Technologies* **23** (8), 3257–3262.

Zhang, C., He, D., Ma, J., Tang, W. & Waite, T. D. 2018 Faradaic reactions in capacitive deionization (CDI) – problems and possibilities: a review. *Water Research* **128**, 314–330.

Zhang, X., Zuo, K., Zhang, X., Zhang, C. & Liang, P. 2020 Selective ion separation by capacitive deionization (CDI) based technologies: a state-of-the-art review. *Environmental Science Water Research & Technology* **6** (2), 243–257.

Zhao, C., Zhang, L., Ge, R., Zhang, A., Zhang, C. & Chen, X. 2019a Treatment of low-level Cu(II) wastewater and regeneration through a novel capacitive deionization-electrodeionization (CDI-EDI) technology. *Chemosphere* **217**, 763–772.

Zhao, C., Ge, R., Zhen, Y., Wang, Y., Li, Z., Shi, Y. & Chen, X. 2019b A hybrid process of coprecipitation-induced crystallization-capacitive deionization-ion exchange process for heavy metals removal from hypersaline ternary precursor wastewater. *Chemical Engineering Journal (Lausanne, Switzerland: 1996)* **378**, 122136.

Zhao, X., Wei, H., Zhao, H., Wang, Y. & Tang, N. 2020 Electrode materials for capacitive deionization: a review. *Journal of Electroanalytical Chemistry* **873**, 114416.

Zuo, K., Kim, J., Jain, A., Wang, T., Verduzco, R. & Li, Q. 2018 Resin-modified capacitive deionization for selective uptake and removal of sulfate. *Abstracts of Papers of the American Chemical Society* **256**, 419.

First received 31 October 2020; accepted in revised form 18 January 2021. Available online 25 February 2021

 © 2021 The Authors

Aggravated biofouling caused by chlorine disinfection in a pilot-scale reverse osmosis treatment system of municipal wastewater

Li-Wei Luo, Yin-Hu Wu, Yun-Hong Wang, Xin Tong, Yuan Bai, Gen-Qiang Chen, Hao-Bin Wang, Nozomu Ikuno and Hong-Ying Hu

ABSTRACT

The reverse osmosis (RO) system is widely applied to produce reclaimed water for high-standard industrial use. Chlorine disinfection is the main biofouling control method in the RO systems for wastewater reclamation. However, researchers reported the adverse effects of chlorine disinfection which aggravated biofouling in laboratory-scale RO systems. In this study, four parallel 4-inch spiral wound RO membranes were used to study the effect of chlorine on biofouling in a pilot-scale RO system. The free chlorine dosages in four experimental groups were 0, 1, 2 and 5 mg/L, respectively. After continuous chlorination and dechlorination, the feed water entered the RO system. It was found that chlorine pretreatment caused a 1.9–36.7% increase in relative feed water pressure of the RO system, suggesting that chlorine aggravated the membrane fouling in the pilot-scale RO system. The microbial community structures of living bacteria in the feed water of the RO system were determined by the PMA (propidium monoazide)-PCR method and showed that the relative abundance of chlorine-resistant bacteria (CRB) was significantly increased after disinfection. Nine major genera which maintained higher relative abundance in experimental groups with high chlorine dosage were considered as possible key species causing membrane fouling, including *Pedobacter*, *Clostridium* and *Bradyrhizobium*.

Key words | biofouling, chlorine disinfection, chlorine-resistant bacteria (CRB), reverse osmosis, wastewater reclamation

Li-Wei Luo
Yin-Hu Wu (corresponding author)
Yun-Hong Wang
Xin Tong
Yuan Bai
Gen-Qiang Chen
Hao-Bin Wang
Hong-Ying Hu
Environmental Simulation and Pollution Control State Key Joint Laboratory, State Environmental Protection Key Laboratory of Microorganism Application and Risk Control (SMARC), School of Environment,
Tsinghua University,
Beijing 100084,
China
and
Beijing Laboratory for Environmental Frontier Technologies,
Beijing 100084,
China
E-mail: *wuyinhu@mail.tsinghua.edu.cn*

Nozomu Ikuno
Kurita Water Industries Ltd,
Nakano-ku,
Tokyo 164-0001,
Japan

Hong-Ying Hu
Shenzhen Environmental Science and New Energy Technology Engineering Laboratory,
Tsinghua-Berkeley Shenzhen Institute,
Shenzhen 518055,
China

HIGHLIGHTS

- Revealed the effects of chlorine disinfection on fouling of a pilot-scale RO system.
- Chlorine disinfection aggravated the membrane fouling in the pilot RO system.
- Analyzed community structure of residual bacteria after disinfection in feed water.
- Identified key chlorine-resistant bacteria (CRB) causing RO membrane fouling.

This is an Open Access article distributed under the terms of the Creative Commons Attribution Licence (CC BY-NC-ND 4.0), which permits copying and redistribution for non-commercial purposes with no derivatives, provided the original work is properly cited (http://creativecommons.org/licenses/by-nc-nd/4.0/).

doi: 10.2166/wrd.2021.108

GRAPHICAL ABSTRACT

Pilot-scale RO system

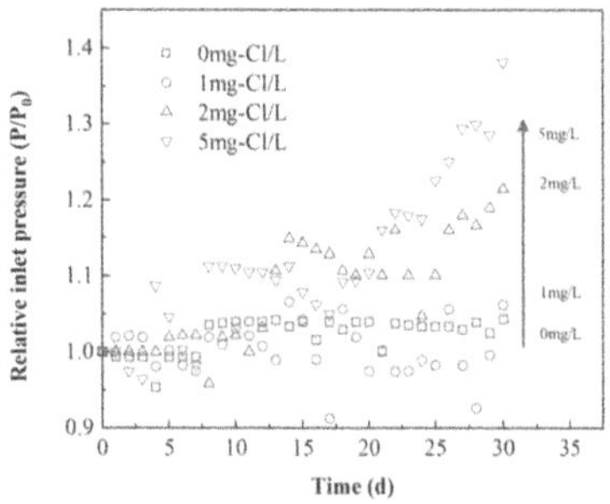

INTRODUCTION

Wastewater reclamation is an attractive approach to addressing water scarcity. The wastewater reclamation reverse osmosis (RO) system was widely applied to produce reclaimed water for municipal or industrial use in water-deficient and economically developed areas, such as Singapore and Beijing, the capital of China (Khan *et al.* 2015; Yen *et al.* 2017; Gu *et al.* 2019; Zhu & Dou 2019; Bai *et al.* 2020).

However, membrane fouling frequently occurred in the wastewater reclamation RO system, which leads to membrane pressure rise and energy consumption increase (Khedr 1998; Khan *et al.* 2015). The attachment and growth of microorganisms on the RO membrane, which was also called biofouling, was one of the main types of membrane fouling (Matin *et al.* 2011; Jiang *et al.* 2017).

Chlorine disinfection as pretreatment and adding biocides were the main strategies to control biofouling (Nguyen *et al.* 2012; Tan *et al.* 2017; Manalo *et al.* 2019). In recent years, some studies have reported the adverse effects of chlorine disinfection, which aggravated biofouling, on the seawater or municipal wastewater desalination RO systems (Khan *et al.* 2015; Wang *et al.* 2019a). The change of community structure and secretory products were considered to be the reasons for the change of RO biofouling. Researchers found that biofouling was more related to extracellular products (EPS) rather than bacterial cells and the fouling potential of EPS from different bacterial strains differed significantly (Ridgway *et al.* 1985; Gomez-Suarez *et al.* 2002; Chen & Ma 2004; Wang *et al.* 2019b). Wang *et al.* (2019a) reported that the abundance of chlorine-resistant bacteria (CRB) and EPS amount on the RO membrane surface significantly increased after chlorine disinfection, and it explained the adverse effects of chlorine disinfection.

However, the adverse effects of chlorine disinfection have only been reported in the laboratory-scale wastewater reclamation RO system with flat RO membranes (Wang *et al.* 2019a), and they have not been verified by pilot-scale experiments with spiral wound RO membranes. This study only analyzed the changes in the bacterial community structure in fouling layers after chlorine pretreatment but did not analyze the influence of chlorine on the community structure in feed water. The changes in community structures in feed water were the direct reason for the changes in community structures in fouling layers.

Therefore, the objectives of this study are to verify the adverse effects of chlorine disinfection to biofouling in the pilot-scale wastewater reclamation spiral wound RO system and reveal the influence of chlorine on the feed water community structure of the RO system.

MATERIALS AND METHODS

The pilot-scale RO system

Water source

Secondary effluent from a wastewater treatment plant in northern China is used as the water source of the pilot-scale

RO system. The treatment process of the wastewater treatment plant includes A^2O, a high-efficiency sedimentation tank and a deep-bed filter.

Process flow

The process flow of the pilot-scale RO system is shown in Figure 1. This system consisted of four parallel experimental groups, and each group consisted of three main parts: pretreatment tank, RO feed tank and RO membrane. For example, A1 pretreatment tank, B1 feed tank and B1 RO membrane constituted an experimental group. In addition, the pretreatment tanks and RO feed tanks were equipped with dosing tanks, respectively. In this experiment, four pretreatment water tanks used the same feed water, the water source mentioned in the section 'Water source' (the secondary effluent from a wastewater treatment plant). Meanwhile, four new 4-inch RO membranes (LP100, Vontron) were installed in the places of B1–B4 RO membranes. The cartridge filters were installed between the pretreatment tanks and RO feed tanks. The filter element size of these cartridge filters was 40-inch, the filter element material was polypropylene fiber, and the filtration accuracy was 1 μm.

Operation conditions and modes of the pilot-scale RO system are also shown in Figure 1. The four experimental groups adopted the same operation mode, but only the dosage of free chlorine (NaClO) and reductants ($NaHSO_3$) was different. Sodium hypochlorite was added to the pretreatment water tanks for disinfection. The volumes of the pretreatment tanks were 1 m^3, the feed water rates were 1 m^3/h, and the hydraulic retention time was 1 h. Excessive reductants $NaHSO_3$ were added into RO feed tanks for dechlorination. The mixed solution of NaClO and excess $NaHSO_3$ were used to adjust the conductivity and ions concentration of each experimental group equal. It can maintain a consistent amount of NaClO and $NaHSO_3$ dose to each of the feed water and exclude the effects of conductivity and ion concentration. The feed water rates of RO feed tanks were 1 m^3/h, and the hydraulic retention time was 1 h. The flow rates of permeate were kept constant at about 0.15 m^3/h by adjusting the feed water flow rates and pressures. The initial feed water flow rates of the RO system were set to 0.8 m^3/h and continuously increased with the aggravation of membrane fouling. The feed water flow rates were adjusted every

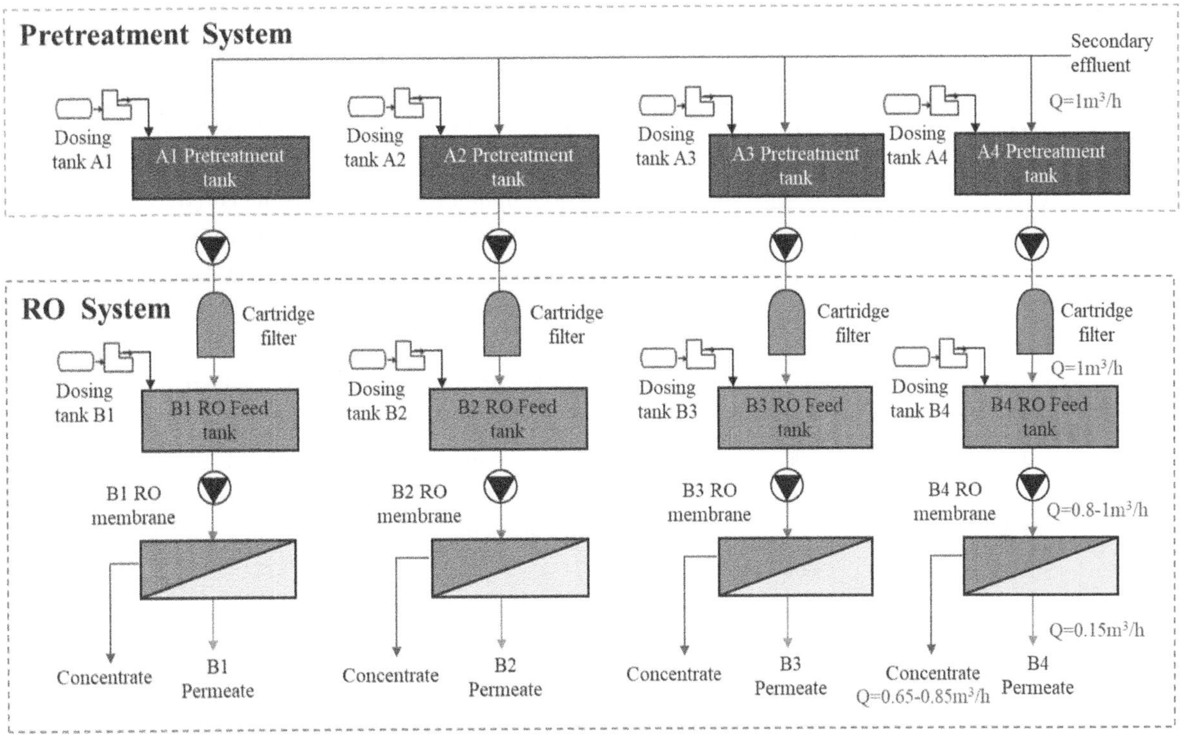

Figure 1 | Process flowchart of the pilot-scale wastewater reclamation RO system.

12 h. All concentrates of RO were discharged without backflow.

Dosing mode

The dosing mode of the pretreatment system is shown in Table 1. The sodium hypochlorite solution was continuously added to the pretreatment tanks from the dosing tanks. The free chlorine concentrations of A1–A4 pretreatment tanks were 0, 1, 2 and 5 mg/L, respectively.

The dosing mode of the RO system is shown in Table 2. The mixing solution of excess $NaHSO_3$ and sodium hypochlorite was continuously added to the RO feed tanks. Sodium hypochlorite was replenished to the B1–B3 feed tanks to make the total amount of sodium hypochlorite in each experimental group equal. The concentration of ions in each group could be comparable by adding sodium hypochlorite. Excess $NaHSO_3$ completely consumed free chlorine in the RO feed tanks to ensure that free chlorine did not damage the RO membranes.

Table 1 | Dosing mode of pretreatment tank

Dosing tank	**A1**	**A2**	**A3**	**A4**
Volume (L)	100	100	100	100
Flow rate (L/d)	100	100	100	100
Free chlorine (g/d)	0	24	48	120
Pretreatment tank	**A1**	**A2**	**A3**	**A4**
Free chlorine (mg/L)	0	1	2	5

Sodium hypochlorite solution was continuously pumped from the A1–A4 dosing tanks to pretreatment tanks.

Table 2 | Dosing mode of RO feed tanks

Dosing tank	**B1**	**B2**	**B3**	**B4**
Volume (L)	100	100	100	100
Flow rate (L/d)	100	100	100	100
Free chlorine (g/d)	120	48	24	0
$NaHSO_3$ (g/d)	600	600	600	600
Feed tank	**B1**	**B2**	**B3**	**B4**
Free chlorine (mg/L)	0	0	0	0

Excessive sodium bisulfite and sodium hypochlorite solution were continuously pumped from the B1–B4 dosing tanks to RO feed tanks to realize dechlorination and salt concentration supplement.

Operational data monitoring

Eleven kinds of operating parameters of the pilot-scale RO system were monitored online and uploaded to the computer in real-time. These operating parameters include: (1) flow rate, pressure, temperature, pH and conductivity of feed water; (2) flow rate, pressure, pH and conductivity of permeates and (3) flow rate, pressure, pH and conductivity of concentrates.

Water quality analysis

TOC (Total Organic Carbon) of secondary effluent (feed water of pretreatment tanks) was measured with a Sievers 5310C analyzer (GE, USA). The concentrations of inorganic elements in the secondary effluent water sample were measured with ICP-AES (Varian, USA) and ICP-MS (Perkin-Elmer, USA). An HPC (heterogeneous plate counties) method (APHA 2012) was used to determine the concentrations of bacteria in water samples of B1–B4 feed tanks.

Water quality parameters of secondary effluent (feed water of pretreatment tanks) are shown in Table 3. The concentration of inorganic ions in secondary effluent (feed water of pretreatment tanks) is shown in Table 4. The concentration of TOC in the water sample was 7.1 ± 0.7 mg/L. The number of bacteria represented by the HPC was $3.2 \pm 0.4 \times 10^4$ CFU/ml. The pH of the water sample was 7.20 ± 0.23. The turbidity of the water sample was lower than the detection limit of the turbidity meter (WZS-186, China). Studies have shown that TOC, pH, salt concentration and other water quality parameters will have an important impact on membrane fouling potential (Cai *et al.* 2019).

Microbial community structure analysis

The community structures of living bacteria of the water samples in B1–B4 feed tanks were determined by the PMA-PCR method. PMA (propidium monoazide) staining

Table 3 | Water quality of secondary effluent (feed water of pretreatment tanks)

TOC (mg/L)	**HPC (CFU/ml)**	**Conductivity (μs/cm)**	**pH**	**Turbidity (-)**
7.1 ± 0.7	$3.2 \pm 0.4 \times 10^4$	$1{,}533.5 \pm 114.4$	7.20 ± 0.23	N.D. < 0.50

Table 4 | Concentration of inorganic ions in secondary effluent (feed water of pretreatment tanks)

TN mg/L	NH_3-N mg/L	TP mg/L	Cl^- mg/L	SO_4^{2-} mg/L	PO_4^{3-} mg/L
6.09 ± 0.33	0.11 ± 0.01	0.17 ± 0.02	313 ± 15	211 ± 11	0.19 ± 0.02
Fe mg/L	**Ca mg/L**	**Na mg/L**	**Mg mg/L**	**K mg/L**	**Zn mg/L**
15.9 ± 0.3	64.1 ± 5.3	266 ± 12	37.0 ± 1.0	20.2 ± 1.7	48.0 ± 4.2

can prevent the amplification of DNA of dead cells in PCR and avoid the interference of dead bacteria on the determination of community structures (Gensberger *et al.* 2014; Li *et al.* 2014).

The experimental method was based on the research of Pang *et al.* (2016) and Gensberger *et al.* (2014). The experimental steps are as follows. (1) Bacteria in 300 ml of the water samples were enriched on the clean 0.1 μm nylon membrane filter (47 mm, Whatman). (2) The ultrafiltration membrane was cut into pieces and placed in a 15 ml tube. Then, 3 ml of sterile phosphate buffer saline (PBS) solution was added to the tube. The tube was shaken for 60 s. The supernatant was separated into another tube. (3) Step (2) was repeated five times to obtain 15 ml of solution with enriched bacteria. (4) The solution with enriched bacteria dewatered by centrifugation (Beckman, USA) at 10,000 rpm for 20 min and that supernatant was discarded. The precipitate (cells) was suspended again with 1 ml of sterile PBS solution. (5) 5 μL of PMAxx dye (Biotium Inc., USA) was added to the solution with cells. After incubation at room temperature for 5 min in the dark, the water sample was irradiated with a PMA-Lite LED photolyzer (Biotium Inc., USA) for 15 min to fully cross-link PAMxx with DNA. (6) DNA was extracted and amplified by PCR. An Illumina high-throughput sequencing method was used to analyze amplified RNA.

RESULTS AND DISCUSSION

The operational conditions of the pilot-scale RO system under different chlorine dosages

The reduction of HPC represented the bacterial inactivation rate of chlorine on water samples in B1–B4 feed tanks, and the results are shown in Figure 2. The log inactivation rate in experimental groups with 1, 2 and 5 mg Cl_2/L were 0.69 ± 0.05, 1.09 ± 0.07, 2.28 ± 0.06 log. Chlorine reduced the concentrations of bacteria in feed water of the RO system significantly.

The operational characteristics of the pilot-scale RO system, namely feed water flow rate, permeate flow rate, feed water pH and salt rejection, are monitored every 12 h and is shown in Figure 3. During the 30-day operation of the pilot-scale RO system, the final feed water flow rates of these experimental groups with 0, 1, 2 and 5 mg Cl_2/L were 0.86, 0.86, 0.96 and 1.09 m^3/h, respectively, as shown in Figure 3(a). Under the low concentration of chlorine (0, 1 mg/L), the feed water flow rates remained unchanged in 30 days. However, the feed water flow rates increased significantly with the high concentration of chlorine (2, 5 mg/L) as pretreatment. The feed water flow rates provided additional pressure to overcome the resistance of the fouling layers.

The permeate flow rate of each experimental group fluctuated around 0.15 m^3/h, as shown in Figure 3(b). The pH of secondary effluent (feed water of pretreatment tanks) was 7.18 ± 0.08 during the 30-day operation. Simultaneously, the salt rejection rate of each group always remained above 98%, as shown in Figure 3(d).

Effect of chlorine disinfection on the membrane fouling of the pilot-scale RO system

The feed water pressures (P) of the RO membrane were normalized based on the initial feed water pressures (P_0). The normalized pressures are called the relative pressure of feed water (P/P_0). It was used to reflect the degree of RO membrane fouling. Higher relative pressure indicated higher resistance of the fouling layer and more serious membrane fouling. In this study, a significant rise of relative feed

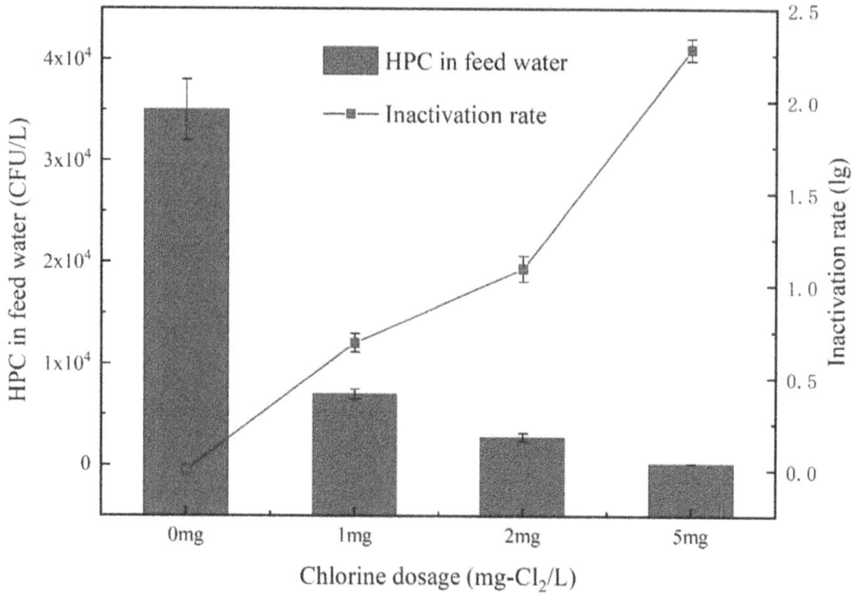

Figure 2 | Bacterial concentrations (CFU/L) and inactivation rates (log) of the water samples in B1–B4 feed tanks measured by the HPC method.

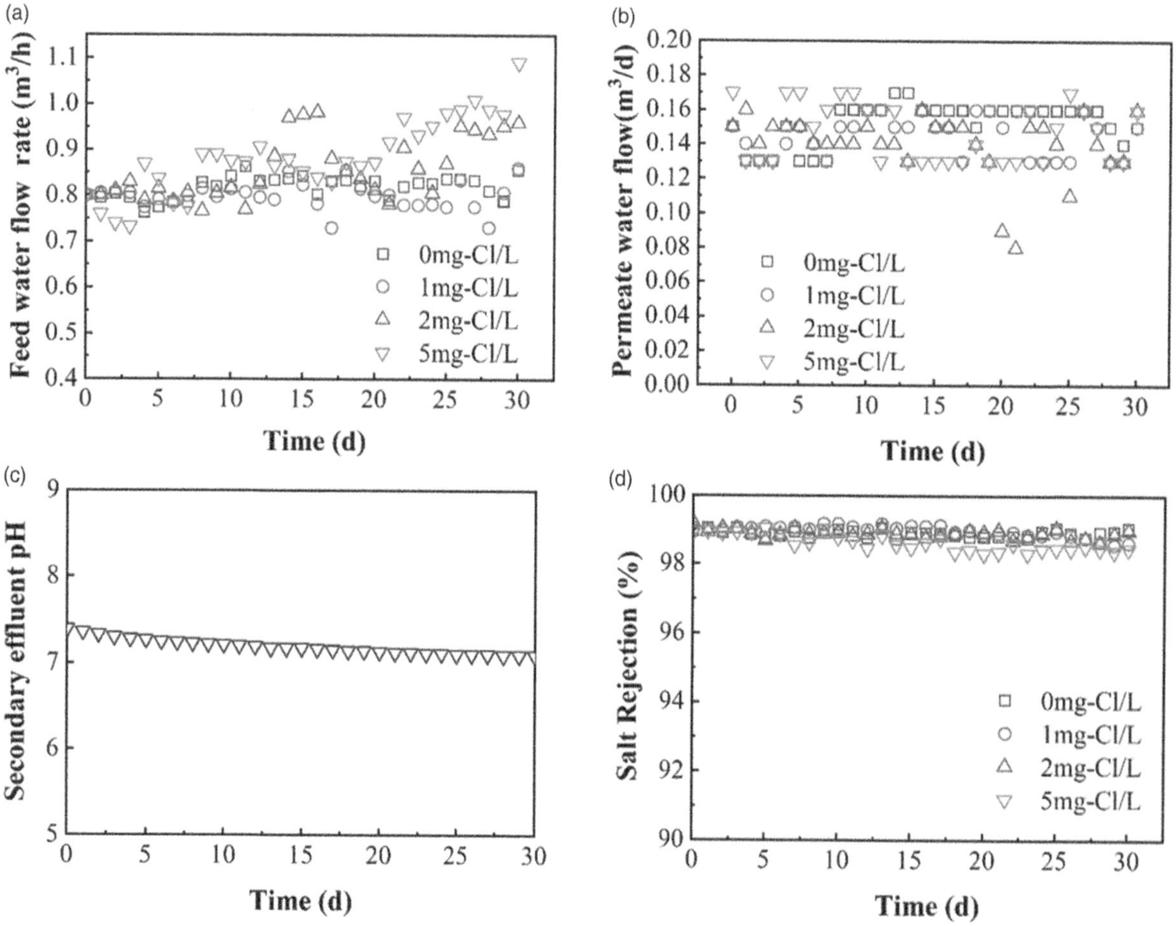

Figure 3 | Operational characteristics of the pilot-scale wastewater reclamation RO system. (a) Feed water flow rate, (b) permeate flow rate, (c) secondary effluent pH and (d) salt rejection.

water pressure after chlorine disinfection was observed. During the 30-day operation, the change of relative feed water pressure of each experimental group is shown in Figure 4. The initial feed water pressures (P_0) of the B1–B4 RO membrane were 867, 844, 856 and 872 kPa, respectively. The relative pressures of the water samples with 0, 1, 2 and 5 mg Cl_2/L as pretreatment finally reached 1.04, 1.06, 1.21 and 1.38, respectively.

Compared with the control group, the final relative feed water pressures of experimental groups with 1, 2 and 5 mg Cl_2/L increased by 1.9, 16.3 and 36.7%, respectively. Chlorine disinfection did not alleviate biofouling but aggravated it in this study.

These results were in accordance with the conclusion obtained by Wang *et al.* (2019a) in a laboratory-scale wastewater reclamation RO system. In addition, researchers also reported some cases, where biofouling could not be controlled by sodium hypochlorite in seawater desalination RO systems (Obaid & Ben Hamida 1998; Khan *et al.* 2015). Wang *et al.* (2019a) found that although chlorine disinfection reduced the number of bacteria in RO feed water, the number of bacteria on the membrane surface did not decrease, but the quality of organic matter on the membrane surface and the thickness of fouling layer increased instead. Since the abundance of CRB in the biofouling layer increased after disinfection, CRB was considered to be the key species leading to the aggravated fouling effect of chlorine (Wang *et al.* 2019a). Studies have

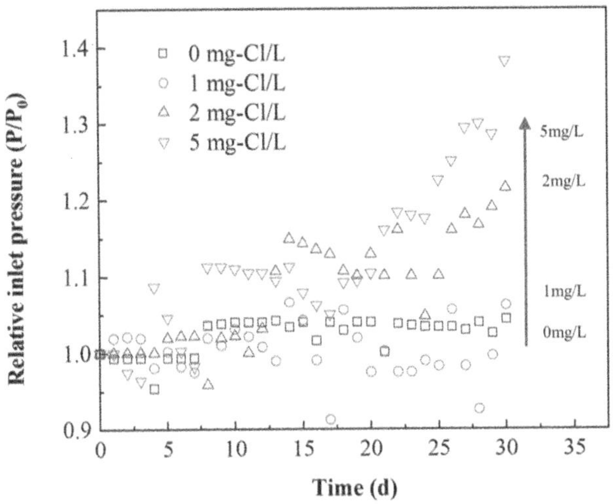

Figure 4 | Relative feed water pressure curves of the pilot-scale wastewater reclamation RO system under different chlorine dosages. The operation time lasted for 30 days.

proved that the EPS production of a strain was positively correlated with the chlorine resistance and the biofouling potential of RO membrane (Tsuneda *et al.* 2003; More *et al.* 2014; Wang *et al.* 2019b). Therefore, the remaining CRB after disinfection have stronger membrane fouling potential.

Researchers reported that chemical cleaning with NaClO causes increases in eDNA and quorum sensing molecules N-acyl homoserine lactones (AHLs) secreted by pure bacteria in the regrowth stage, which eventually lead to the aggravation of biofouling of ultrafiltration membrane (Wang *et al.* 2020). After disinfection, the increase of eDNA and AHLs secretion may also lead to the increase of fouling of the RO membrane.

The aggravated fouling phenomenon of disinfection showed that the limited inactivation of bacteria might not effectively alleviate biofouling of the RO membrane. Because the size of bacteria (0.5–5 μm) is much larger than the pore size (0.1–0.7 nm) of RO membrane, bacteria could not directly block the pores. The process of RO biofouling can be described in three stages: adhesion, *in situ* regrowth and biofilm diffusion (Matin *et al.* 2011; Liu *et al.* 2020). Biofilm diffusion means that the bacteria in the mature biofilm fall off and enter the liquid phase, and then, the bacteria adhere, grow and form a new biofilm in other places (Matin *et al.* 2011). The effects of chlorine on the three stages of biofouling formation were not completely clear. So far, there is no research about the effects of chlorine on the adhesion ability of a single bacterium from a microscopic perspective. Wang *et al.* (2019a) studied the effects of chlorine on the regrowth of disinfection residual bacteria and found that chlorine increased the competitive advantage of CRB on the RO membranes. In addition, chlorine might promote the regrowth of bacteria by affecting the water quality. Liu *et al.* (2002) found that chlorine disinfection could improve the level of assimilable organic carbon (AOC) of the water samples. The rise of AOC reduced biological stability and promoted bacterial regrowth.

Effect of chlorine disinfection on the community structure of the feed water

The microbial community structures of the water samples in B1–B4 feed tanks were analyzed by 16S rRNA gene sequencing, and the results are shown in Figure 5.

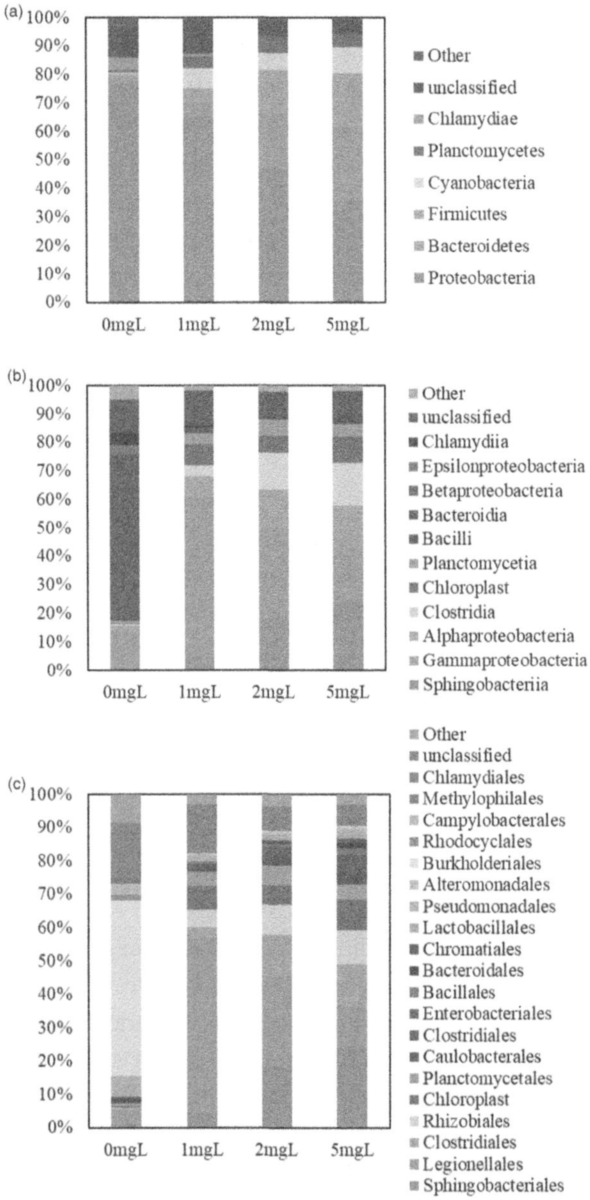

Figure 5 | Effect of chlorine disinfection on the community structure of the water samples in B1–B4 feed tanks at the (a) phylum level, (b) class level and (c) family level. 'Others' in the legend included bacteria whose abundances were lower than 1%.

On the phylum level, with the increase in chlorine concentration, the relative abundance of Bacteroidetes and Firmicutes increased and that of Proteobacteria decreased obviously. In the water samples with 0 and 5 mg Cl_2/L as pretreatment, the relative abundance of Bacteroidetes was 2.42 and 25.8%, that of Firmicutes was 0.77 and 18.9% and that of Proteobacteria was 76.9 and 35.6%, respectively.

On the class level, Betaproteobacteria, the dominant bacteria in the control group (with 0 mg Cl_2/L), almost completely inactivated after disinfection. Gammaproteobacteria, Sphingobacterium, Alphaproteobacteria and Clostridia became the dominant bacteria in the microbial community after disinfection. With the increase in chlorine concentration, the relative abundance of Gammaproteobacteria increased firstly and then decreased, while that of Sphingobacterium, Alphaproteobacteria and Clostridia increased continuously. The relative abundance of Gammaproteobacteria, Sphingobacterium, Alphaproteobacteria and Clostridia was 19.1, 23.9, 14.5 and 15.2%, respectively, at the experimental group with 5 mg Cl_2/L.

The community structure on the order level and the class level were compared. Betaproteobacteria mainly include Burkholderiales. Gammaproteobacteria mainly include Legionellales. Sphingobacteria mainly include Sphingobacteriales. Clostridia mainly include Clostridiales.

The differences of the water samples on the genus level were shown with the heat map (Figure 6). After chlorine disinfection, the relative abundance of some bacteria increased, showing different levels of chlorine resistance between different genera. The relative abundance of *Acidovorax*, *Acinetobacter*, *Arcobacter* and *Undibacterium* in the control group (0 mg Cl_2/L) was 47.5, 4.83, 2.15 and 2.12%, respectively. They were dominant bacteria in the control group. After disinfection, their relative abundance decreased significantly to less than 1%.

In this study, the aggravated fouling of chlorine disinfection only occurred when the chlorine concentrations in the pretreatment stage was relatively high (2, 5 mg Cl_2/L). Therefore, the dominant species in these two experimental groups with 2, 5 mg Cl_2/L might be more closely related to the aggravated fouling. On the genus level, nine major genera maintained higher relative abundance in experimental groups with 2, 5 mg Cl_2/L than the groups with 0, 1 mg Cl_2/L. The relative abundances of these major genera are shown in Figure 7. Among them, *Clostridium*, *Citrobacter* and *Bacillus* were reported as CRB in many other researches (Khan *et al.* 2016; Owoseni & Okoh 2017; Roy & Ghosh 2017; Luo *et al.* 2020). *Legionella* was also CRB and showed higher abundance in all three disinfection groups (with 1, 2 and 5 mg Cl_2/L)

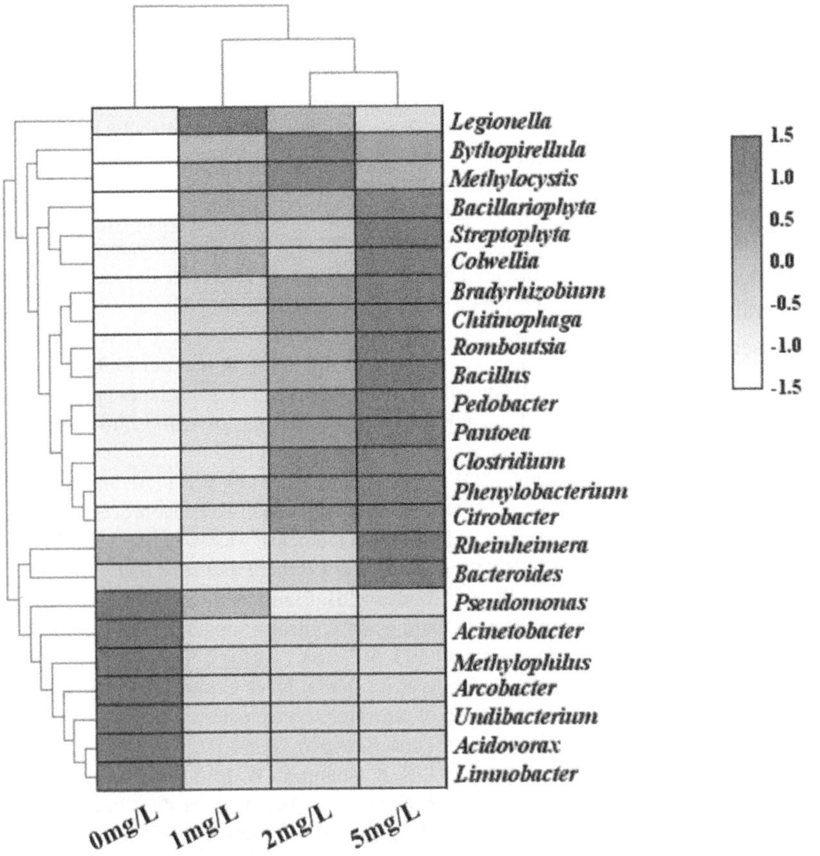

Figure 6 | Heat maps of microbial communities of the water samples in B1–B4 feed tanks on the genus level. Genera whose abundances were higher than 1% were shown and counted.

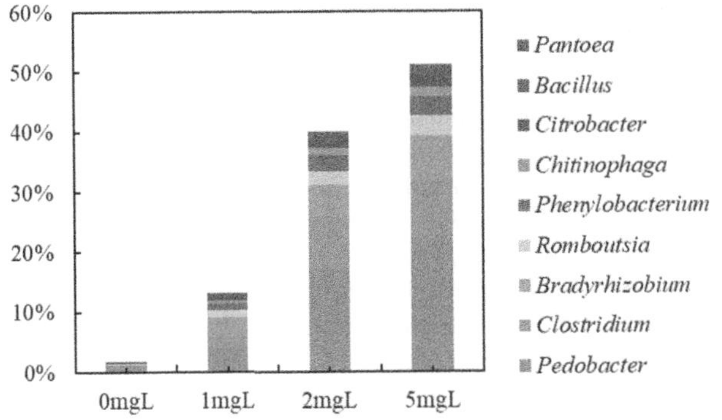

Figure 7 | Nine species with higher abundance under the high concentration of chlorine (2, 5 mg/L) than the low concentration of chlorine (0, 1 mg/L).

than the control group (with 0 mg Cl_2/L). Since *Legionella* was pathogenic bacteria, many studies focused on its chlorine resistance (Kuchta *et al.* 1983; Miyamoto *et al.* 2000). However, the relative abundance of *Legionella* in the experimental group with 1 mg Cl_2/L (53.2%) was higher than that with 2 and 5 mg Cl_2/L chlorine (27.9 and 13.0%). This trend suggested that *Legionella* might not be the cause of the aggravated biofouling after disinfection. These nine bacteria with abundance advantages in the experimental groups with 2 and 5 mg Cl_2/L were more likely to be the key species causing the increased fouling potential of RO.

It is noted that the abundances change of predominant CRB were not linear with increasing chlorine concentration. As mentioned previously, the low concentration of chlorine increased the relative abundance of *Legionella*, but the high concentration decreased it. In addition, some strong CRB did not dominate in the low concentration of chlorine. Therefore, the change of fouling potential caused by the change of the community structure was not linear, either. This might explain the sudden and significant increase in fouling potential at the experimental groups with 2, 5 mg Cl_2/L.

CONCLUSION

The pilot-scale RO system verified the conclusion of the laboratory-scale trial that chlorine disinfection in the wastewater reclamation RO system could aggravate RO membrane fouling. Chlorine pretreatment caused a 1.9, 16.3–36.7% increase in the relative feed water pressure of the RO system in this study. Analysis of the bacterial community structure in the feed water showed that the relative abundance of CRB was significantly increased after disinfection. The changing trend of dominant species with the increase in chlorine concentration was analyzed, and nine strains were considered as possible key species causing membrane fouling, including *Pedobacter*, *Clostridium*, *Bradyrhizobium* and *Romboutsia*.

ACKNOWLEDGEMENTS

This study was supported by the National Natural Science Foundation of China (No. 52000114) and the Key Program of the National Natural Science Foundation of China (No. 51738005).

DATA AVAILABILITY STATEMENT

All relevant data are included in the paper or its Supplementary Information.

REFERENCES

American Public Health Association 2012 *Standard Methods for the Examination of Water and Wastewater*, 22nd edn. American Public Health Association, Washington, DC, USA.

Bai, Y., Shan, F., Zhu, Y.-y., Xu, J.-y., Wu, Y.-s., Luo, X.-g., Wu, Y.-h., Hu, H.-Y. & Zhang, B.-l. 2020 Long-term performance and economic evaluation of full-scale MF and RO process – A case study of the changi NEWater Project Phase 2 in Singapore. *Water Cycle* **1**, 128–135.

Cai, B. J., Baudin, I. & Ng, H. Y. 2019 A modified fouling index (MFI40) and fouling predicting approach for ultrafiltration of secondary effluents. *Journal of Water Reuse and Desalination* **9** (1), 67–82.

Chen, V. J. & Ma, P. X. 2004 Nano-fibrous poly(L-lactic acid) scaffolds with interconnected spherical macropores. *Biomaterials* **25** (11), 2065–2073.

Gensberger, E. T., Polt, M., Konrad-Koeszler, M., Kinner, P., Sessitsch, A. & Kostic, T. 2014 Evaluation of quantitative PCR combined with PMA treatment for molecular assessment of microbial. *Water Research* **67**, 367–376.

Gomez-Suarez, C., Pasma, J., van der Borden, A. J., Wingender, J., Flemming, H. C., Busscher, H. J. & van der Mei, H. C. 2002 Influence of extracellular polymeric substances on deposition and redeposition of *Pseudomonas aeruginosa* to surfaces. *Microbiology-SGM* **148**, 1161–1169.

Gu, J., Liu, H., Wang, S., Zhang, M. & Liu, Y. 2019 An innovative anaerobic MBR-reverse osmosis-ion exchange process for energy-efficient reclamation of municipal wastewater to NEWater-like product water. *Journal of Cleaner Production* **230**, 1287–1293.

Jiang, S., Li, Y. & Ladewig, B. P. 2017 A review of reverse osmosis membrane fouling and control strategies. *Science of the Total Environment* **595**, 567–583.

Khan, M. T., Hong, P.-Y., Nada, N. & Croue, J. P. 2015 Does chlorination of seawater reverse osmosis membranes control biofouling? *Water Research* **78**, 84–97.

Khan, S., Beattie, T. K. & Knapp, C. W. 2016 Relationship between antibiotic- and disinfectant-resistance profiles in bacteria harvested from tap water. *Chemosphere* **152**, 132–141.

Khedr, M. G. 1998 A case study of RO plant failure due to membrane fouling, analysis and diagnosis. *Desalination* **120** (1–2), 107–113.

Kuchta, J. M., States, S. J., McNamara, A. M., Wadowsky, R. M. & Yee, R. B. 1983 Susceptibility of legionella-pneumophila to chlorine in tap water. *Applied and Environmental Microbiology* **46** (5), 1134–1139.

Li, D., Tong, T., Zeng, S., Lin, Y., Wu, S. & He, M. 2014 Quantification of viable bacteria in wastewater treatment plants by using propidium monoazide combined with quantitative PCR (PMA-qPCR). *Journal of Environmental Sciences* **26** (2), 299–306.

Liu, W., Wu, H., Wang, Z., Ong, S. L., Hu, J. Y. & Ng, W. J. 2002 Investigation of assimilable organic carbon (AOC) and bacterial regrowth in drinking water distribution system. *Water Research* **36** (4), 891–898.

Liu, Y.-N., Lv, Z.-T., Lv, W.-L. & Liu, X.-W. 2020 Plasmonic probing of the adhesion strength of single microbial cells. *Proceedings of the National Academy of Sciences of the United States of America* **117** (44), 27148–27153.

Luo, L.-W., Wu, Y.-H., Yu, T., Wang, Y.-H., Chen, G.-Q., Tong, X., Bai, Y., Xu, C., Wang, H.-B., Ikuno, N. & Hu, H.-Y. 2020 Evaluating method and potential risks of chlorine-resistant bacteria (CRB): a review. *Water Research* **188**, 116474.

Manalo, C. V., Ohno, M., Nishimoto, S., Okuda, T., Nakai, S. & Nishijima, W. 2019 Long-term pilot plant study using direct chlorination for biofouling control of a chlorine-resistant polyamide reverse osmosis membrane. *Desalination and Water Treatment* **138**, 57–67.

Matin, A., Khan, Z., Zaidi, S. M. J. & Boyce, M. C. 2011 Biofouling in reverse osmosis membranes for seawater desalination: phenomena and prevention. *Desalination* **281**, 1–16.

Miyamoto, M., Yamaguchi, Y. & Sasatsu, M. 2000 Disinfectant effects of hot water, ultraviolet light, silver ions and chlorine on strains of Legionella and nontuberculous mycobacteria. *Microbios* **101** (398), 7–13.

More, T. T., Yadav, J. S. S., Yan, S., Tyagi, R. D. & Surampalli, R. Y. 2014 Extracellular polymeric substances of bacteria and their potential environmental applications. *Journal of Environmental Management* **144**, 1–25.

Nguyen, T., Roddick, F. A. & Fan, L. 2012 Biofouling of water treatment membranes: a review of the underlying causes, monitoring techniques and control measures. *Membranes* **2** (4), 804–840.

Obaid, M. & Ben Hamida, A. 1998 Practical solutions to problems experienced in open seawater RO plants operating on the Arabian Gulf. *Desalination* **120** (1–2), 137–142.

Owoseni, M. & Okoh, A. 2017 Assessment of chlorine tolerance profile of *Citrobacter* species recovered from wastewater treatment plants in Eastern Cape, South Africa. *Environmental Monitoring and Assessment* **189** (4), 201.

Pang, Y.-C., Xi, J.-Y., Xu, Y., Huo, Z.-Y. & Hu, H.-Y. 2016 Shifts of live bacterial community in secondary effluent by chlorine disinfection revealed by Miseq high-throughput sequencing combined with propidium monoazide treatment. *Applied Microbiology and Biotechnology* **100** (14), 6435–6446.

Ridgway, H. F., Rigby, M. G. & Argo, D. G. 1985 Bacterial adhesion and fouling of reverse-osmosis membranes. *Journal American Water Works Association* **77** (7), 97–106.

Roy, P. K. & Ghosh, M. 2017 Chlorine resistant bacteria isolated from drinking water treatment plants in West Bengal. *Desalination and Water Treatment* **79**, 103–107.

Tan, Y.-J., Sun, L.-J., Li, B.-T., Zhao, X.-H., Yu, T., Ikuno, N., Ishii, K. & Hu, H.-Y. 2017 Fouling characteristics and fouling control of reverse osmosis membranes for desalination of dyeing wastewater with high chemical oxygen demand. *Desalination* **419**, 1–7.

Tsuneda, S., Aikawa, H. & Hayashi, H. 2003 Extracellular polymeric substances responsible for bacterial adhesion onto solid surface. *FEMS Microbiology Letters* **223** (2), 287.

Wang, Y.-H., Wu, Y.-H., Tong, X., Yu, T., Peng, L., Bai, Y., Zhao, X.-H., Huo, Z.-Y., Ikuno, N. & Hu, H.-Y. 2019a Chlorine disinfection significantly aggravated the biofouling of reverse osmosis membrane used for municipal wastewater reclamation. *Water Research* **154**, 246–257.

Wang, Y.-H., Wu, Y.-H., Yu, T., Zhao, X.-H., Tong, X., Bai, Y., Huo, Z.-Y. & Hu, H.-Y. 2019b Effects of chlorine disinfection on the membrane fouling potential of bacterial strains isolated from fouled reverse osmosis membranes. *Science of the Total Environment* **693**, 133579.

Wang, X., Ma, J., Wu, Z. & Wang, Z. 2020 Stimulatory effects on bacteria induced by chemical cleaning cause severe biofouling of membranes. *Journal of Water Reuse and Desalination* **10** (1), 82–94.

Yen, F.-C., You, S.-J. & Chang, T.-C. 2017 Performance of electrodialysis reversal and reverse osmosis for reclaiming wastewater from high-tech industrial parks in Taiwan: a pilot-scale study. *Journal of Environmental Management* **187**, 393–400.

Zhu, Z. & Dou, J. 2019 Current status of reclaimed water in China: an overview. *Journal of Water Reuse and Desalination* **9** (3), 338–338.

First received 30 October 2020; accepted in revised form 4 February 2021. Available online 24 March 2021

 © 2021 The Authors

The removal efficiencies and mechanism of aniline degradation by peroxydisulfate activated with magnetic Fe-Mn oxides composite

Lin Qiao, Yu Shi, Qingli Cheng, Bingtao Liu and Jing Liu

ABSTRACT

The Fe-Mn oxides composite prepared by a chemical co-precipitation method was used as a heterogenous peroxydisulfate catalyst for the decomposition of aniline. This study investigated the mechanism of aniline degradation by PDS activated with catalyst. Reactive species resulting in the degradation of aniline was investigated via radical quenching experiments with different scavengers, including methanol, tert-butyl alcohol, EDTA and sodium azide. Based on the experiments made here, it is speculated that the predominant reactive species responsible for the degradation of aniline may be holes and singlet molecular oxygen rather than $SO_4^{\cdot -}$ and ·OH radicals. The degradation of compounds in catalyst/peroxydisulfate system was put forward. The three possible intermediates were speculated by high performance liquid chromatography-mass spectrometry, and two possible degradation pathways were proposed.

Key words | Fe-Mn oxides, intermediate, oxidation, peroxydisulfate, reaction mechanism

Lin Qiao
Yu Shi
Qingli Cheng
Bingtao Liu (corresponding author)
Jing Liu
School of Environment and Municipal Engineering, North China University of Water Resources and Electric Power,
Zhengzhou, Henan 450011,
China
E-mail: *liubingtao@ncwu.edu.cn*

Qingli Cheng
Bingtao Liu
Henan Key Laboratory of water Environment Simulation and Treatment,
Zhengzhou, Henan, 450011,
China

HIGHLIGHTS

- Magnetic Fe-Mn particles based on spinel-type ferrite $MnFe_2O_4$ were synthesized.
- Diagrammatic sketch of aniline degradation mechanism by peroxydisulfate in the presence of $MnFe_2O_4$ system was demonstrated.
- Three main intermediate products and two possible paths were studied.

INTRODUCTION

As one of the most widely used sewage treatment technologies, advanced oxidation processes (AOPs) are applied in the oxidation of non-biodegradable, recalcitrant and toxic organic pollutants on account of the generation of highly reactive species under ambient conditions in recent years (Hodges *et al.* 2018). In previous studies, traditional AOPs based on the generation of hydroxyl radical (·OH), e.g. Fenton reaction (Fe^{2+}/H_2O_2) (Verma & Haritash 2019), electron-Fenton (Yong *et al.* 2017), UV/H_2O_2 (Benito *et al.* 2017), O_3/H_2O_2 (Can & Cakir 2010), have received popular attention owing to producing the strong oxidant ·OH (oxidation potential is 1.8–2.7 V vs. normal hydrogen electrode (NHE)) in mild reaction conditions. However, several weaknesses, such as narrow and limited pHs (2–4) under optimal conditions and the chemical instability and high-cost of hydrogen peroxide, were gradually discovered in the continuous research of the AOPs, leading to tremendously placed restrictions on its practical application (Duran *et al.* 2018; Huang *et al.* 2020).

In recent years, AOPs based on the generation of sulfate radical ($SO_4^{\cdot -}$) have been the focus of numerous studies, and

This is an Open Access article distributed under the terms of the Creative Commons Attribution Licence (CC BY 4.0), which permits copying, adaptation and redistribution, provided the original work is properly cited (http://creativecommons.org/licenses/by/4.0/).

doi: 10.2166/wrd.2021.102

peroxymonosulfate (PMS, HSO_5^-) and peroxydisulfate (PDS, $S_2O_8^{2-}$) have been proposed as candidates for such a purpose (Wacławek *et al.* 2017). Compared with ·OH, sulfate radical (SO_4^-) not only carries forward the same advantages due to the higher reactive oxidation (2.5–3.1 V vs. NHE) (Zhu *et al.* 2019), but also solve the shortcomings resulting from its stability and wider working pH range (Hu & Long 2016). SO_4^- is generated by activating the PDS and PMS through various methods such as heat (Ji *et al.* 2015), UV light (Khan *et al.* 2013), alkaline (Lominchar *et al.* 2018), metal ions (Wang *et al.* 2019a), and metal oxide (Du *et al.* 2017; Zhang *et al.* 2018) though the cleavage of their O–O bonds. They are both very stable in the solid state, but PDS is a cheaper oxidant than PMS (considered as oxone salt ($2KHSO_5·KHSO_4·K_2SO_4$) and H_2O_2 and have a higher oxidation potential than PMS (Zhou *et al.* 2019). Consequently, PDS has shown to be a promising application prospect for the decomposition of organic pollutants.

Among the catalytic methods in AOPs based on PDS, metal ion and metal oxide catalysis are gaining greater attention and have wider application rather than sonocatalysis, photocatalyst is and thermocatalysis because of the lower energy consumption and low price. Metal oxide can be reused and regenerated in the catalytic process, which demonstrates that metal oxide catalysis can further reduce the total cost compared with other methods (Zhou *et al.* 2019). In addition, the separation between catalysts and liquid in the waste water treated by the method is easy to be implemented in subsequent processes. Therefore, it is highly significant to research metal oxide catalysis in AOPs based on PDS.

Spinel-type ferrite oxides, mainly including $CuFe_2O_4$, $MgFe_2O_4$, $NiFe_2O_4$, $CoFe_2O_4$ and $MnFe_2O_4$ etc., have attracted attention due to their chemical stability and photochemical properties (Junlabhut *et al.* 2018). Recently, spinel-type ferrite oxides were used as the catalytic agent in AOPs. Wang *et al.* (2019b) report that 67% of levofloxacin was degraded within 1 h with the reaction condition of 0.6 g/L magnetic nickel ferrite/carbon sphere composite and 1.8 g/L PDS, which is described as a potentially useful catalyst in AOPs. Pan *et al.* (2017) reported that UV-filter benzophenone-3 in aqueous solution is removed in 6 h under the condition of $[\text{benzophenone-3}]_0$: $[\text{PDS}]_0 = 1:1{,}000$, catalyst load = 500 mg L^{-1}. Generally, spinel-type ferrite oxides have gained more attention in catalytic oxidation.

Although $SO_4·^-$ is the main production of most persulfate processes, there are some investigations which indicate that partial persulfate activation processes involve a non-radical mechanism. Some investigations have found a non-radical mechanism for the degradation of organic contaminants. Zhang *et al.* (2014) reported that copper oxide (CuO) could efficiently activate PDS without producing $SO_4·^-$. Meanwhile, Lee *et al.* (2015) reported that the persulfates bind onto the surface of CNTs, forming reactive complexes that are immediately decomposed upon reaction with organic compounds. Wu *et al.* (2018) discovered that the reaction of non-radical mechanism was observed in the degradation of aniline by rice straw biochar RSBC/PDS system and the holes were responsible for the reaction process. However, the research about the mechanism of non-radicals has not been greatly reported in current reports.

Single oxygen ($^1O_2$), which is one of the typical nonradical reactive oxygen species, has been widely used in the degradation of organic pollutants, particularly in photocatalytic degradation processes (Lee *et al.* 2011). The $^1O_2$ in the process of degradation has strong selectivity to organic pollutants owing to its electrophilic nature (Cheng *et al.* 2017). Some studies reported that the activation process of PDS can produce $^1O_2$ in some systems. Ma *et al.* (2018) reported that PDS is activated by nitrogen-doped mesoporous carbon to degrade the phenol through $^1O_2$ and electron transfer. Shi *et al.* (2020) reported that $^1O_2$ was one of the reactive oxidative species involved in the degradation of sulfamethoxazole through activation of persulfate by EDTA-2 K-derived nitrogen-doped porous carbons.

In this study, we investigated the activation mechanism of PDS by $MnFe_2O_4$ for the degradation of aniline. Radical scavengers identified the main reactive species responsible for aniline decomposed. Additionally, degradation mechanisms were proposed based on experimental results from spectroscopy measurement.

MATERIALS AND METHODS

Reagents and materials

All chemicals used in this study were at least of analytical grade and used as received without further purification, except as noted. Aniline was provided by Sigma-Aldrich Chemical Co., Ltd (Shanghai, China). Potassium persulfate was provided by ANPEL Laboratory Technologies Reagent Inc. (Shanghai, China). N-(1-naphthyl) ethylenediamine dihydrochloride, iron (III) chloride hexahydrate ($FeCl_3$. $\cdot 6H_2O$) and manganese sulfate monohydrate ($FeSO_4 \cdot 6H_2O$) were supplied by Kermel Chemical Reagent (Tianjin, China). Methanol anhydrous (MeOH) and tert-Butanol (*t*-BuOH), serving the scavengers of radicals, were obtained from Macklin Biochemical Co., Ltd (Shanghai, China). Ethylenediaminetetraacetic acid disodium salt (EDTA-2Na) used as the hole scavenger and sodium azide (NaN_3) used as singlet molecular oxygen ($^1O_2$) scavenger, was bought from Guangfu Technology Development Co., Ltd (Chengdu, China). Deionized water was used throughout the experiments.

Synthesis of $MnFe_2O_4$

The $MnFe_2O_4$ catalyst was synthesized using the chemical coprecipitation method (Liu *et al.* 2020). First, 5.07 g $MnSO_4 \cdot H_2O$ and 16.5 g $FeCl_3 \cdot 6H_2O$ were dissolved into 300 mL deionized water (Mn^{2+}:Fe^{3+} = 1:2 at a molar ratio) with continuous mechanical stirring at 62°. Secondly, 8 M NaOH solution was added slowly to the mixture solution until the pH value of the solution reached approximately 11. Then, stirring was stopped and the mixture solution was maintained at 75° for 30 min. When cooling down to room temperature, the mixture solution was filtered to obtain the suspension which was then washed with deionized water and ethanol at least three times. Finally, the composite was calcined at 300° for 4 h in the muffle furnace and ground to powder for further use.

Characterization of $MnFe_2O_4$

The morphological representation and internal structure of $MnFe_2O_4$ was collected on a field emission scanning electron microscope (FE-SEM, JSM-6700F) equipped with energy-dispersive X-ray spectrometer and transmission electron microscopy (TEM, JEM-200CX, JEOL, Japan). Powder X-ray diffraction spectrum was recorded at 0.02°/s and operated at 40 kV and 15 mA current with Cu kα1 radiation (λ = 0.15406 nm) diffractometer (XRD, Dmax 2,500 V). The various functional groups of $MnFe_2O_4$ were detected by Fourier-transform infrared (FTIR, Nicolet380).

Degradation experiments

In a typical run, conical flasks containing 50 mL reaction solution were kept at a controlled temperature in an air bath shaker (HNY-2102C, Honour, China) to conduct the tests. The initial concentration of aniline was 20 g/L. After 30 min of adsorption equilibrium between aniline and $MnFe_2O_4$, the oxidation reaction was initiated by adding a desired amount of PDS. The pH in the solution was adjusted by H_2SO_4 or NaOH solution. After 0, 30, 60, 90, 120, 180 and 240 min, respectively, 1 mL of reaction solution was withdrawn and filtered through a 0.45 mm membrane into a liquid sampling bottle for further analysis.

Analytical methods

The concentration of aniline was analyzed with the colorimetric method using N-(1-naphthyl) ethylenediamine as a chromogenic agent. An N5000 UV-VIS spectrometer (Shanghai Youke, China) was used to measure the absorbance in a 1 cm cell at 545 nm. The intermediates were identified on a Waters W2489-QDa (USA) high performance liquid chromatography (HPLC) equipped with a Waters QDa mass spectrometer with a reversed-phase C18 column (4.7 × 250 mm). The mobile phase was also the mixture of acetonitrile-water (55:45, *V:V*) and the flow rate was set at 0.50 mL/min. The mass spectrometer was operated under electrospray positive ionization at a fragmentor voltage of 800 V with a mass scan range of 50–600 amu. The ionization source conditions were set at the drying gas flow at 600 °C. The metal valance state was identified by X-ray photoelectron spectroscopy (XPS, Al Kα, AXIS Supra spectrometer, Shimadzu, Japan).

RESULTS AND DISCUSSION

Characterization of catalysts

The XRD patterns were used to analyze the information of material composition, structure or morphology of atoms or molecules in materials. X-ray diffraction analysis is the main method to research the phase and crystal structure of materials. Thus, the X-ray diffraction of catalyst is shown in Figure 1 to determine the composition and crystallinity.

In the range of 2θ from 20 to 80°, it was noticed that there were several significant peaks in the X-ray diffraction

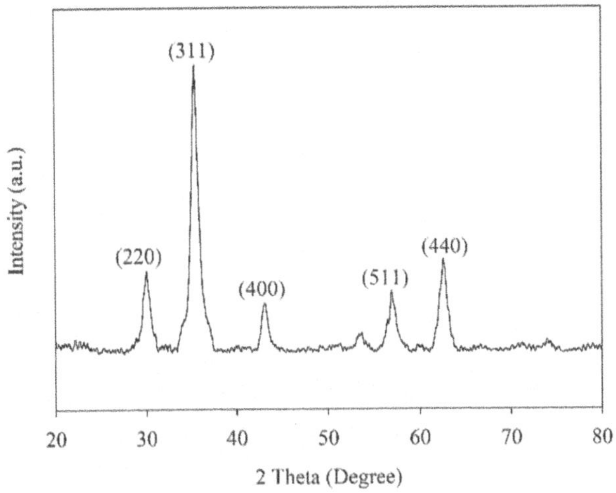

Figure 1 | XRD spectrum of $MnFe_2O_4$.

of the material. The diffraction peaks at 2θ of 29.71, 34.98, 42.53, 56.20 and 61.66° mainly corresponded to the typical reflections of (220), (311), (400), (511) and (440) planes, respectively. The results mean that the material was indexed into the cubic spinel-type $MnFe_2O_4$ (JCPDS No. 38-0430). Wang *et al.* (2019b) also reported similar peaks in the XRD spectrum of $MnFe_2O_4$. Therefore, the pure $MnFe_2O_4$ was efficiently produced by using the chemical coprecipitation method in the experiment.

The transmission electron microscopy (TEM) and scanning electron microscope (SEM) images, where surface morphology and particle size was obtained, is demonstrated in Figure 2. It could be seen from TEM images that $MnFe_2O_4$ particles have spherical structures, while catalyst particles in micron scale with a diameter of 0.1–0.5 μm were confirmed by the SEM images.

The various functional groups of $MnFe_2O_4$ before and after reaction were confirmed in the FTIR spectra (Figure 3). The appearance of absorption peak at 3,424 cm^{-1} was due to the stretching vibrations of -OH moieties in carboxyl and hydroxyl groups (Yang *et al.* 2019). The corresponding functional groups with wave numbers of 2,853 and 2,923 cm^{-1}

Figure 2 | SEM and TEM spectrum of $MnFe_2O_4$.

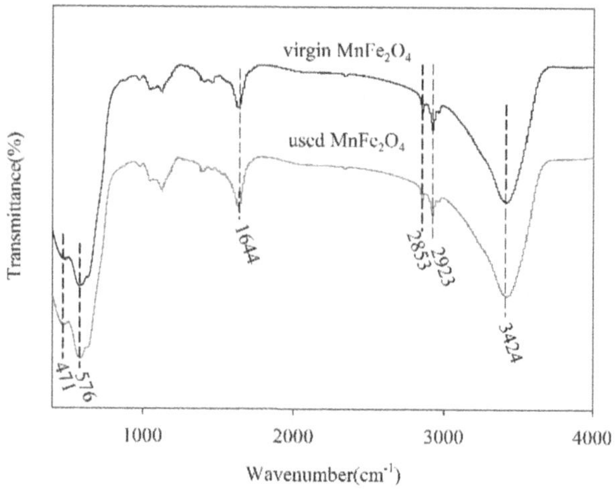

Figure 3 | FT-IR spectra of $MnFe_2O_4$.

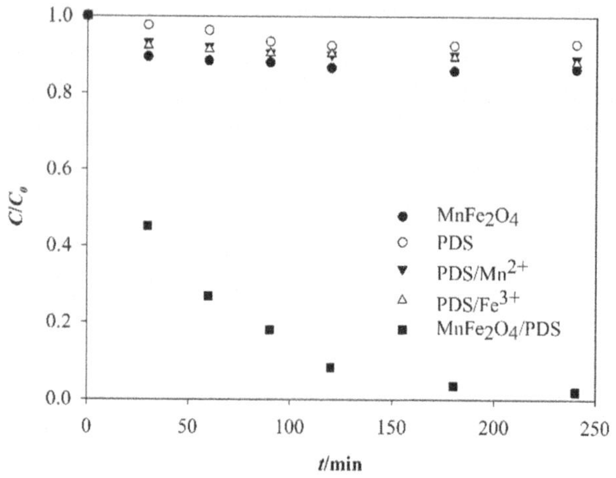

Figure 4 | Effect of different systems on degradation of aniline by PDS catalyzed with $MnFe_2O_4$. Experimental conditions: aniline = 20 mg/L, PDS = 2.4 mM, $MnFe_2O_4$ = 1.3 g/L, Fe^{3+} = Mn^{2+} = 5.6 mM.

were associated to C-H stretching vibrations (Chakraborty *et al.* 2017). The peak at 1,644 cm^{-1} could be assigned to the bending vibrations of H_2O molecules on the surface of $MnFe_2O_4$ (Zhu *et al.* 2019). Furthermore, there were two strongly characteristic peaks in the wave numbers of 471 and 576 cm^{-1} which is related to Mn-O and Fe-O vibrations in $MnFe_2O_4$ respectively. The lower wave number of 471 cm^{-1} was assigned to the octahedral coordination Mn^{2+}, and the higher wave number of 576 cm^{-1} was related to the tetrahedral coordination Fe^{3+} (Kafshgari *et al.* 2017). It was observed that there were no differences in the FTIR spectra of the used and the virgin $MnFe_2O_4$. The results demonstrated the $MnFe_2O_4$ used as the catalyst was stable in the system $MnFe_2O_4$/PDS (Liu *et al.* 2020).

Degradation experiment in different systems

To confirm the $MnFe_2O_4$ accessed to activate PDS, Figure 4 shows the degradation of aniline by different systems. It can be seen that the maximum adsorption capacity of $MnFe_2O_4$ for aniline is 2.1 mg/L and aniline degradation by 2.4 mM PDS is only 7.1% at 240 min, which shows that they both had a negligible contribution to adsorb or oxidize the compound. Aniline removal efficiency in the $MnFe_2O_4$/PDS system was nearly 100% at 240 min. It is noteworthy that a pseudo-first-order degradation pattern ($R^2 > 0.90$) was suitable for the aniline removal in PDS activation by $MnFe_2O_4$ and the fitting reaction rate constant (k_{app}) was 0.016. Besides, aniline degradation efficiencies in PDS/Fe^{3+} and PDS/Mn^{2+} system was 11 and 12% at 240 min, respectively, demonstrating that the heterogeneous activation was the dominant oxidation in the $MnFe_2O_4$/PDS system. These results demonstrate that $MnFe_2O_4$ could effectively activate PDS. Similarly, Deng *et al.* (2017) reported that the degradation of bisphenol A by homogeneous Mn^{2+}/PMS, and Fe^{3+}/PMS activation was less than 10% compared with the degradation of 90% by $MnFe_2O_4$/PMS system. Abroshan *et al.* (2018) reported that $MnFe_2O_4$ had a certain function in catalyzing PDS for the removal of levofloxacin in aqueous solution.

Effect of initial pH and reaction temperture

The influence of initial pH ranging from 3.0 to 11.0 on aniline degradation is illustrated in Figure 5. It is clear that the aniline degradation in the $MnFe_2O_4$/PDS system showed high efficiencies over a wide pH range of 3.0–9.0 and strongly alkaline pH could not benefit the degradation of aniline. Specially, when the pH value increased from 3.0 to 9.0, the minimal removal efficiencies of aniline achieved 95% within 4 h, and the corresponding k_{obs} value was 0.0113 min^{-1}. The aniline degradation had a significant reduction with the increasing of initial pH value to 11 and the relevant k_{obs} was found to be 0.0019 min^{-1}. The pH_{pzc} of $MnFe_2O_4$, the pK_a of aniline, and the pH value of solution should all be taken into consideration when discussing the

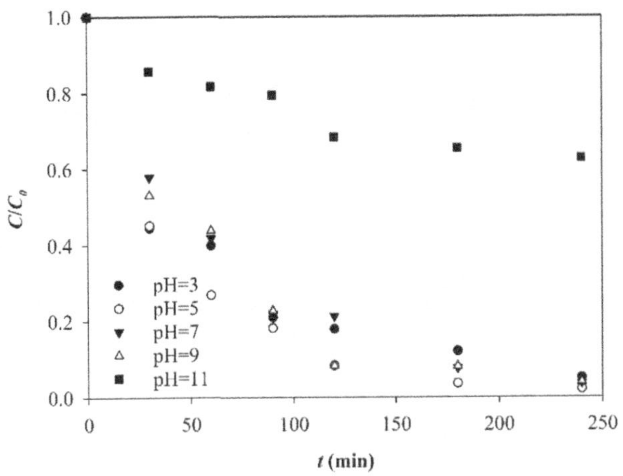

Figure 5 | Effect of solution initial pH on degradation of aniline by PDS catalyzed with $MnFe_2O_4$. Experimental conditions: initial pH = 5.00 ± 0.2 and T = 25°, initial concentration of aniline, PDS and $MnFe_2O_4$ were 20 mg/L, 2.4 mM, and 1.3 g/L, respectively.

influence of initial pH. It has been reported that the pK_a value of aniline is 4.6, which means that aniline exists mainly in cationic forms in solution at pH 3 and mainly as neutral in a pH range of 5–11. The pH_{pzc} of $MnFe_2O_4$ was determined to be 4.81. It demonstrated that the surface of the particle had a positive charge due to pH less than pH_{pzc} and would be negatively charged owing to pH larger than pH_{pzc}. To further determine the change of pH value in the system, variations of solution pH in these processes are also confirmed and it is clear that the values of pH from 30 to 240 min are controlled in a certain range from 4.50 to 3.50 when the initial pH was from 3.0 to 9.0 and the solution pH in 240 min tardily decreased to 10 when the initial pH was 11. Thus, as the initial pH value was under 10, the surface on $MnFe_2O_4$ would engender the electrostatic attraction force with PDS (anion) instead of the repulsion force with aniline. It could be speculated that the reaction between PDS and $MnFe_2O_4$ took precedence over the oxidative degradation of aniline in the PDS/$MnFe_2O_4$ system. As the initial pH value reached 11, the electrostatic attraction force between PDS and $MnFe_2O_4$ vanished ($pH > pH_{ZPC}$), leading to a decrease in the aniline degradation. Above all, it was likely that the $MnFe_2O_4$/PDS system had broad application prospects as part of wastewater treatment except in the strongly alkaline condition. What is more, to reach such prospects, the leaching of metal ion in this system is one of the questions of common interest. The leaching concentration of ferric ion after 4 h reaction was detected to be 0.14, 0.15, 0.1, 0.17 and 0.45 mg/L with the initial pH value increasing from 3.0 to 11.0 g/L, respectively. Meanwhile, the leaching concentration of manganese ion was 30.30,17.09, 13.40, 14.24 and 0.57 mg/L, respectively. The results show that further treatment in practical applications needs to be carried out for decreasing the concentration level of manganese in the solution.

The effect of reaction temperature in the $MnFe_2O_4$/PDS system was confirmed. Specially, the increasing of the corresponding k_{obs} values is from 0.0055 min^{-1} at 10 °C to 0.018 min^{-1} at 30 °C. The results showed reaction temperature is an essential factor for this system, thus it is hard for it to be widely used in some lower temperature areas.

Mechanism of aniline degradation by $MnFe_2O_4$/PDS systems

In terms of previous reports, the predominant reactive species in PDS activation were $SO_4^{\cdot-}$ and ·OH (Nie *et al.* 2014). It was reported that MeOH could quench both $SO_4^{\cdot-}$ ($k_{SO_4^{\cdot-}} = 1 \times 10^9\ M^{-1}\ s^{-1}$) and ·OH ($k_{\cdot OH} = 1 \times 10^7\ M^{-1}\ s^{-1}$). *t*-BuOH showed a high reaction rate in quenching ·OH ($k_{\cdot OH} = 5.2 \times 10^8\ M^{-1}\ s^{-1}$), yet it reacted slower with $SO_4^{\cdot-}$ ($k_{SO_4^{\cdot-}} = 8.4 \times 10^5\ M^{-1}\ s^{-1}$) than MeOH (Guan *et al.* 2018). Thus, MeOH and *t*-BuOH were used as scavengers to distinguish the existence of $SO_4^{\cdot-}$ and ·OH in $MnFe_2O_4$/PDS systems separately. As Figure 6 shows, the degradation of aniline by the $MnFe_2O_4$/PDS system was examined in the presence of excess radical scavengers. Both the introduction of MeOH and *t*-BuOH in PDS/$MnFe_2O_4$ system had little influence on the degradation of aniline even if the scavenger was 1,000 times the molar ratio of aniline. This suggested that the mechanism of $SO_4^{\cdot-}$ generation probably has little influence on the aniline decomposition by the $MnFe_2O_4$/PDS system. Similarly, Shi *et al.* (2020) reported that the removal rate of sulfamethoxazole through the PDS activated by EDTA-2 K-derived nitrogen-doped porous carbons only reduced by 23% as the addition of MeOH which demonstrated that $SO_4^{\cdot-}$ and ·OH were not dominantly responsible in the reaction system. However, $SO_4^{\cdot-}$ and ·OH were still generated in the system.

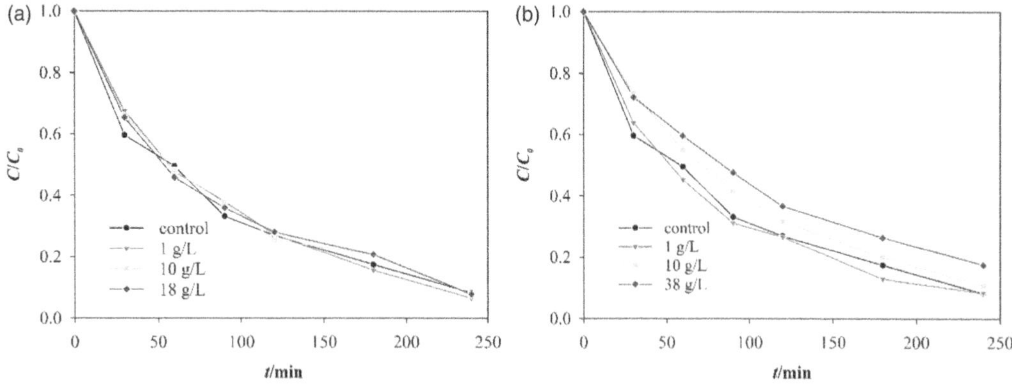

Figure 6 | Effect of different MeOH(a) and *t*-BuOH (b) on degradation of aniline by PDS catalyzed with $MnFe_2O_4$. Experimental conditions: initial concentration of aniline, PS and $MnFe_2O_4$ were 20 mg/L, 2.4 mM, and 0.9 g/L, respectively.

To verify whether the main reactive species in $MnFe_2O_4$/PDS system responsible for the degradation of aniline was the hole which was similar to photogenerated active sites, we tested the radical trapping experiments, where EDTA was added into the reaction system as the hole radical scavenger (Figure 7(a)). Specifically, the removal efficiencies of aniline after 4 h significantly decreased from 90.1 to 47.0% with the concentrate of EDTA increasing from 0.01 to 1.00 g/L, respectively. This result showed the possible involvement of the hole in the degradation of the aniline. However, Wu *et al.* (2018) reported that the degradation of aniline in RSBC/PDS was completely inhibited when the addition of EDTA reached 1.00 g/L. This demonstrated that the $MnFe_2O_4$/PDS system may produce other reactive species except holes.

One of the main photogenerated reactive species was $^1O_2$ (Zhang *et al.* 2020), which might have an effect on the facilitated degradation of aniline. To prove the hypothesis, NaN_3, which was used as the scavengers of $^1O_2$ to inhibit the oxidation process, was fed into the $MnFe_2O_4$/PDS system (Figure 7(b)). When the concentrate of NaN_3 increased to 1.00 g/L, the degradation of the aniline in the $MnFe_2O_4$/PDS system was drastically decreased to over half of the control system. This indicated that the generation of $^1O_2$ may be responsible for the degradation of aniline in the $MnFe_2O_4$/PDS system.

TOC removal efficiencies in $MnFe_2O_4$/PDS system

As shown in Figure 8, the removal efficiencies of TOC during the oxidation of aniline with different PDS concentrations were quantified. It can be seen that the TOC degradation degree of the PDS/$MnFe_2O_4$ system increased slowly with the reaction time. In particular, the TOC

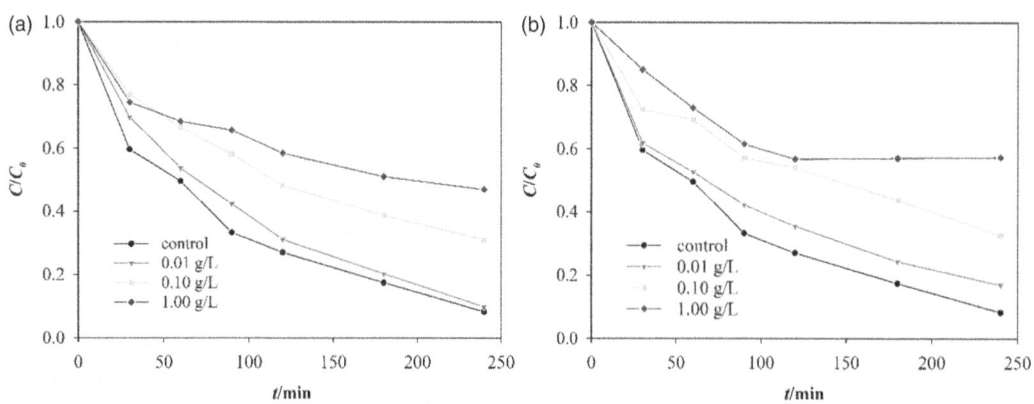

Figure 7 | Effect of different EDTA(a) and NaN_3 (b) concentration on degradation of aniline by PDS catalyzed with $MnFe_2O_4$. Experimental conditions: initial concentration of aniline, PS and $MnFe_2O_4$ were 20 mg/L, 2.4 mM, and 0.9 g/L, respectively.

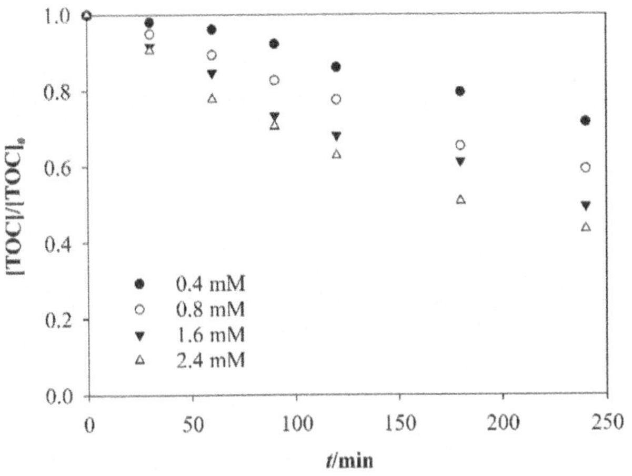

Figure 8 | Effect of different PDS concentrations on the removal efficiencies of TOC by PDS catalyzed with $MnFe_2O_4$. Experimental conditions: initial concentration of aniline and $MnFe_2O_4$ were 20 mg/L and 0.5 g/L, respectively.

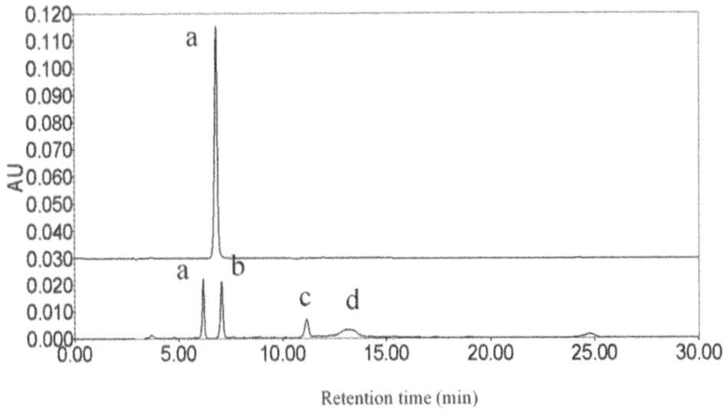

Figure 9 | HPLC typical chromatograms of intermediates (a–d) for aniline solutions collected after 0 and 4 h in MnFe2O4/PDS system.

removal efficiencies after 4 h were 71, 59, 49 and 43% for 0.4, 0.8, 1.6 and 2.4 mM PDS, respectively. These results indicated that some organic oxidation products were formed and remained in the reaction solution.

Identification of intermediates and transformation pathways

To further explore the mechanism of aniline degradation in the $MnFe_2O_4$/PDS system, high-performance liquid chromatography-mass spectrometry (HPLC-MS) was introduced to identify the intermediates during the $MnFe_2O_4$/PDS system. The detected samples collected at the reaction time of 0 and 4 h were extracted with n-Hexane (the extraction ratio was 1:1) to analyze in HPLC-MS (Xie *et al.* 2012), respectively. The three main intermediates in the system are shown in Figure 9. All the peaks in the chromatogram were marked to respond with the analytical results in mass spectrometry.

As shown in mass spectrometry (Figure 10), the peak of a, b, c and d corresponded to m/z values of 94.07, 201.16, 124.01 and 184.04, respectively. The chemical formulas of products in the system was proposed by both the mass spectrometry and the retention time. It can be observed that there are three main byproducts which have the chromatographic peak areas with m/z 201.16, 124.01 and 184.04.

Based on the above discussion, two possible degradation pathways of aniline degradation in the system were proposed and demonstrated in Figure 11. In pathway **I**, through the hydroxylation of aniline, the aromatic ring was first attacked by ·OH and $SO_4^{\cdot-}$ and the deprotonation of aniline accomplished, leading to the formation of imino radical. Next, imino radical was further oxidized by holes and $^1O_2$ into nitrobenzene. Then, a certain amount of nitrobenzene was possibly cleaved to small molecules such as the maleic acid and oxalic acid, and finally to CO_2. The other may react with imino radical resulting in converting into 4–4′-diaminodiphenyl, which was oxidized to small molecular organics by the reactive species. In pathway II, the intermediate dianiline resulting from the bimolecular reaction occurred. Further, the reactive species 'attacked', inducing benzene ring opening of dianiline to phenylsuccinic acid and maleic acid.

Through the analysis of XPS, which was carried out to elucidate the charge of chemical valence, the element composition of Mn and Fe species is further confirmed, see Figure 12. The C 1s peak was located at 284.8 eV in order to calibrate this spectrum (Figure 12(a)), and it was detected that the main chemical components including Fe, Mn and O continuously existed in $MnFe_2O_4$ during the reaction. Figure 12(b) showed that the two peaks locating at binding energy of 710 and 724 eV corresponded to Fe $2p_{3/2}$ and Fe $2p_{1/2}$ in spectrum of virgin $MnFe_2O_4$ (Lai *et al.* 2020), respectively, which strongly confirmed that only Fe(III) species existed in $MnFe_2O_4$ rather than the Fe(II). As shown in Figure 12(c), it can also be observed that the peaks centered at 641.8 eV (Mn $2p_{3/2}$) and 653 eV (Mn $2p_{1/2}$) suggesting that Mn(II) was the entire state of Mn species (Deng *et al.* 2018). After the

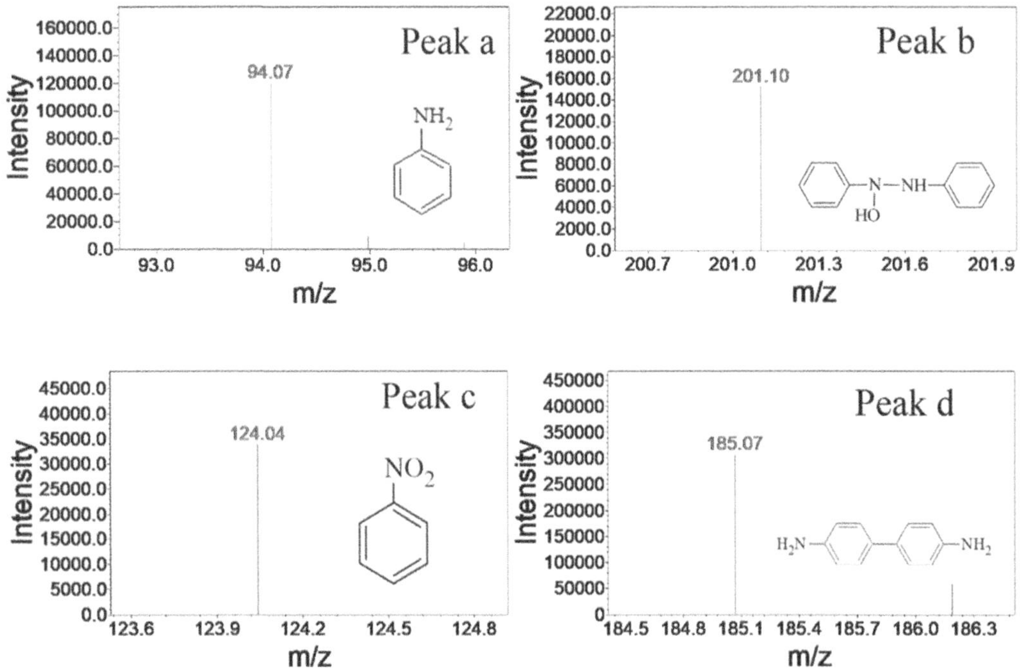

Figure 10 | Proposed intermediates and mass spectra of the degradation products.

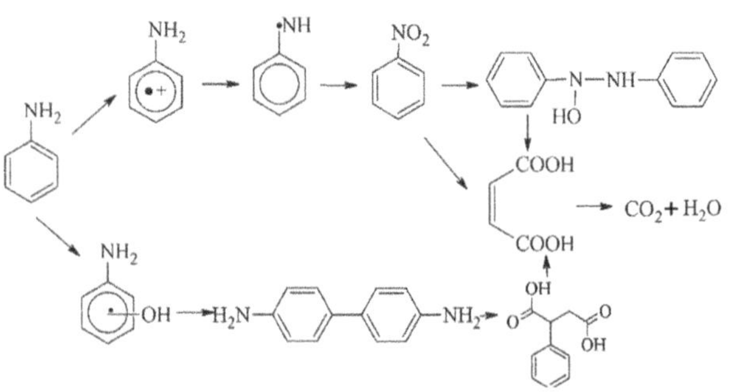

Figure 11 | Possible pathways of aniline mineralization by PDS/$MnFe_2O_4$.

reaction process, the peaks of Fe $2p_{3/2}$ located at 712.7 eV and Mn $2p_{3/2}$ located at 645.7 eV corresponded to Fe(II) and Mn(III), and deconvolution of these two peaks represents that multivalent states of Fe(II)/Fe(III) and Mn(II)/Mn(III) coexisted on the surface. The proportion of Fe(II)/(Fe(III) + Fe(II)) and Mn(II)/(Mn(III) + Mn(II)) increased to 33.02 and 51.57%, which confirmed that the chemical transformations between these multivalent states occurred on the surface of specimen. Figure 12(d) presented the O 1s spectrum of used and virgin $MnFe_2O_4$, which both had three peaks, one locating at 529.9 eV indicating the form of surface lattice oxygen (O_{latt}) and others locating at 531.7 and 533.2 eV derived from surface adsorbed oxygen (O_{ads},) (Lai *et al.* 2021). Due to its higher mobility than O_{latt}, O_{ads} could be integrated with Fe(III) and Mn(II) and then take part in the chain reaction in AOPs to the degradation of pollution (Kim & Shim 2010; Deng *et al.* 2017), and may cause the surface of $MnFe_2O_4$ to have more surface oxygen vacancies, which may further result in the generation of holes. Consequently, the proportion of O_{ads} in $MnFe_2O_4$ may affect the catalyst activity. Specially, the $MnFe_2O_4$ corresponding area ratio of $O_{ads}/(O_{ads} + O_{latt})$ was increased by 13.98% after the reaction, which demonstrated that the catalytic activity of $MnFe_2O_4$ after use was promoted to a certain extent.

According to the results above, a schematic diagram of the possible oxidation mechanism of aniline by PDS/$MnFe_2O_4$ is shown in Figure 13. It speculated that the degradation of compounds in the $MnFe_2O_4$/PDS system mainly contained three results: (1) the adsorption between the $S_2O_8^{2-}$ and the surface of $MnFe_2O_4$ occurred first; (2) then $S_2O_8^{2-}$ was converted by $MnFe_2O_4$ which was used as the electron donor into the production of not only holes and $^1O_2$ but also $SO_4^{\cdot-}$ and ·OH; (3) finally, reaction activation species reacted with compounds.

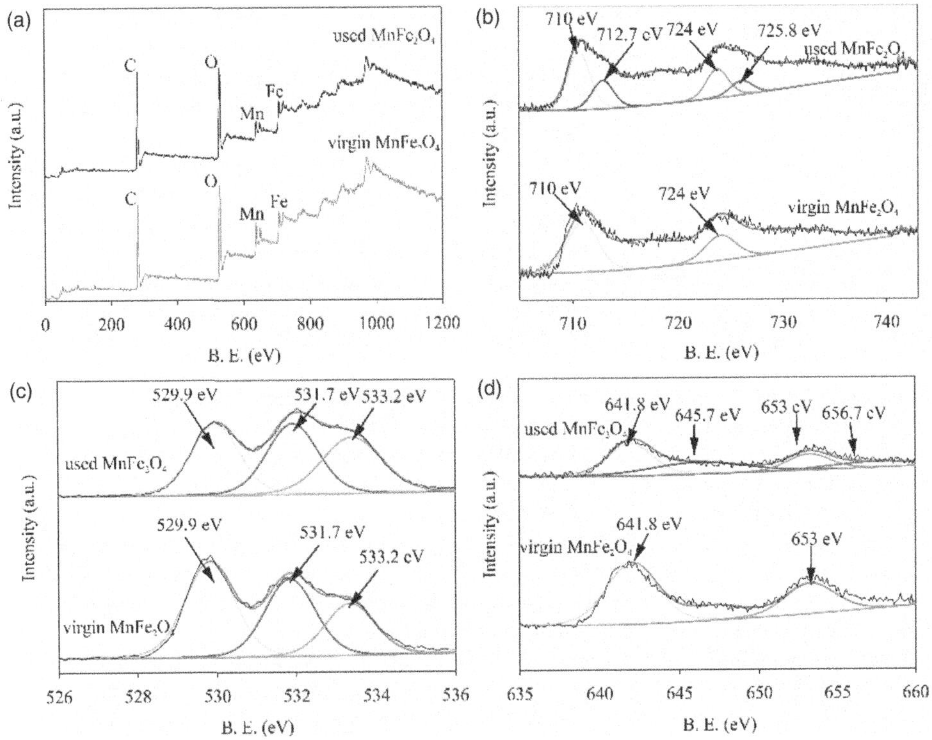

Figure 12 | XPS spectra of the virgin and used $MnFe_2O_4$: (a) full-range scan, (b) Fe 2p core level, (c) Mn core level O, and (d) 1 s core level.

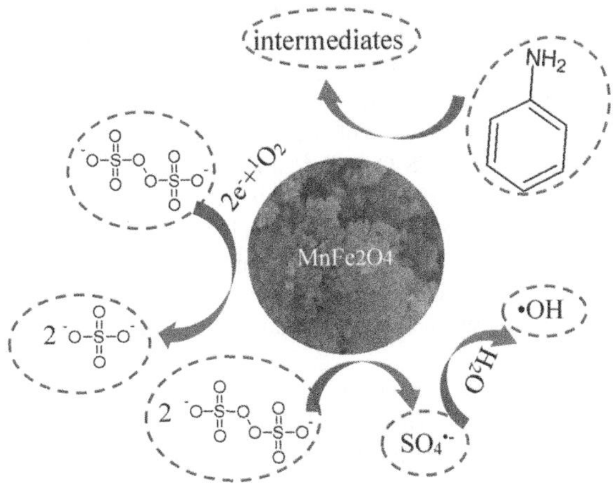

Figure 13 | Schematic diagram of the possible oxidation mechanism of aniline by PDS/ $MnFe_2O_4$.

CONCLUSIONS

In this study, $MnFe_2O_4$ used as catalyst was synthesized by a chemical co-precipitation method and the heterogeneous catalytic oxidation of aniline with PDS was investigated; 97.98% of aniline was degraded in 4 h by $MnFe_2O_4$/PDS system. According to radical scavenging tests, both sulfate and hydroxyl radicals have little influence on the degradation of aniline in the $MnFe_2O_4$/PDS system. However, the degradation of aniline mainly depended on the reactive species of holes and $^1O_2$. There were several intermediate productions, including 4–4′-diaminodiphenyl, dianiline and nitrobenzene for the decomposition of aniline in the $MnFe_2O_4$/PDS system. All these productions were further broken into small molecular substances and finally into CO_2 and H_2O.

ACKNOWLEDGEMENTS

This work was financially supported by the Science and Technology Department of Henan (161100310700), and Key Research and Extension Project of Henan Province (192102310225).

DATA AVAILABILITY STATEMENT

Data cannot be made publicly available; readers should contact the corresponding author for details.

REFERENCES

Abroshan, E., Farhadi, S. & Zabardasti, A. 2018 Novel magnetically separable $Ag_3PO_4/MnFe_2O_4$ nanocomposite and its high photocatalytic degradation performance for organic dyes under solar-light irradiation. *Solar Energy Materials and Solar Cells* **178**, 154–163. doi:10.1016/j.solmat.2018.01.026.

Benito, A., Penades, A., Lliberia, J. L. & Gonzalez-Olmos, R. 2017 Degradation pathways of aniline in aqueous solutions during electro-oxidation with BDD electrodes and UV/H_2O_2 treatment. *Chemosphere* **166**, 230–237. doi:10.1016/j.chemosphere.2016.09.105.

Can, Z. S. & Cakir, E. 2010 Treatability of organic constituents in the pasakoy wastewater treatment plant effluent by O_3 and O_3/H_2O_2. *Ozone: Science & Engineering* **32** (3), 209–214. doi:10.1080/01919511003796087.

Chakraborty, I., Majumder, D., Talukdar, S., Roy, S. & Mandal, K. 2017 Surface engineered magneto fluorescent $MnFe_2O_4$ nanoparticles in the realm of biomedical applications. *Surfaces and Interfaces* **9**, 154–159. doi:10.1016/j.surfin.2017.09.005.

Cheng, X., Guo, H. G., Zhang, Y. L., Wu, X. & Liu, Y. 2017 Non-photochemical production of singlet oxygen via activation of persulfate by carbon nanotubes. *Water Research* **113**, 80–88. doi:10.1016/j.watres.2017.02.016.

Deng, J., Feng, S. F., Zhang, K. J., Li, J., Wang, H. Y., Zhang, T. Q. & Ma, X. Y. 2017 Heterogeneous activation of peroxymonosulfate using ordered mesoporous co3o4 for the degradation of chloramphenicol at neutral pH. *Chemical Engineering Journal* **308**, 505–515. doi:10.1016/j.cej.2016.09.075.

Deng, J., Xu, M., Qiu, C., Chen, Y., Ma, X., Gao, N. & Li, X. 2018 Magnetic $MnFe_2O_4$ activated peroxymonosulfate processes for degradation of bisphenol A: performance, mechanism and application feasibility. *Applied Surface Science* **459**, 138–147. doi:10.1016/j.apsusc.2018.07.198.

Du, X., Zhang, Y., Hussain, I., Huang, S. & Huang, W. 2017 Insight into reactive oxygen species in persulfate activation with copper oxide: activated persulfate and trace radicals. *Chemical Engineering Journal* **313**, 1023–1032. doi:10.1016/j.cej.2016.10.138.

Duran, A., Monteagudo, J. M., San Martin, I. & Merino, S. 2018 Photocatalytic degradation of aniline using an autonomous rotating drum reactor with both solar and UV-C artificial radiation. *Journal of Environmental Management* **210**, 122–130. doi:10.1016/j.jenvman.2018.01.012.

Guan, R., Yuan, X., Wu, Z., Wang, H., Jiang, L., Zhang, J., Li, Y., Zeng, G. & Mo, D. 2018 Accelerated tetracycline degradation by persulfate activated with heterogeneous magnetic $Ni_xFe_{3-x}O_4$ catalysts. *Chemical Engineering Journal* **350**, 573–584. doi:10.1016/j.cej.2018.05.195.

Hodges, B. C., Cates, E. L. & Kim, J. H. 2018 Challenges and prospects of advanced oxidation water treatment processes using catalytic nanomaterials. *Nature Nanotechnology* **13** (8), 642–650. doi:10.1038/s41565-018-0216-x.

Hu, P. & Long, M. 2016 Cobalt-catalyzed sulfate radical-based advanced oxidation: a review on heterogeneous catalysts and applications. *Applied Catalysis B: Environmental* **181**, 103–117. doi:10.1016/j.apcatb.2015.07.024.

Huang, D., Zhang, Q., Zhang, C., Wang, R., Deng, R., Luo, H., Li, T., Li, J., Chen, S. & Liu, C. 2020 Mn doped magnetic biochar as persulfate activator for the degradation of tetracycline. *Chemical Engineering Journal* **391**, 123532. doi:10.1016/j.cej.2019.123532.

Ji, Y., Dong, C., Kong, D., Lu, J. & Zhou, Q. 2015 Heat-activated persulfate oxidation of atrazine: implications for remediation of groundwater contaminated by herbicides. *Chemical Engineering Journal* **263**, 45–54.

Junlabhut, P., Nuthongkum, P. & Pechrapa, W. 2018 Influences of calcination temperature on structural properties of mnfe2o4 nanopowders synthesized by co-precipitation method for reusable absorbent materials. *Materials Today: Proceedings* **5** (6), 13857–13864. doi:10.1016/j.matpr.2018.02.028.

Kafshgari, L. A., Ghorbani, M. & Azizi, A. 2017 Fabrication and investigation of mnfe2o4/MWCNTs nanocomposite by hydrothermal technique and adsorption of cationic and anionic dyes. *Applied Surface Science* **419**, 70–83. doi:10.1016/j.apsusc.2017.05.019.

Khan, J. A., He, X., Khan, H. M., Shah, N. S. & Dionysiou, D. D. 2013 Oxidative degradation of atrazine in aqueous solution by $UV/H_2O_2/Fe^{2+}$, $UV/S_2O_82-/Fe^{2+}$ and $UV/HSO_5-/Fe^{2+}$ processes: a comparative study. *Chemical Engineering Journal* **218**, 376–383. doi:10.1016/j.cej.2012.12.055.

Kim, S. C. & Shim, W. G. 2010 Catalytic combustion of VOCs over a series of manganese oxide catalysts. *Applied Catalysis B – Environmental* **98** (3–4), 180–185. doi:10.1016/j.apcatb.2010.05.027.

Lai, L. D., Zhou, H. Y., Zhang, H., Ao, Z. M., Pan, Z. C., Chen, Q. X., Xiong, Z. K., Yao, G. & Lai, B. 2020 Activation of peroxydisulfate by natural titanomagnetite for atrazine removal via free radicals and high-valent iron-oxo species. *Chemical Engineering Journal* **387**, 124165. doi:10.1016/j.cej.2020.124165.

Lai, L., Ji, H., Zhang, H., Liu, R., Zhou, C., Liu, W., Ao, Z., Li, N., Liu, C., Yao, G. & Lai, B. 2021 Activation of peroxydisulfate by V-Fe concentrate ore for enhanced degradation of carbamazepine: surface equivalent to V(III) and equivalent to V(IV) as electron donors promoted the regeneration of equivalent to Fe(II). *Applied Catalysis B – Environmental* **282**, 119559. doi:10.1016/j.apcatb.2020.119559.

Lee, J., Hong, S., Mackeyev, Y., Lee, C., Chung, E., Wilson, L. J., Kim, J. H. & Alvarez, P. J. J. 2011 Photosensitized oxidation of emerging organic pollutants by tetrakis C-60 aminofullerene-derivatized silica under visible light irradiation. *Environmental Science & Technology* **45** (24), 10598–10604. doi:10.1021/es2029944.

Lee, H., Lee, H. J., Jeong, J., Lee, J., Park, N. B. & Lee, C. 2015 Activation of persulfates by carbon nanotubes: oxidation of organic compounds by nonradical mechanism. *Chemical Engineering Journal* **266**, 28–33. doi:10.1016/j.cej.2014.12.065.

Liu, J., Qiao, L., Wang, Y., Li, G. & Liu, B. 2020 Aniline degradation by peroxydisulfate activated with magnetic Fe-Mn oxides composite: efficiency, stability, and mechanism. *Reaction Kinetics Mechanisms and Catalysis* **131** (2), 567–582. doi:10.1007/s11144-020-01861-1.

Lominchar, M. A., Santos, A., de Miguel, E. & Romero, A. 2018 Remediation of aged diesel contaminated soil by alkaline activated persulfate. *Science of the Total Environment* **622–623**, 41–48. doi:10.1016/j.scitotenv.2017.11.263.

Ma, W. J., Wang, N., Fan, Y. A., Tong, T. Z., Han, X. J. & Du, Y. C. 2018 Non-radical-dominated catalytic degradation of bisphenol A by ZIF-67 derived nitrogen-doped carbon nanotubes frameworks in the presence of peroxymonosulfate. *Chemical Engineering Journal* **336**, 721–731. doi:10.1016/j.cej.2017.11.164.

Nie, M., Yang, Y., Zhang, Z., Yan, C., Wang, X., Li, H. & Dong, W. 2014 Degradation of chloramphenicol by thermally activated persulfate in aqueous solution. *Chemical Engineering Journal* **246**, 373–382.

Pan, X., Yan, L., Li, C., Qu, R. & Wang, Z. 2017 Degradation of UV-filter benzophenone-3 in aqueous solution using persulfate catalyzed by cobalt ferrite. *Chemical Engineering Journal* **326**, 1197–1209. doi:10.1016/j.cej.2017.06.068.

Shi, Y. W., Zhu, J. D., Yuan, G., Liu, G. Z., Wang, Q. F., Sun, W. J., Zhao, B., Wang, L. & Zhang, H. W. 2020 Activation of persulfate by EDTA-2 K-derived nitrogen-doped porous carbons for organic contaminant removal: radical and non-radical pathways. *Chemical Engineering Journal* **386**, 124009. doi:10.1016/j.cej.2019.124009.

Verma, M. & Haritash, A. K. 2019 Degradation of amoxicillin by Fenton and Fenton-integrated hybrid oxidation processes. *Journal of Environmental Chemical Engineering* **7** (1), 102886. doi:10.1016/j.jece.2019,102886.

Wacławek, S., Lutze, H. V., Grübel, K., Padil, V. V. T., Černík, M. & Dionysiou, D. D. 2017 Chemistry of persulfates in water and wastewater treatment: a review. *Chemical Engineering Journal* **330**, 44–62. doi:10.1016/j.cej.2017.07.132.

Wang, S., Wu, J., Lu, X., Xu, W., Gong, Q., Ding, J., Dan, B. & Xie, P. 2019a Removal of acetaminophen in the Fe^{2+}/persulfate system: kinetic model and degradation pathways. *Chemical Engineering Journal* **358**, 1091–1100. doi:10.1016/j.cej.2018.09.145.

Wang, Z. J., Zhang, X. Y., Zhang, H. X., Zhu, G. X., Gao, Y. J., Cheng, Q. F. & Cheng, X. W. 2019b Synthesis of magnetic nickel ferrite/carbon sphere composite for levofloxacin elimination by activation of persulfate. *Separation and Purification Technology* **215**, 528–539. doi:10.1016/j.seppur.2019.01.063.

Wu, Y., Guo, J., Han, Y., Zhu, J., Zhou, L. & Lan, Y. 2018 Insights into the mechanism of persulfate activated by rice straw biochar for the degradation of aniline. *Chemosphere* **200**, 373–379. doi:10.1016/j.chemosphere.2018.02.110.

Xie, X., Zhang, Y., Huang, W. & Huang, S. 2012 Degradation kinetics and mechanism of aniline by heat-assisted persulfate oxidation. *Journal of Environmental Sciences* **24** (5), 821–826. doi:10.1016/s1001-0742(11)60844-9.

Yang, W. C., Jiang, Z., Hu, X. X., Li, X. Y., Wang, H. Y. & Xiao, R. Y. 2019 Enhanced activation of persulfate by nitric acid/annealing modified multi-walled carbon nanotubes via non-radical process. *Chemosphere* **220**, 514–522. doi:10.1016/j.chemosphere.2018.12.136.

Yong, X. Y., Gu, D. Y., Wu, Y. D., Yan, Z. Y., Zhou, J., Wu, X. Y., Wei, P., Jia, H. H., Zheng, T. & Yong, Y. C. 2017 Bio-Electron-Fenton (BEF) process driven by microbial fuel cells for triphenyltin chloride (TPTC) degradation. *Journal of Hazardous Materials* **324**, 178–183. doi:10.1016/j.jhazmat.2016.10.047.

Zhang, T., Chen, Y., Wang, Y., Le Roux, J., Yang, Y. & Croue, J. P. 2014 Efficient peroxydisulfate activation process not relying on sulfate radical generation for water pollutant degradation. *Environmental Science & Technology* **48** (10), 5868–5875. doi:10.1021/es501218f.

Zhang, Y., Zhang, Q., Dong, Z., Wu, L. & Hong, J. 2018 Degradation of acetaminophen with ferrous/copperoxide activate persulfate: synergism of iron and copper. *Water Research* **146**, 232–243. doi:10.1016/j.watres.2018.09.028.

Zhang, B. T., Wang, Q., Zhang, Y., Teng, Y. G. & Fan, M. H. 2020 Degradation of ibuprofen in the carbon dots/Fe_3O_4 carbon sphere pomegranate-like composites activated persulfate system. *Separation and Purification Technology* **242**, 116820. doi:10.1016/j.seppur.2020.116820.

Zhou, Z., Liu, X. T., Sun, K., Lin, C. Y., Ma, J., He, M. C. & Ouyang, W. 2019 Persulfate-based advanced oxidation processes (AOPs) for organic-contaminated soil remediation: a review. *Chemical Engineering Journal* **372**, 836–851. doi:10.1016/j.cej.2019.04.213.

Zhu, J., Chen, C., Li, Y., Zhou, L. & Lan, Y. 2019 Rapid degradation of aniline by peroxydisulfate activated with copper-nickel binary oxysulfide. *Separation and Purification Technology* **209**, 1007–1015. doi:10.1016/j.seppur.2018.09.055.

First received 31 October 2020; accepted in revised form 20 January 2021. Available online 11 March 2021

© 2021 The Authors

Photocatalytic degradation of hydrocarbons and methylene blue using floatable titanium dioxide catalysts in contaminated water

T. Schnabel, N. Jautzus, S. Mehling, C. Springer and J. Londong

ABSTRACT

Photocatalytic disintegration is a novel approach to eliminate pollution. The method utilizes the semiconductor titanium dioxide to degrade organic molecules in the presence of ultraviolet (UV) light. In this study, it is shown how the capabilities of several types of catalyst designs degrade the non-polar substance diesel fuel and the polar substance methylene blue. The floating design of foam glass coated with titanium dioxide could reduce the concentration of diesel fuel by 329 mg/L in 16 h; the submerged designs for coated glass fiber and coated steel grit could reduce methylene blue concentration by 96.6% after 4 h and 99.1% after 6 h, respectively. It could be shown that photocatalysis is a promising cost- and energy-efficient method for managing air and water pollution. It can be established as a low-technology method without requiring the use of a conventional source of energy, given an adequate amount of sun hours, or as an additional cleaning stage in water treatment plants using UV-LEDs.

Key words | clean-up, mineral oil, petrol-derived hydrocarbons, photocatalysis, titanium dioxide

T. Schnabel (corresponding author)
Department for Construction and Environmental Chemistry,
Laboratory for Material Testing and Research (MFPA) Weimar,
Weimar,
Germany
E-mail: *tobias.schnabel@mfpa.de*

N. Jautzus
S. Mehling
J. Londong
Chair for Urban Water Management and Sanitation,
Bauhaus-Universität Weimar,
Weimar,
Germany

C. Springer
Chair for Urban Water Management and Environmental Technology,
Erfurt University of Applied Sciences,
Erfurt,
Germany

HIGHLIGHTS

- Floating photocatalysts for solar photocatalysis.
- Low tech application for the treatment of surface water.
- Inexpensive material based on titanium dioxide.

This is an Open Access article distributed under the terms of the Creative Commons Attribution Licence (CC BY 4.0), which permits copying, adaptation and redistribution, provided the original work is properly cited (http://creativecommons.org/licenses/by/4.0/).

doi: 10.2166/wrd.2021.118

GRAPHICAL ABSTRACT

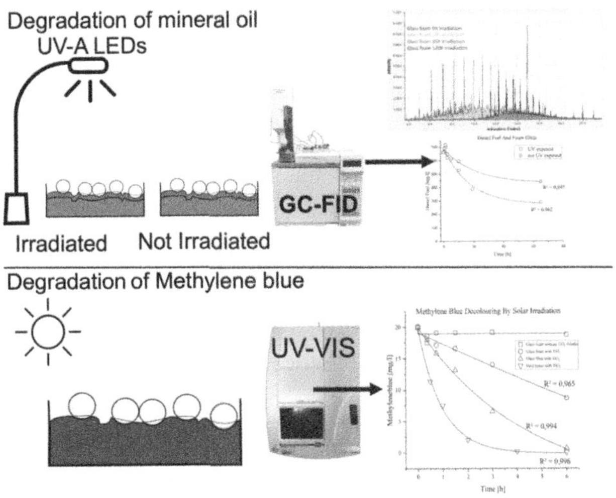

INTRODUCTION

Mineral oil-based pollution of water bodies is a problem worldwide today. The sources of pollution are diverse and include roadways, accidents, pollution from shipping, etc. Polluted water can be found in rainwater retention areas, docks, and, ultimately, in the natural water resources. Additionally, anthropogenic trace substances, such as pharmaceuticals, additives from personal care products, industrial chemicals, and pesticides, have been the focus of urban water management research for some time (Bergmann *et al.* 2011). Particularly in the context of further wastewater treatment, these substances and the process-technical possibilities for their elimination from water are considered.

Separation and/or degradation of contaminants from water is possible through physical, chemical, and biological means. While biological approaches rely on natural self-purification capabilities of the water body, this may take a considerable amount of time and results may be unsatisfactory (Weichgrebe *et al.* 1991). Physical and chemical methods tend to be more cost-intensive but offer faster results than biological methods (Pinnekamp *et al.* 2010). Precipitation, flocculation, disinfection, radiation, reverse osmosis, electro-coagulation, and absorption by activated charcoal are conventional methods (Pinnekamp *et al.* 2010).

Flexible and decentralized solutions are especially required in the case of accidents involving the release of large quantities of mineral oil hydrocarbons. Furthermore, increasing demand and parallel scarcity of freshwater and increasing pollution of water bodies, especially in developing countries and areas with insufficient infrastructure, create the need for a sustainable low-cost alternative for water clean-up. Particularly in warm regions, which receive high solar radiation throughout the year, a possible answer could be the use of solar-powered photocatalytic purification.

The novel approach of solar photocatalysis presented here relies on ultraviolet (UV) radiation from the sun that triggers the semiconductor characteristics of titanium dioxide (Pitre *et al.* 2017). The activation of the semiconductor is shown in Equation (1) (Nosaka & Nosaka 2013):

$$TiO_2 + h_v \rightarrow e^-_{cb} + h^+_{vb} \tag{1}$$

$$h^+_{vb} + R \rightarrow R^{\bullet +} \tag{2}$$

$$h^+_{vb} + H_2O \rightarrow OH^\bullet + H^+ \tag{3}$$

$$h^+_{vb} + OH^- \rightarrow OH^\bullet \tag{4}$$

$$e^-_{cb} + O_2 \rightarrow O_2^{\bullet -} \tag{5}$$

When irradiated by sunlight, titanium dioxide forms highly reactive oxygen species that can disintegrate organic molecules to their mineral components (Pedanekar *et al.* 2020) and are potentially lethal to microbes and can even destroy viruses (Cabiscol 2000). Several important features for the heterogeneous photocatalysis have extended its feasible applications in water treatment, such as ambient operating temperature and pressure, complete mineralization of organic substances, their derivate and decomposition products without secondary pollution, as well as low operating costs (Chong *et al.* 2010). It is well-documented (Mehrjouei *et al.* 2015) that highly reactive oxygen species cause cell lysis, generated as a result of the photo-induced charge separation on titanium dioxide (TiO_2) surfaces, which leads to microbial inactivation and organic mineralization without creating secondary pollution. The most important reactive oxygen species are hydroxyl radicals, which are formed by the reaction of hydroxide-ions and water molecules with the electron defects in the semiconductor (Equations (3) and (4)) (Irawaty *et al.* 2014). Superoxide anion radicals can be formed as further oxidizing agents in the electron band (Equation (5)). The electron defects themselves are also strong oxidizing agents, which can react directly with the organic compounds to be broken down (Equation (2)) (Vargas & Núñez 2010).

The degradation takes place through the reaction between C–H bonds with the reactive oxygen species under the formation of radical residuals of the hydrocarbons, which are so unstable that they disintegrate further on their own (Fox & Dulay 1993). The principal reaction between the hydroxyl radicals and the organic pollutant is shown in Equation (2). The well-known kinetics of the heterogeneous catalysis defines the step of the pollutant adsorption as the step of the kinetics which decides the velocity of the reaction. Therefore, a rough surface structure of the titanium dioxide coating will help to accelerate the disintegration of the pollutant (Didier *et al.* 2013).

Nevertheless, there are still a series of technical challenges for the application of TiO_2 catalysts to water treatment. The post-separation of the semiconductor TiO_2 catalyst after water treatment remains the major obstacle to its practicality as an industrial process (Barka *et al.* 2014). The fine particle size of the TiO_2 along with their large surface area-to-volume ratio and surface energy creates a strong tendency for catalyst agglomeration during the operation. Such agglomeration of the particles is highly detrimental given the particles' size preservation, surface-area reduction, and its reusable lifespan (Chong *et al.* 2010).

Using the catalyst designs on a fixed surface has the advantage that only a small amount of TiO_2 powder is necessary and only a small fraction ends up in the water column, thus not requiring elaborate separation. The reduction of free nanoparticles is a notable ecological benefit over the use of loose TiO_2 powder nanoparticles, which harm aquatic life forms (Adams *et al.* 2006; Haynes *et al.* 2017; Li *et al.* 2017). Photocatalytic disintegration with carrier-bound titanium dioxide catalysts provides a practical, low-cost approach for dealing with organic pollution originating from petrol-derived hydrocarbons (Cazoir *et al.* 2012), pharmaceuticals and their derivates (Schnabel *et al.* 2020), and even pathogens while possessing self-cleaning capabilities (Cabiscol 2000). The catalysts can be left on the water, utilizing a floating or fixed frame to hold them in place. Constant and predictable solar radiation is required, and no further energy use is needed. This simple low-technology method does not need trained personnel and is advantageous in acquisition and operation. These floating catalysts could be used to break down various types of pollution on the water surface with the help of solar radiation. Mineral oil-based pollution in particular that floats on the surface is a worthwhile cleaning goal for this approach.

Therefore, the goal of this study is to design and present floatable titanium dioxide-catalyst materials, which have a high surface contact with non-polar hydrocarbons that also float on the surface of the waterbody. The idea of a floatable sun driven catalyst (Qiu *et al.* 2019) was already described with bismuth-based materials (Qiu *et al.* 2019; Nasir *et al.* 2020). In this study, the goal was to investigate the suitability of three different types of floatable titanium dioxide catalysts (foam glass, glass fiber, and steel grid), provided by the Laboratories for Material Testing and Research (MFPA) Weimar, to disintegrate polar and non-polar substances in water. Ideally, these catalysts can be placed on polluted water bodies (such as water reservoirs and rain-water basins) or used as an additional cleaning stage in sewage treatment plants (Schnabel 2016; Mayer 2017), purifying the water by removing organic pollutants like drugs and their derivates and pathogenic germs (Cabiscol 2000). In the experiments, diesel fuel served as a substitute for non-polar pollutants such as petrol-derived hydrocarbons, while the pigment methylene blue was used to investigate the degradability of polar organic substances.

EXPERIMENTAL

A series of experiments was conducted for each type of catalyst. To simulate solar radiation, an actively cooled array of UV-LEDs manufactured by Synantik (Ohrdruf, Thuringia, Germany) was used. The array has a power rating of 8 W with an efficiency of approximately 8% at 385 nm from electrical to radiometric power. Conventional 200 mL aluminum trays from the food industry with a resulting water surface area of 70 cm^2 were used as containers for each experiment. Therefore, the specific irradiation power was 91.4 W/m^2. Each aluminum tray was filled with 100 mL demineralized water, into which 100 μL (0.082 g) of diesel fuel was injected beneath the water surface. Subsequently, the fuel rose to the surface and spread evenly. The coated glass foam was put in the tray until the water surface was covered. The other photocatalytic materials (coated steel grid and glass fiber tissue) were attached to wood pieces, which helped the material float on the surface. The UV-array was placed over one of the trays at 1.5 cm on a UV light-transmissive glass panel. During the experiments, the reactors were not magnetically stirred. Various experiments with different radiation times were executed, and a new preparation for every single experiment was applied. Within the closed reactor system, there was a gas phase present, so that evaporation of volatile components occurred. To exclude the effect of evaporation, two setups were used. One was exposed to UV light, while the other was shielded from it. This allowed us to estimate the influence of evaporation and distinguish it from the effect of photocatalytic disintegration. Figures 1–3 show the floating materials.

The extraction of the remaining hydrocarbons was performed using cyclohexane. A total of 20 mL of eluent was poured into the tray, and a sample was drawn after 15 min of stirring to be analyzed via GC-FID (Agilent ChemStation 6890N and Agilent 7683B Injector). This resulted in a complete sampling of all hydrocarbons within the reactor system including adsorbed and undissolved components. The experiments using coated glass fiber and steel grit were conducted in the same manner, with the exception of the glass fibers, which were provided with a floating frame. The grits were folded such that they would remain approximately 1 mm under the water surface.

Depletion of methylene blue with natural solar radiation

Field tests were conducted in Weimar, Germany (50°N, 11°E) at 28 °C and under sunny weather on the 9 August 2017 to show the performance of the three catalysts for polar organic compounds and to include real sunlight. The experiments were performed to compare the activity of the

Figure 1 | Uncoated (left) and coated (right) glass foams.

Figure 2 | Uncoated (left) and coated (right) glass fiber tissue.

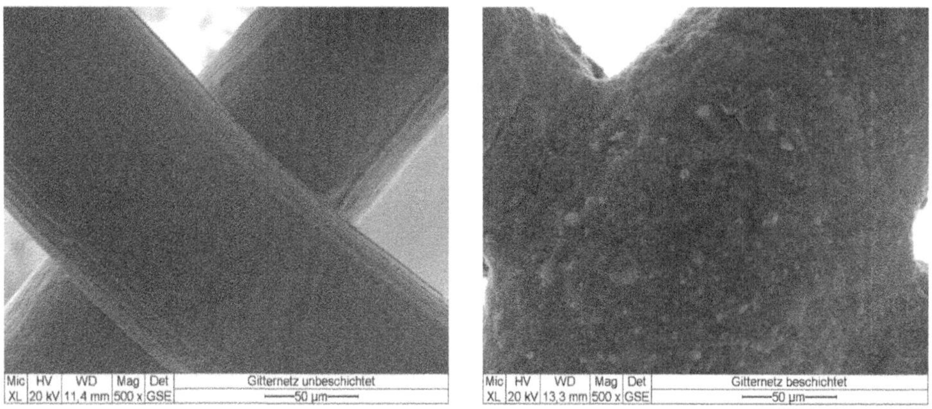

Figure 3 | Uncoated (left) and coated (right) glass steel tissue.

different materials in a homogenous solution. Methylene blue (20 µmol/L) was used as a representative of water-soluble polar substances. Discoloration and thus disintegration were quantified by a 'Merck SQ 118' photometer. The extinction was measured with a 1 cm quartz glass cuvette at a wavelength of 620 nm. The extinction has a linear relation to the methylene blue concentration, so the concentration was calculated using a 10-point calibration curve between 1 and 20 µmol/L methylene blue.

Kinetic and modeling

The photocatalytic degradation of dissolved organic substances can be described by the Langmuir–Hinshelwood equation, which is shown in the following equation (Lazar *et al.* 2012):

$$r = -\frac{dC_{eq}}{dt} = -k_{LH}\frac{K_L C_{eq}}{1 + K_L C_{eq}} \tag{6}$$

where C_{eq} is the concentration of dissolved reactant, t is the reaction time, k_{LH} is the reaction rate constant, and K_L is the adsorption coefficient of the reactant. For $K_L * \mathrm{C_{eq}} \ll 1$, Equation (6) can be simplified and the reaction rate follows a pseudo-first-order decay rate:

$$r = \frac{dC_{eq}}{dt} = -k_{LH} * \mathrm{K_L} * \mathrm{C_{eq}} = \mathrm{k} * \mathrm{C_{eq}} \tag{7}$$

This applies for small reactant concentrations or low levels of adsorption of the reactant on the catalyst surface.

For high concentrations of the reactant or strong adsorption effect, the reaction rate is no longer influenced by the reactant concentration. Equation (6) can be converted to a pseudo-zeroth-order reaction rate:

$$r = \frac{dC_{eq}}{dt} = -k_{LH} \tag{8}$$

The degradation of mixtures of mineral oil hydrocarbons, which are present as a separate phase, is further significantly influenced by the specific solubility and vapor pressures of their individual compounds. Mineral oil hydrocarbons are not miscible with water but are soluble in water to a small extent. Diesel, as a representative of the mineral oil hydrocarbons, belongs to the middle distillates, which are characterized by a boiling range of 160–390 °C and a chain length of 9–24 C atoms (Cho *et al.* 2006). The total water solubility of diesel as a mixture of mineral oil hydrocarbons is determined by the respective water solubility of the individual components. Furthermore, in the case of mineral oil mixtures, there is a mutual influence of the individual compounds, which reduces their individual solubility. With increasing chain length, a decrease of the solubility can be recognized. With the same number of C atoms, a decrease in solubility is evident in the following series: aromatics > alkenes > cycloalkanes > *i*-alkanes > *n*-alkanes. The total solubility of diesel can be classified as low and amounts to 5–20 mg/L (Speight 2020). The volatility of the individual compounds behaves analogously, whereby a decrease in volatility occurs to a greater extent with increasing molecule size compared with solubility. According to Speight (2020), the total vapor pressure of diesel mixtures is 0.05 kPa, whereby mixtures of substances above 0.01 kPa are classified as volatile.

For the photocatalytic degradation of mineral oil hydrocarbons, the degradation rate of the respective individual compound is essentially determined by its solubility. As long as the respective components are present in a separate phase, their concentration in the water phase is equal to the maximum water solubility (with the assumption of a continuous stirred-tank reactor). Due to a constant concentration, the degradation kinetics initially follows a zero-order reaction. As soon as the mineral oil phase of the respective components is no longer present due to progressive degradation, the concentration in the liquid phase is successively reduced so that the kinetics follows a first-order reaction. Parallel to this, an evaporation especially of shorter hydrocarbons takes place. As the chain length increases, the vapor pressure decreases so that the evaporation rate decreases. Meanwhile, a decrease of the solubility takes place more slowly, so that hydrocarbons of medium chain length accumulate in the water phase and are primarily decomposed photocatalytically. Longer-chain hydrocarbons continue to show a strong reduction in their solubility, so that these mineral oil components can only be decomposed very slowly by photocatalysis. For the total degradation of diesel by photocatalytic processes, the following formula was chosen to describe the reaction kinetics:

$$c_t = c_{non\ disintegrated} + c_0 * e^{k_0 * t} \tag{9}$$

In this equation, $c_{non\ disintegrated}$ is the amount of mineral oil which could not be disintegrated further during the time of the experiment. c_0 is the maximum concentration which could be disintegrated, k_0 is the exponential disintegration factor, and t is the time in hours. The kinetics show that the percentile disintegration does not depend on the concentration, but the reaction velocity depends on the concentration (Equation (9)). Within this formula, the total decrease in diesel is described by an exponential disintegration factor. As a result, both photocatalytic degradation and evaporation are represented by the disintegration factor. The influence of the evaporation was covered by an analogous data processing of the test series without the use of a catalyst. Furthermore, the photocatalytic degradation within the experiments is influenced by the mass transfer within the reactor system as well as between mineral oil and water phase, since no devices for the reactor stirring were included. For methylene blue as a water-soluble substance, the term $c_{non\ disintegrated}$ equals zero.

PHOTOCATALYTIC MATERIALS USED AND THEIR PREPARATION

Three different photocatalytic materials were used for the experiments. All three materials were coated with a layer of titanium dioxide nanoparticles, which were sinter-fused on the hot material surface out of a waterish aerosol suspension. The load of titan dioxide was calculated so that it would be 30 g/m^2 in all experiments. The three materials used were glass foam with a diameter between 1 and 2 mm and glass fiber tissue and stainless-steel tissue with 200 µm mesh. The steel grit is made of V4A high-grade stainless steel (316 L/1.4404). An aqueous nanoparticle suspension made from anatase-modified titanium dioxide nanoparticles of 14 nm in size was used as the coating. The anatase content of the titanium dioxide was 90%, and the BET surface area of the TiO_2 particles was 50 m^2/g. The aqueous suspension was sprayed onto the heated carrier material (150–200 °C surface temperature). The fast evaporation of the solvent creates a fractal coating structure with a high inner surface (Figures 1–3). The concentration of the aqueous suspension was 10 m% titanium dioxide. The scanning electron microscope (SEM) images (Figures 1–3) show the different materials before and after the coating with the nanoparticle suspension. As can be seen in the transmission electron microscope (TEM) images, the coating of the glass tissue and the steel grit is very homogeneous while the coating of the glass foam is not as even and needs to be improved in the future to reach better photocatalytic properties. The following images show the floatability of the glass foam pearls (Figure 4(a) and 4(b)) and the other materials with the wooden floaters (Figure 4(c)).

Figure 4 | (a) Glass foam, (b) floating glass foam, and (c) glass fiber with floaters.

RESULTS AND DISCUSSION

In this study, the disintegration induced by the photocatalysis of organic substances with titanium dioxide in the presence of UV light was investigated. The disintegration of UV light-exposed and non-exposed diesel fuel with different forms of catalyst was measured via gas chromatography as well as photometer measurements for methylene blue decoloring. For the evaluation of the measurements, an exponential curve fit was using the described kinetics for the disintegration of the mineral oil and methylene blue.

With the floating foam glass design, the concentration of diesel fuel could be lowered in 68 h from 688 to 271 mg/L; this equals the disintegration of 417 mg/L or 60.6%. In the experiment without UV light, the concentration decreased through evaporation of the volatile mineral oil compounds from 688 to 418 mg/L over the same time; this equals evaporation of 270 mg/L or 39%. Figure 5 shows the results as a concentration of mineral oil hydrocarbons over time. The contrast between radiated and non-radiated experiments is

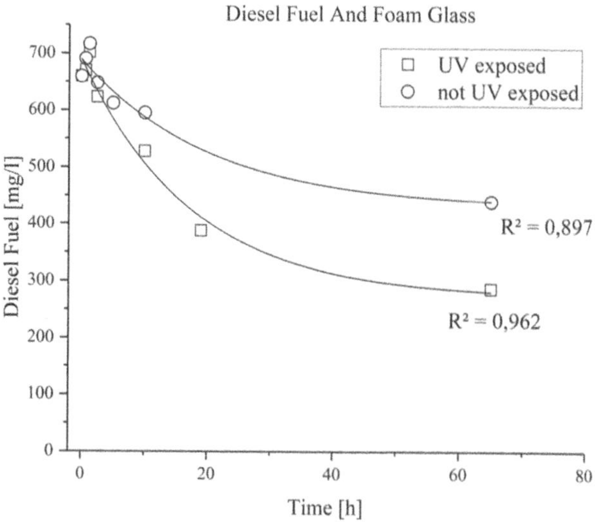

Figure 5 | Diesel fuel disintegration over time with UV light and titanium dioxide-coated foam glass. Squares represent exposure to UV; circles stand for experiments not exposed to UV. Each item represents an independent experiment. The curves were fitted with an exponential decay first order.

clearly visible throughout the tests: it is always higher in the non-radiated experiments. The calculated difference of non-radiated and radiated values in the arrangements indicates the effect of photocatalytic disintegration.

The experiments using diesel fuel were conducted with artificial UV light from LEDs because the effect of evaporation proved to be influential in an open-air setup with natural solar radiation and heating, thus corrupting the outcome. The analysis of the FID chromatograms in Figure 6 shows that the smaller compounds disintegrate faster by hydroxyl radicals than the long-chained aliphatic compounds of the mineral oil. This is due to lower volatility, solubility, and possibly lower photocatalytic reaction rates of longer-chain components. These effects overlap within the selected experimental setup, which is why this question cannot be solved completely by the existing results.

This was also shown in the fluorescent behavior of the mineral oil in the experiments after 2 h of irradiation that the mineral oil film under the photocatalytic materials was not fluorescent anymore. This provides strong evidence that especially the short aromatic compounds in the mineral oil (benzene and toluene) disintegrate faster than others. Figure 7 shows the irradiated and unirradiated samples after 2 h.

The glass fiber material has, in comparison, the highest surface area. The problem remains that this material does not float by itself and needs floaters of wood to stay close to the surface of the water. Otherwise, the contact of the catalyst and the mineral oil is not given, which may decrease reaction rate for low turbulence applications. Regardless of those problems, a higher and faster disintegration of mineral oil was achieved in comparison to the experiment without UV light (Figure 8). In 16 h, the evaporation was 189 mg/L (23% of the initial concentration). In the experiment with UV light, a disintegration of 591 mg/L (71% of the initial concentration) was achieved. The same effect regarding fluorescing of the mineral oil like with the glass foam could be observed. Additionally, the chromatograms were equal.

The experiment with the steel grid materials showed the smallest difference between evaporation and photocatalytic disintegration (Figure 9). The experiments showed higher variation between values, which is reflected by R^2. Presumably, insufficient adsorption of diesel fuel on the catalyst

Figure 7 | Irradiated (left) and unirradiated (right) sample after 2 h using glass foam catalysts.

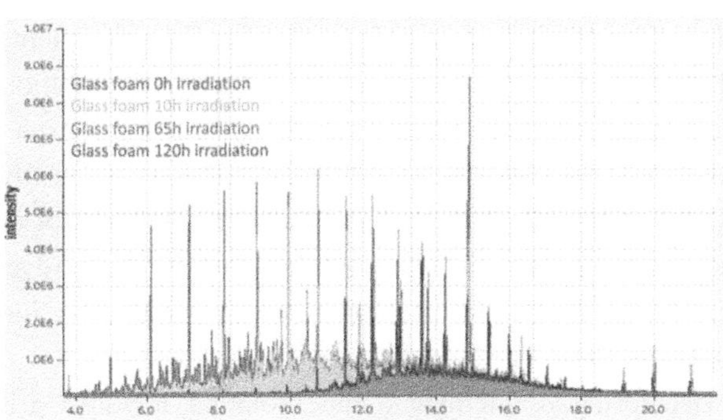

Figure 6 | Chromatogram of the analysis of hydrocarbons after different irradiation times.

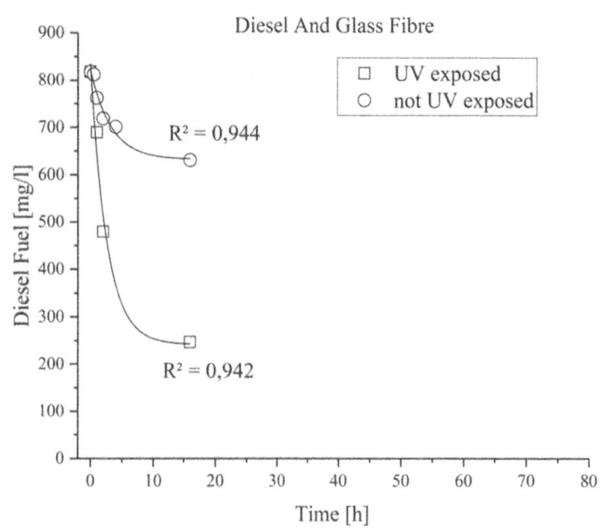

Figure 8 | Diesel fuel disintegration over time with UV light and titanium dioxide-coated glass fiber. Squares represent UV exposed; circles stand for non-UV exposed experiments. Each item represents an independent experiment. The curves were fitted with an exponential decay first order.

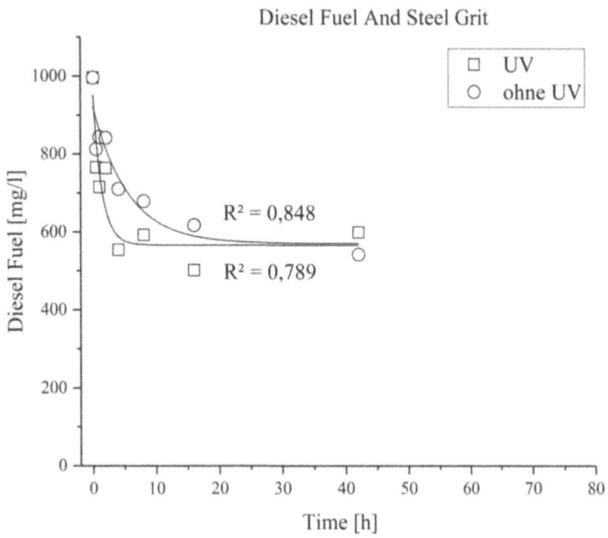

Figure 9 | Diesel fuel disintegration over time with UV light and titanium dioxide-coated steel grit. Squares represent UV exposed; circles stand for non-UV exposed experiments. Each item represents an independent experiment. The curves were fitted with an exponential decay first order.

surface led to inconsistent measurements. One hypothesis is that this effect occurs due to the greater distance of the catalyst surface from the mineral oil phase. However, glass wool, which is also a non-floatable material, showed a very high degradation rate for diesel fuel. It seems more likely that the adsorption of petroleum hydrocarbons is significantly reduced due to the electrochemical properties of steel-based catalyst carriers.

To compare all experimental setups, Table 1 provides the functions and a combination of evaporation and disintegration after 16 h. The results of the comparison showed that the difference between evaporation and photocatalytic experiment is 329 mg/L for the glass foam, 401 mg/L for the glass fiber materials, and 55 mg/L for the steel grid. This means that the glass fiber material achieved the highest photocatalytic disintegration rate, and the steel grid achieved almost no photocatalytic activity with the disintegration of mineral oil.

Thus, the decomposition of diesel fuel with coated foam glass and glass fiber proved to be promising, while the design of the coated steel grid was less efficient. Using immobilized TiO_2, a solar reactor with collectors and groundwater samples with similar hydrocarbon concentrations but a mixture of different hydrocarbons (Cho *et al.* 2006) achieved only a degradation of 10% after 4 h. When

Table 1 | Comparison of the different experiments with mineral oil

Experiment	Function	Evaporation or disintegration after 16 h
Glass foam with UV	$c_t = 271$ mg/L + 418 mg/L* $e^{(-0.057*t)}$	470 mg/L
Glass foam without UV	$c_t = 430$ mg/L + 258 mg/L* $e^{(-0.049*t)}$	141 mg/L
Glass fiber with UV	$c_t = 242$ mg/L + 591 mg/L* $e^{(-0.38*t)}$	589 mg/L
Glass fiber without UV	$c_t = 632$ mg/L + 189 mg/L* $e^{(-0.31*t)}$	188 mg/L
Steel grit with UV	$c_t = 566$ mg/L + 385 mg/L* $e^{(-0.72*t)}$	385 mg/L
Steel grit without UV	$c_t = 570$ mg/L + 349 mg/L* $e^{(-0.18*t)}$	330 mg/L

switching to 0.2% TiO_2 slurry 80% of the hydrocarbons could be degraded after 4 h. In these experiments, the measured solar irradiation power per surface was 16 W/m^2, which is lower than the value chosen in this paper. However, due to the use the reactor designs, a higher specific irradiation surface of 0.55 m^2/L was created. The volume-specific irradiation power of 8.8 W/L is of a similar order of magnitude to the selected experimental design (6.37 W/L). The higher degradation rate of the hydrocarbons (Cho *et al.* 2006) can be explained by the use of groundwater samples which only contain the soluble contents of mineral oil hydrocarbons. In conclusion, it is shown that the floating, carrier-bound catalysts can achieve nearly the same reaction velocity for the degradation of mineral oil as particulate material.

The experiments with methylene blue depletion were conducted with real sunlight in sunny weather conditions, as direct solar radiation had – in contrast to the experiments with diesel fuel – no observable effect on dye concentration via evaporation. As shown in Figure 10, glass foam and steel grit catalysts could disintegrate the pigment more efficiently and in a shorter time than the glass foam. The control of uncoated glass foam had no observable effect on the dye. The attainable degradation was much higher than in the experiments with diesel fuel.

Decoloring of methylene blue solution for steel grit yielded up to 99.1% decoloring after 4 h, for glass fiber 96.6% after 6 h, and for glass foam 56.4% after 6 h. The

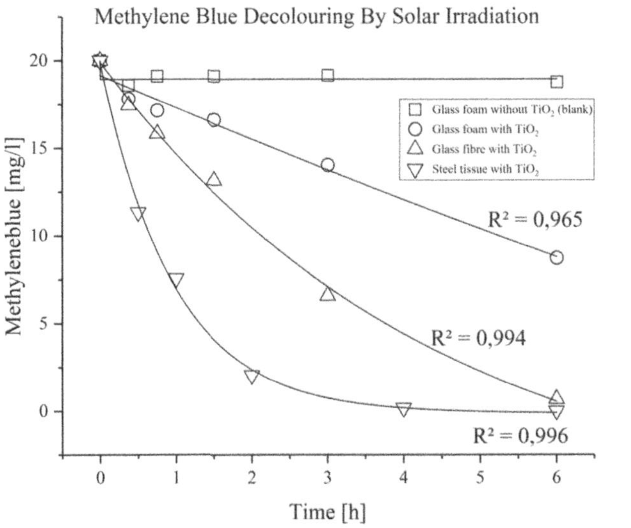

Figure 10 | Methylene blue disintegration over time with natural sunlight and foam glass coated with titanium dioxide, glass fiber, and steel grit as well as uncoated foam glass. Each item represents an independent experiment.

trend line fits well with the values measured with exponential decay of the first order (except foam glass without TiO_2), which diverges from it only to a small extent as shown by its correlation (R^2). The catalysts investigated showed a higher effect on the polar substance methylene blue than on the diesel fuel. Foam glass could disintegrate over 50% of methylene blue after 6 h, while only foam glass was able to mineralize 47% of diesel fuel concentration after 16 h. Glass fiber and steel grit could diminish methylene blue concentration by 96.6% after 4 h 99.1% after 6 h, respectively. Given that the experiments with diesel and glass fiber/steel grit produced inconsistent results, the outcome of the methylene blue examinations, in contrast, showed that the capability of the catalysts relies mostly on adequate adsorption and contact of the organic compounds to make them available for photocatalysis. The non-polar nature of hydrocarbons and the polarity of dissolved methylene blue may also play an important role during adsorption. Table 2 provides the exponential functions of the methylene blue experiment as well as the half-life time for the different materials.

Table 2 | Comparison of the methylene blue discoloration with sunlight

Experiment	Function	Half-life time
Glass foam	$c_t = -1.72\ (\mu mol/L)/h{*}t$	5.8 h
Glass fiber	$c_t = c_0\ e^{(-0.22{*}t)}$	2.4 h
Steel grit	$c_t = c_0\ e^{(-1.03{*}t)}$	0.6 h

CONCLUSION

Within the scope of this work, it could be demonstrated that carrier-bound TiO_2 catalysts can achieve similar degradation rates in their application as floating catalysts to those achieved by particulate catalyst material. However, the investigated catalysts show a slightly reduced reaction rate, which in practical applications should be compensated by a much easier handling. The treatment of mineral oil phases in larger water bodies represents a promising application of this technology. In the carried out experiments, the floating design of foam glass coated with titanium dioxide could reduce the concentration of diesel fuel by 329 mg/L in 16 h; the submerged designs for coated glass fiber and coated steel grit could reduce methylene blue concentration by 96.6% after 4 h and 99.1% after 6 h, respectively. In relation to an application for mineral oil degradation, the question arises as to how the behavior of dissolved mineral oil hydrocarbons in large water bodies with high dilution potentials and higher turbulence is influenced by the positioning of the catalysts. It is assumed that the distance between catalyst and mineral oil phase has a significant influence on the degradation rate. Given this context, the use of self-floatable catalysts, such as glass foam as a carrier material, seems to be promising. The conducted experiments have shown that this material is generally suitable but shows a lower activity. An optimization of the coating process is, therefore, recommended.

For the degradation of diesel fuel, the experiments further showed a limitation to the maximum amount accessible to photocatalytic degradation. Photocatalytic degradation requires a dissolved state for the compounds to be oxidized. For longer-chain mineral oil components, this is only given to a very limited extent. It should be noted that highly insoluble and correspondingly less mobile substances usually have an equally low ecotoxicological impact potential. It would also be feasible to counteract this effect by using solubilizers (e.g. surfactants). In addition, a loss of function is to be feared in the permanent reactor operation due to the retention of a highly viscous, almost inert mineral oil phase. With regard to the development of an effective technical application, the investigation of permanent modes of operation represents an essential step and should possibly be taken into account within future research projects.

A comparison of the tests with mineral oil and methylene blue revealed significant differences in the activity of the catalysts investigated. The catalysts with steel grids as carrier material showed almost no activity for diesel fuel, while they degraded methylene blue at a very high rate. The catalysts with glass foam and glass fiber as carrier material showed the same ranking in activity for both groups of substances. In addition to varying solubility, the investigated substance groups differed substantially with respect to their molecular charge. Catalysts fixed on conductive stainless steel appear to show an affinity for the degradation of polar substances. An increased positive potential on the catalyst would cause an increased adsorption of polar substances and thus explain their higher reaction rate according to the Langmuir–Hinshelwood kinetic. Furthermore, by discharging excited electrons to a counter electrode, the recombination rate of the electron-hole sites is reduced, which can increase the quantum efficiency of the catalyst. Increased degradation rates of photoelectric catalysis using an external voltage source are described in Song *et al.* (2019); Collivignarelli *et al.* (2020); Wang *et al.* (2020). It is also known that electrical potentials are generated by excitation of the semiconductor material titanium dioxide (Sclafani & Herrmann 1996; Li & Shen 2006; Suhadolnik *et al.* 2016). The influence of the carrier material on this electrochemical system has not yet been sufficiently investigated for carrier-based photocatalytic applications. Within this work, overlapping effects of solubility and electric charge were found, which is why an exact quantification of each process is not possible. Therefore, there is a need for further research on the influence of the carrier materials on the obtainable reaction rates for different groups of substances.

Apart from the main objective of the study, it can be stated that carrier-bound solar photocatalysis offers an energy-neutral and low-maintenance option for the oxidation of dissolved substances. While in most applications it is not possible to achieve a continuously operating treatment stage due to varying solar radiation, this technology could provide a supporting purification stage which, in the case of solar irradiation, allows energy and more cost-intensive conventional processes to be relieved. For these applications, the floatability of the catalysts only influences the absorption of solar radiation by the water layer above the catalyst. Due to the higher activity, the use of glass wool or steel grit catalysts, equipped with floating frames or fixed installations, seems to be reasonable. An application idea would be to use the free surfaces of existing post clarification plants for solar photocatalysis as a supplementary measure for the elimination of micro pollutants. The use of carrier-bound catalysts as a sole purification stage for micro pollutants or other organic compounds is also possible. In this case, an artificial irradiation source is required to achieve a permanent and efficient system operation. UV-LEDs have proven to be an efficient source of UV light and electricity consumption.

DATA AVAILABILITY STATEMENT

All relevant data are included in the paper or its Supplementary Information.

REFERENCES

Adams, L. K., Lyon, D. Y. & Alvarez, P. J. J. 2006 Comparative ecotoxicity of nanoscale TiO_2, SiO_2, and ZnO water suspensions. *Water Research* **40** (19), 3527–3532. https://doi.org/10.1016/j.watres.2006.08.004.

Barka, N., Abdennouri, M., Boussaoud, A., Galadi, A., Baâlala, M., Bensitel, M., Sahibed-Dine, A., Nohair, K. & Sadiq, M. 2014 Full factorial experimental design applied to oxalic acid photocatalytic degradation in TiO_2 aqueous suspension. *Arabian Journal of Chemistry* **7** (5), 752–757. https://doi.org/10.1016/j.arabjc.2010.12.015.

Bergmann, A., Fohrmann, R. & Weber, F. A. 2011 *Zusammenstellung von Monitoringdaten zu Umweltkonzentrationen von Arzneimitteln (Compilation of Monitoring Data on Environmental Concentrations of Pharmaceuticals)*. Text 66/2012, Umweltbundesamt, Dessau-Roßlau.

Cabiscol, E. 2000 Oxidative stress in bacteria protein damage by reactive oxygen species. *International Microbiology* **3** (1), 3–8.

Cazoir, D., Fine, L., Ferronato, C. & Chovelon, J.-M. 2012 Hydrocarbon removal from bilgewater by a combination of air-stripping and photocatalysis. *Journal of Hazardous Materials* **235–236**, 159–168. https://doi.org/10.1016/j.jhazmat.2012.07.037.

Cho, I.-H., Kim, L.-H., Zoh, K.-D., Park, J.-H. & Kim, H.-Y. 2006 Solar photocatalytic degradation of groundwater contaminated with petroleum hydrocarbons. *Environmental Progress* **25** (2), 99–109. https://doi.org/10.1002/ep.10124.

Chong, M. N., Jin, B., Chow, C. W. K. & Saint, C. 2010 Recent developments in photocatalytic water treatment technology: a review. *Water Research* **44** (10), 2997–3027. https://doi.org/10.1016/j.watres.2010.02.039.

Collivignarelli, M. C., Abbà, A., Carnevale Miino, M., Arab, H., Bestetti, M. & Franz, S. 2020 Decolorization and biodegradability of a real pharmaceutical wastewater treated by H_2O_2-assisted photoelectrocatalysis on TiO_2 meshes. *Journal of Hazardous Materials* **387**, 121668.

Didier, R., Valerie, K. & Nicolas, K. 2013 Immobilization of a semiconductor photocatalyst on solid supports: methods, material, and application. In: *Photocatalysis and Water Purification, Chapter 6* (P. Pichat & M. Lu, eds). Wiley-VCH Verlag GmbH Co.KGaA, Weinheim, pp. 145–178.

Fox, M. A. & Dulay, M. T. 1993 Heterogeneous photocatalysis. *Chemical Reviews* **93**, 341–357.

Haynes, V. N., Evan Ward, J., Russell, B. J. & Agrios, A. G. 2017 Photocatalytic effects of titanium dioxide nanoparticles on aquatic organisms – current knowledge and suggestions for future research. *Aquatic Toxicology* **185**, 138–148. https://doi.org/10.1016/j.aquatox.2017.02.012.

Irawaty, W., Soetaredjo, F. E. & Ayucitra, A. 2014 Understanding the relationship between organic structure and mineralization rate of TiO_2-mediated photocatalysis. *Procedia Chemistry* **9**, 131–138. https://doi.org/10.1016/j.proche.2014.05.016.

Lazar, M., Varghese, S. & Nair, S. 2012 Photocatalytic water treatment by titanium dioxide: recent updates. *Catalyst* 572–601. https://doi.org/10.3390/catal2040572.

Li, M. C. & Shen, J. N. 2006 Photoelectrochemical oxidation behavior of organic substances on TiO_2 thin-film electrodes. *Journal of Solid State Electrochemistry* **10** (12), 980–986. https://doi.org/10.1007/s10008-005-0043-5.

Li, Z., Wang, X., Ma, B., Wang, S., Zheng, D., She, Z. & Guo, L. 2017 Long-term impacts of titanium dioxide nanoparticles (TiO_2 NPs) on performance and microbial community of activated sludge. *Bioresource Technology* **238**, 361–368. https://doi.org/10.1016/j.biortech.2017.04.069.

Mayer, M. 2017 *Photokatalytische Ozonierung zur Spurenstoffelimination aus Kommunalem Abwasser (Reduction of Trace Pollutants in Waste Water by Photocatalytic Ozonation)*. Master's Thesis, Bauhaus-Universität Weimar, Weimar.

Mehrjouei, M., Müller, S. & Möller, D. 2015 A review on photocatalytic ozonation used for the treatment of water and wastewater. *Chemical Engineering Journal* **263**, 209–219.

Nasir, A. M., Jaafar, J., Aziz, F., Yusof, N., Salleh, W. N. W., Ismail, A. F. & Aziz, M. 2020 A review on floating nanocomposite photocatalyst: fabrication and applications for wastewater treatment. *Journal of Water Process Engineering* **36**, 101300. https://doi.org/10.1016/j.jwpe.2020.101300.

Nosaka, Y. & Nosaka, A. Y. 2013 Identification and roles of the active species generated on various photocatalysts. In: *Photocatalysis and Water Purification, Chapter 1* (P. Pichat & M. Lu, eds). Wiley-VCH Verlag GmbH Co.KGaA, Weinheim, pp. 3–24.

Pedanekar, R. S., Shaikh, S. K. & Rajpure, K. Y. 2020 Thin film photocatalysis for environmental remediation: a status review. *Current Applied Physics* **20** (8), 931–952. https://doi.org/10.1016/j.cap.2020.04.006.

Pinnekamp, J., Keysers, C., Montag, K. & Veltmann, D. 2010 *Elimination von Mikroschadstoffen – Stand der Wissenschaft (Elimination of Micropollutants – State of the Art)*. Gewässerschutz-Wasser-Abwasser.

Pitre, S. P., Yoon, T. & Scaiano, T. 2017 Titanium dioxide visible light photocatalysis: surface association enables photocatalysis with visible light irradiation. *Chemical Communications* **31** (53), 4335–4338.

Qiu, H., Hu, J., Zhang, R., Gong, W., Yu, Y. & Gao, H. 2019 The photocatalytic degradation of diesel by solar light-driven floating BiOI/EP composites. *Colloids and Surfaces A: Physicochemical and Engineering Aspects* **583**, 123996. https://doi.org/10.1016/j.colsurfa.2019.123996.

Schnabel, T. 2016 *Photokatalytischer Abbau von Medikamentenrückständen als Möglichkeit für eine 4.Reinigungsstufe in kommunalen Kläranlagen (Photocatalytic Breakdown of Drug Residues as a Possibility for a 4th Purification Stage in Municipal Sewage Treatment Plants)*. Master theses, Bauhaus-Universität Weimar.

Schnabel, T., Mehling, S. & Londong, J. 2020 Hydrogen-peroxide-assisted photocatalytic water treatment for the removal of anthropogenic trace substances from the effluent of wastewater treatment plants. *Water Science and Technology* **82** (10), 2019–2028. https://doi.org/10.2166/wst.2020.481.

Sclafani, A. & Herrmann, J. M. 1996 Comparison of the photoelectronic and photocatalytic activities of various anatase and rutile forms of titania in pure liquid organic phases and in aqueous solutions. *The Journal of Physical Chemistry* **100** (32), 13655–13661.

Song, T., Li, R., Li, N. & Gao, Y. 2019 Research progress on the application of nanometer TiO_2 photoelectrocatalysis technology in wastewater treatment. *Science of Advanced Materials* **11** (2), 158–165. https://doi.org/10.1166/sam.2019.3411.

Speight, J. G. 2020 *Handbook of Industrial Hydrocarbon Processes*, 2nd edn. Gulf Professional Publishing, an imprint of Elsevier, Cambridge, MA.

Suhadolnik, L., Pohar, A., Likozar, B. & Čeh, M. 2016 Mechanism and kinetics of phenol photocatalytic, electrocatalytic and photoelectrocatalytic degradation in a TiO_2-nanotube fixed-bed microreactor. *Chemical Engineering Journal* **303**, 292–301.

Vargas, R. & Núñez, O. 2010 Photocatalytic degradation of oil industry hydrocarbons models at laboratory and at pilot-plant scale. *Solar Energy* **84** (2), 345–351. https://doi.org/10.1016/j.solener.2009.12.005.

Wang, Y., Zu, M., Zhou, X., Lin, H., Peng, F. & Zhang, S. 2020 Designing efficient TiO_2-based photoelectrocatalysis systems for chemical engineering and sensing. *Chemical Engineering Journal* **381**, 122605.

Weichgrebe, D., Geißen, S. U. & Vogelpohl, A. 1991 *Deponiesickerwasserbehandlung in Biologischen Systemen (Landfill Leachate Treatment in Biological Systems)*. Tagungsband der GVC-Jahrestagung, (Kurzfassung der Vorträge und Poster), Aachen.

First received 30 November 2020; accepted in revised form 20 March 2021. Available online 30 March 2021

 © 2021 The Authors

Application of an integrated dissolved ozone flotation process in centralised fracturing wastewater treatment plant

Zhuodi Yao, Xin Jin, Jingwen Liang, Rui Wang, Pengkang Jin IWA,
Guangping Li, Liushuan Yao and Zhe Wang

ABSTRACT

To solve the problems of unstable chemical oxygen demand (COD), turbidity and suspended solid (SS) removal for the electrocatalytic process and unstable operation of the subsequent ultrafiltration membrane–reverse osmosis membrane in a centralised fracturing wastewater treatment plant in Inner Mongolia, the integrated dissolved ozone flotation (DOF) process was proposed to replace the original electrocatalytic process. Multiple processes, such as ozonation, flotation, coagulation and decolourisation, can be achieved in one integrated DOF reactor. The results showed that the removal efficiency of COD, colour, turbidity and SS in the DOF process could reach 25.4, 49.9, 95 and 96%, respectively. Meanwhile, the treatment cost was reduced by 47% (i.e., 1.77 RMB/m^3 for the DOF process) compared with the electrocatalytic process.

Key words | advanced treatment, dissolved ozone flotation, fracturing wastewater, reconstruction project

Zhuodi Yao
Xin Jin
Jingwen Liang
Rui Wang
Pengkang Jin IWA (corresponding author)
Key Lab of Northwest Water Resource, Environment and Ecology, MOE,
Xi'an University of Architecture and Technology,
Xi'an 710055,
China
E-mail: *pkjin@hotmail.com*

Guangping Li
Liushuan Yao
Zhe Wang
Inner Mongolia Hengsheng Environmental Technology Engineering Co., Ltd,
Ordos 017399,
China

HIGHLIGHTS

- Multiple physio-chemical processes can be achieved in one integrated DOF reactor.
- The DOF exhibited higher removal efficiency compared with the electrocatalytic process.
- The DOF has stable removal efficiency during drastical influent quality variation.
- The treatment cost was reduced by 47% compared with the electrocatalytic process.

This is an Open Access article distributed under the terms of the Creative Commons Attribution Licence (CC BY 4.0), which permits copying, adaptation and redistribution, provided the original work is properly cited (http://creativecommons.org/licenses/by/4.0/).

doi: 10.2166/wrd.2021.096

GRAPHICAL ABSTRACT

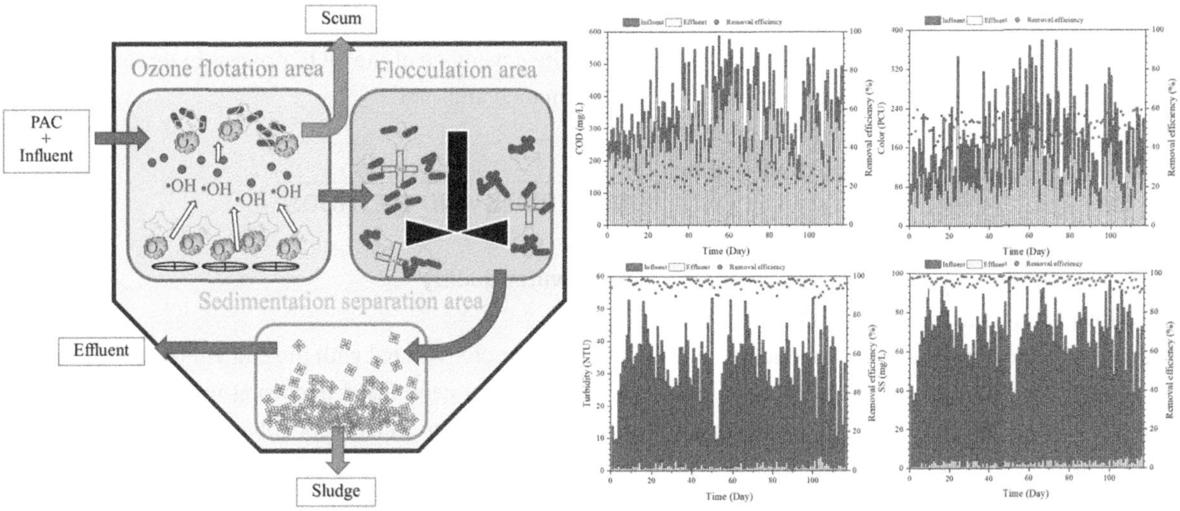

INTRODUCTION

The fracturing wastewater (FWW) comes mainly from the fracturing fluid that flows back to the surface from the wellbore during the fracturing process (Shih *et al.* 2015). Hydraulic fracturing is a technique used to extract oil or natural gas from impermeable host rocks, which involves horizontal drilling into rock formations and injection of high-pressure fracturing fluid (Williams *et al.* 2017; Schultz *et al.* 2020). In the hydraulic fracturing process, up to 24,500 m^3 of water-based fluid can be injected into a single well, and 70% of the fracturing fluid is returned to the surface for harmless disposal (Wang *et al.* 2018a). FWW is a mixture of a variety of compositions, including fracturing fluid additives, as well as geological inorganic and organic substances and their conversion products (Barati & Liang 2014; Luek & Gonsior 2017). FWW has the characteristics of high salt, high organic matter and high viscosity (Yost *et al.* 2016; Stringfellow *et al.* 2017).

In the FWW treatment area, no single treatment process can meet the requirements for stable effluent quality, and two or more combined processes can ensure stable effluent characteristics (Fakhru'l-Razi *et al.* 2009). These independent processes that can be combined include physical, chemical, and biological treatments. Extensive research has shown that: (1) the physical treatment method has the problems of high initial investment and sensitivity to the influent flow; (2) chemical treatment has the problems of high operating cost and disposal of waste sludge; and (3) the biological treatment is sensitive to influent organic matter and salt concentration (Gordalla *et al.* 2013; Olsson *et al.* 2013; Lester *et al.* 2015). Choosing the most suitable FWW treatment technology will ultimately depend on the pollutant characteristics and treatment volume of the single processing unit (Estrada & Bhamidimarri 2016).

A centralised fracturing wastewater treatment plant (CFWTP) is located in Ordos, Inner Mongolia, with an FWW treatment capacity of 5,000 m^3/d. The CFWTP applies microelectrolysis, Fenton oxidation, coagulation and chemical precipitation for the pre-treatment of FWW to initially reduce the chemical oxygen demand (COD) and Ca^{2+} and Mg^{2+} concentrations while increasing the biodegradability of the effluent (Wang *et al.* 2006; He *et al.* 2015; Zhang 2017). The effluent from the pre-treatment stage enters the biological anaerobic–anoxic–oxic (AAO) process for treatment. This stage further reduces the COD in the water while reducing the concentrations of total nitrogen and total phosphorus (Ding *et al.* 2016; Zhang *et al.* 2016). Because FWW has the characteristics of high salt, high organic matter and high viscosity, the effluent from the biological treatment cannot meet the reuse water standard (GB/T18920-2002), which can be used as a compound drilling fluid and for miscellaneous municipal purposes (Lira-Barragán *et al.* 2016). If the AAO effluent can be used for miscellaneous municipal purposes, it must pass through an ultrafiltration membrane and a reverse osmosis membrane (Qurie *et al.* 2013). Therefore, advanced treatment is necessary for the AAO effluent to meet the requirements of reclaimed water.

The existing advanced treatment process for this CFWTP is an electrocatalytic system (ECS) and dual-media filtration. However, due to the large number of chloride ions in the raw water during the actual operation, the chloride ions in the water were oxidised to chlorine gas under the action of the electric current, and the chlorine gas further reacted with water to form hypochlorous acid and hydrochloric acid (Zodi *et al.* 2009; Gao *et al.* 2010). This reaction not only causes a waste of electric energy but also generates oxidising substances that need to be neutralised by adding reducing substances; otherwise, the presence of these miscellaneous will cause irreversible pollution in the subsequent processing system (Dixon *et al.* 2013; Jin *et al.* 2019a). At the same time, partially generated chlorine gas emissions to the air would also threaten the health of on-site operators (Clark *et al.* 2013). To solve the above problems and maintain favourable dissolved organic matter removal efficiency, the proposed integrated DOF process was applied in this study (Jin *et al.* 2006). In this process, ozonation and flocculation are carried out simultaneously during the air flotation process, and the flocs are attached to the microbubbles discharged with the scum (Jin *et al.* 2015). Our previous study indicated that the metal coagulants in the integrated DOF process can act as catalysts to promote the generation of hydroxyl radicals (Jin *et al.* 2016).

The DOF process has been applied in many fields such as urban wastewater treatment plants, printing and dyeing wastewater, milk industry wastewater and cosmetic wastewater (Lee *et al.* 2008; Bogacki *et al.* 2017; Wang *et al.* 2018b; Pereira *et al.* 2020). Nevertheless, there are few reports on the application of advanced FWW treatment. Therefore, it is necessary to carry out long-term experimental research on the application of the DOF process to FWW advanced treatment to evaluate the feasibility of the process. The aim of this study was to investigate the removal performance of the DOF process under long-term operating conditions. In addition, a comparison between the DOF and the original ECS process was conducted to better understand the advantage of the DOF process in terms of removal performance and cost.

MATERIALS AND METHODS

Raw water quality

The DOF reactor was fed with the effluent from the sedimentation tank in a centralised fracturing wastewater treatment plant (CFWTP) in Inner Mongolia, China. The CFWTP mainly treats pre-treated FWW with a biological anaerobic–anoxic–oxic (AAO) treatment process. The main operating parameters of the oxic section are as follows: total hardness = 469.6 ± 187.3 mg/L, total alkalinity = 770.1 ± 386.5 mg/L, total phosphorus = 5.8 ± 1.4 mg/L, total nitrogen = 21.1 ± 6.8 mg/L, SV30 = 48.5 ± 19.5%, SVI = 23.0 ± 10% and sludge concentration = 19,600 ± 4,600 mg/L. The raw water possessed the following characteristics: COD = 396.3 ± 192.7 mg/L, colour = 223.8 ± 155.8 PCU, turbidity = 31.45 ± 22.99 NTU, SS = 67.80 ± 32.72 and pH = 7.89 ± 0.16.

DOF experimental set-up

Figure 1 presents the experimental set-up of the DOF reactor, which contains four main parts: the water inlet system, DOF unit, dosing system and ozone generation system. The DOF unit is composed of two cylindrical reaction vessels, which are independent of each other. A single reactor is divided into three portions, consisting of inner, middle and outer cylinders. The outer cylinder is the ozone flotation area, the middle cylinder is the flocculation area, and the inner cylinder is the sedimentation separation region. The inner diameter of a single reactor column is 3,400 mm, the total height of the main body is 7,000 mm, and the effective volume is 50 m^3.

Table 1 shows the operating parameters of the DOF reactor. Treatment capacity refers to the total influent to the DOF unit. The influent of a single cylindrical reaction

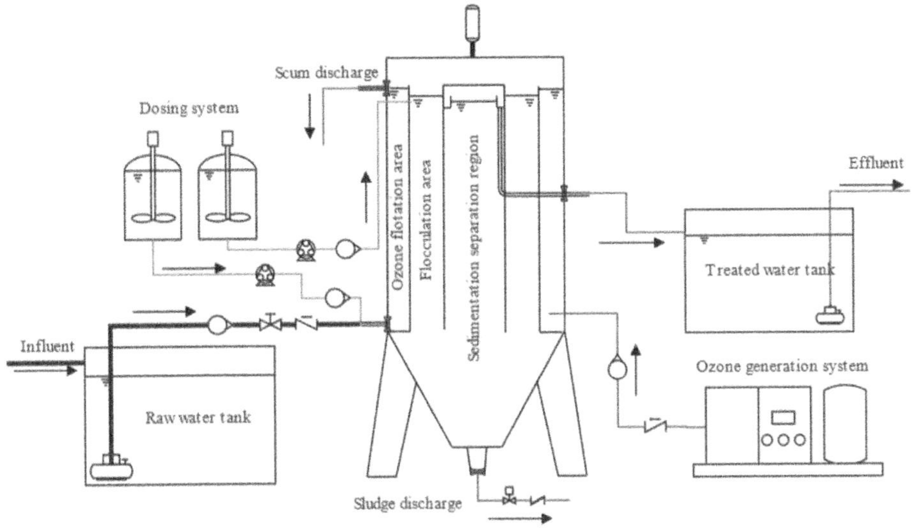

Figure 1 | Experimental set-up of the DOF reactor.

Table 1 | DOF process operating parameters

Project	Value
Treatment capacity	50 m^3/h
PAC dosage	550 mg/L
PAM dosage	2.5 mg/L
Ozone concentration	35 mg O_3/L_{liquid}
Total hydraulic retention time	120 min

vessel is 25 m^3/h, and the remaining items refer to the operating parameters of a single cylindrical reaction vessel.

The raw water is pumped into the DOF reactor, mixed with the poly-aluminium chloride (PAC) in the pipeline mixer and enters the outer cylinder area. Ozone, PAC and water pollutants are in full contact and mixed in this region. On the one hand, organic matter reacts under the combined action of ozone and PAC; on the other hand, the microbubbles generated during the aeration process adsorb suspended pollutants to float on the liquid surface in the form of scum (Jin *et al.* 2016). The scum formed after ozone flotation is removed by the slag scraper, then the water flows to the middle cylinder area through the overflow port, and polyacrylamide (PAM) is added to this area. In the middle cylinder area, the stirring device installed on the top of the processing unit performs slow stirring to make the water fully mixed, and the flocs produced by flocculation enter the inner cylinder area with the water flow. In the inner cylinder area, the sludge-water separation efficiency is improved by the inclined plate, and the clean water is discharged through the outlet weir. The sludge is collected at the bottom of the tank and regularly discharged out of the tank.

Analytical methods

This study measures the COD, colour, turbidity and SS of the water. To avoid the interference caused by the high sodium chloride salinity in the secondary effluent, COD adopts the low-range multiple dilution method in the 'Determination of Water Quality Chemical Oxygen Demand (HJ828-2017)'. Turbidity is measured by a turbidity meter (HACH-2100N, Shanghai, China), and colour and SS are measured by a water quality analyser (HACH-DR900, Qingdao, China). The three-dimensional fluorescence was analysed using a fluorescence spectrophotometer (FP-6500, Jasco, Japan); the three-dimensional fluorescence spectrum was measured at an excitation wavelength of 200–400 nm and an emission wavelength of 280–500 nm.

Data analysis

In this study, two software programmes, origin (OriginPro 2019b) and SPSS (IBM SPSS Statistics 26), were used for data analysis. Between these programmes, Origin is mainly used to analyse the long-term operation data of the DOF-integrated reactor, and the Mann–Whitney tests in SPSS are used to comparatively analyse the treatment effects between DOF and ECS. The Mann–Whitney test is used to test whether two independent samples originate from the same population, that is, to judge whether the mean of two populations is significantly different (Mikocka-Walus & Andrews 2016; Ye & Ahammed 2020).

For the Mann–Whitney test, the data are recoded according to the process: 1 = DOF process and 2 = ECS process. The new sample is called 'Group'. The COD, colour, turbidity and SS removal efficiency for processes 1 and 2 were used as dependent variables and analysed together with the independent variable 'Group'. SPSS itself generates the hypothesis. Null hypothesis, H_0: The distribution of COD (or colour, turbidity and SS) removal efficiency is the same across categories of Group. Alternative hypothesis H_A: The distribution of COD (or colour, turbidity and SS) removal efficiency is not the same across categories of Group.

RESULTS AND DISCUSSION

Organic matter removal performance of the DOF process

COD and colour removal performance

The COD variation in the influent and effluent of the DOF reactor is shown in Figure 2(a). The influent COD of the DOF reactor is 204–588 mg/L, and the treated water COD is 143–457 mg/L. The average COD removal efficiency is 25.4%, and the highest removal efficiency can reach 34%. The influent COD of the DOF reactor varies drastically. An explanation for this variation might be that the raw water of the CFWTP comes from different gas wells as well as different fracturing operation stages. FWW quantity and composition changes significantly in the early stages (e.g., high flow, high COD and low TDS) to the later stages (e.g., low flow, low COD and high TDS from the second week) (Hayes 2009; Slutz *et al.* 2012). Furthermore, during AAO treatment, a high TDS concentration can hinder biological activity and affect COD effluent, so that the quality of the biotreated water is not stable (Lester *et al.* 2014).

Another important finding is that the removal efficiency of the DOF reactor remains relatively stable when the influent COD changes. This finding also agrees with our previous observations, which showed that the organic matter removal performance remained constant when the ozone dosage was greater than 0.8 mg/L (Jin *et al.* 2006; Jin *et al.* 2015). This consistent phenomenon could be attributed to the interactions between ozone and coagulants. Due to the addition of metal coagulants, ozone reacts with the hydroxyl groups on the surface of the hydrolysed metal coagulants in the aqueous solution through electrostatic adsorption and hydrogen bonding force (Jin *et al.* 2018). Studies have found that the carboxyl group on the surface of the metal

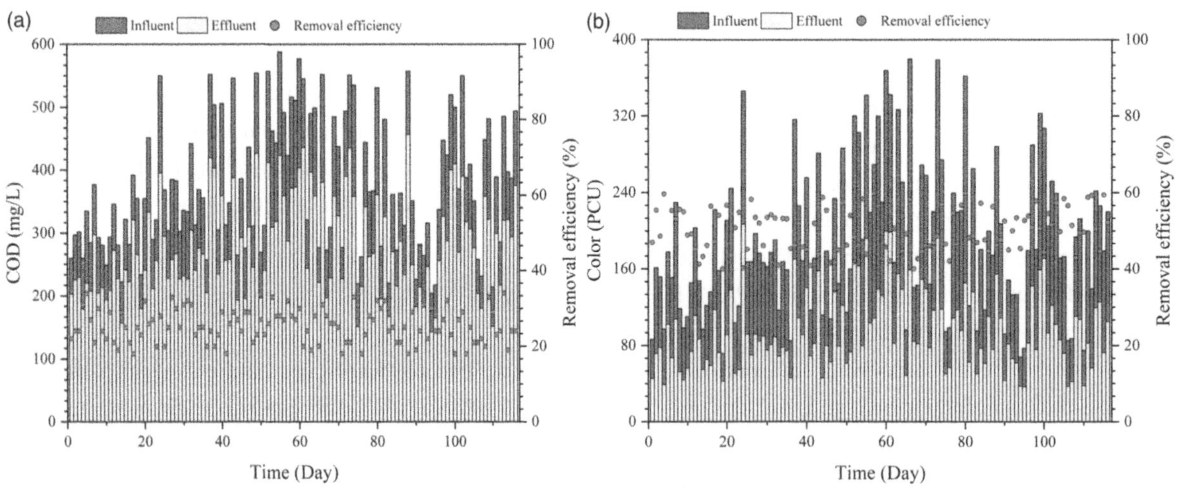

Figure 2 | (a) The removal of COD by the DOF device. (b) The removal of colour by the DOF device.

catalyst will cause the decomposition of ozone to form $^{\cdot}O_2H$, $^{\cdot}O_3H$ and $^{\cdot}O_4H$ and then generate $^{\cdot}OH$ through a chain reaction (Jin *et al.* 2017). In this study, aluminium-based coagulants will form a large number of hydroxyl functional groups on the surface through hydration (Song *et al.* 2019). In addition to the direct reaction of ozone with OH^- in the water to initiate a chain reaction to generate $^{\cdot}OH$, the hydroxyl groups on the surface of the coagulant can also react with ozone to initiate a chain reaction and produce a large amount of $^{\cdot}OH$ (Jin *et al.* 2019b). The simultaneous existence of the above two effects makes the DOF reactor more efficient for organic matter removal.

Figure 2(b) shows the colour removal performance of the DOF process, which shows that the influent colour of the DOF reactor is between 68 and 379 times, with an average of 191 times. The removal of colour by the device is relatively stable. The average removal efficiency is 49.9%, and the highest removal efficiency can reach 59.8%. In accordance with the present results, previous studies have demonstrated that coagulation and ozonation have decolourising effects on wastewater (Collivignarelli *et al.* 2019). Several reports have shown that coagulation can make the wastewater colour removal efficiency reach more than 40% through three processes of coagulation, flocculation and sedimentation (Malik 2004; Ahmad *et al.* 2006; Zonoozi *et al.* 2008; Erkanlı *et al.* 2017). Ozonation has been reported to be able to destroy chromogenic and auxochrome groups such as azo, carbon–carbon double bonds and benzene rings in water to achieve decolourisation (Zhou *et al.* 2013).

Fluorescence characteristics of dissolved effluent organic matter

Figure 3(a) presents the three-dimensional fluorescence spectrum of dissolved organic matter in the biochemical effluent, i.e., the influent of the DOF process. Figure 3(b)–3(d)

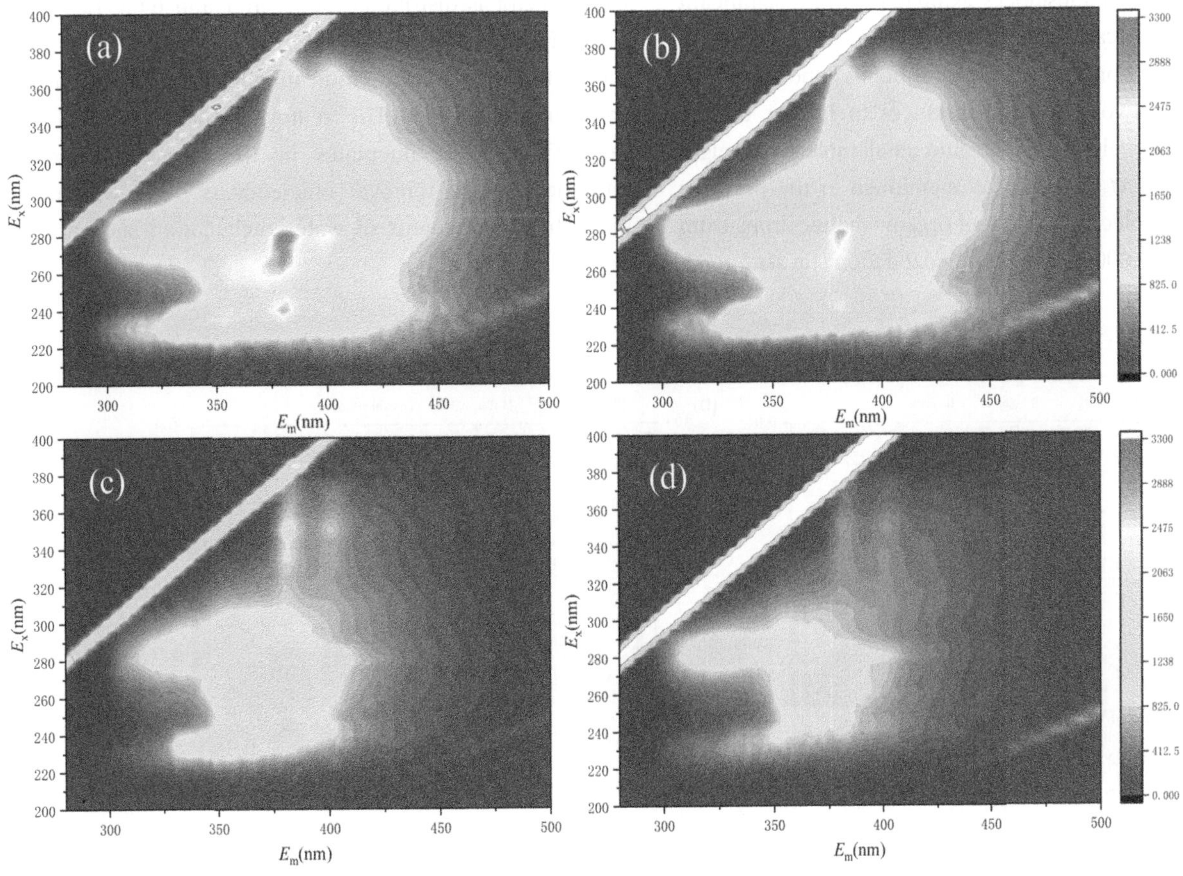

Figure 3 | Three-dimensional fluorescence spectra of (a) raw water and effluent at ozone dosages of (b) 15 mg/L, (c) 25 mg/L and (d) 35 mg/L.

presents the three-dimensional fluorescence spectra of the dissolved organic matter in the effluent of the DOF process at ozone dosages of 15, 25 and 35 mg/L.

There are three main fluorescence peaks in Figure 3(a): the fluorescence peaks of proteins or phenols containing tyrosine aromatic amino acids (emission wavelength 300–340 nm, excitation wavelength 260–290 nm), fulvic acid-like fluorescence peaks of amino acids (emission wavelength 370–440 nm, excitation wavelength 240–270 nm) and fulvic acid-like fluorescence peaks (emission wavelength 370–450 nm, excitation wavelength 310–360 nm) (Komatsu *et al.* 2008). The soluble organic matter in the biochemical effluent is attributed mainly to proteins and fulvic acid, which is consistent with the results of Wang *et al.* (2017). Figure 3(b)–3(d) presents the three-dimensional fluorescence spectra of the effluent with the addition of ozone at 15, 25 and 35 mg/L, respectively. Figure 3 shows that with an increasing ozone concentration, the fluorescence intensity of these two types of substances obviously decreased, and the fluorescence peak of fulvic acid substances exhibited a more significant decrease. Ozone can react with functional groups such as unsaturated carbon–carbon double bonds and aromatic rings of these two types of substances to further oxidise large-molecular-weight organics into small-molecular-weight organics (Lucas *et al.* 2010). As mentioned in the previous research, the molecular weight of organic matter drops from 2,000–6,000 to 2,000–3,000 Da after ozonation (Jin *et al.* 2015).

Suspended pollutant removal performance in the DOF process

Figure 4(a) shows the turbidity removal performance of the DOF process. Figure 4(a) shows that the turbidity of the influent water fluctuates in the range of 10–54 NTU, and the quality of the influent water changes drastically. The turbidity of the effluent of the device is stable within the range of 0–5 NTU. Nevertheless, the effluent of the DOF process can remain stable regardless of the variation in the raw water quality. The turbidity removal efficiency is above 98%, and the average removal efficiency can reach 95%, indicating that the device exhibited excellent turbidity removal ability and can adapt to the variation in water quality.

Figure 4(b) shows the suspended solid (SS) removal performance of the DOF process. Figure 4(b) shows that the influent SS of the DOF reactor fluctuates in the range of 35–100 mg/L. However, the effluent SS of the DOF reactor is distributed between 1 and 6 mg/L, and in 50% of the water samples during the operation period, SS cannot be detected, indicating the excellent removal performance for SS. The removal efficiency for SS by the DOF reactor fluctuates in the range of 90–99%, and the average removal efficiency is 96%, which can meet the requirements of water reclamation such as drilling fluid preparation.

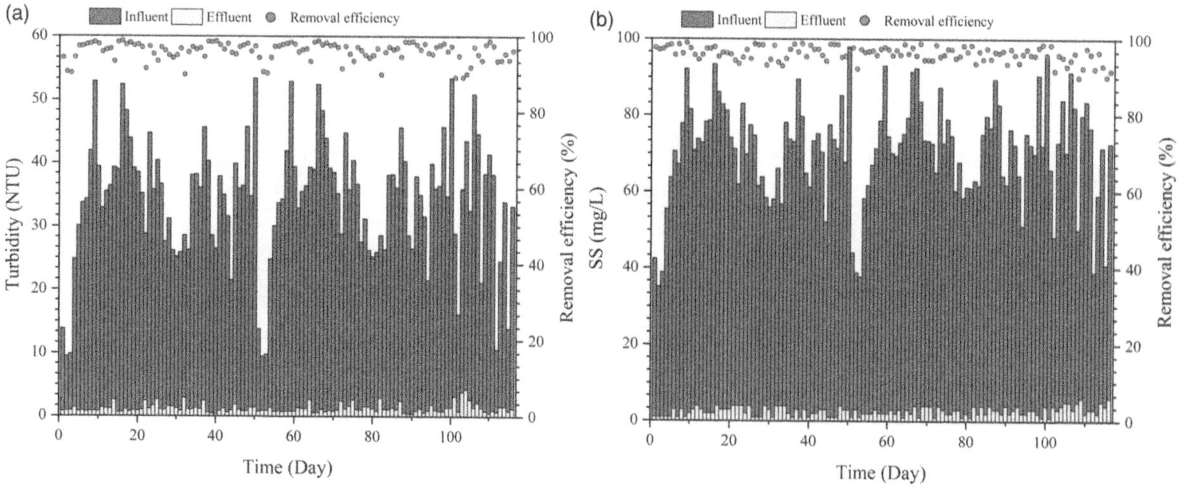

Figure 4 | (a) The removal of turbidity by the DOF device. (b) The removal of SS by the DOF device.

Comparison between the DOF and ECS processes

Removal performance

Table 2 shows the SPSS outputs for the Mann–Whitney tests. Table 2 shows that the significance level is less than 0.05 ($p = 0.000$ for COD, colour, turbidity and SS). Therefore, the test results are statistically significant. The decisions from the hypothesis testing were that the null hypotheses were rejected for all four cases. Therefore, the distribution of the above four detection indicators in the effluent of the DOF and ECS processes is significantly different, showing that the advanced treatment effluent of CFWTP has a significant impact due to changes in the process.

Figure 5 shows the distribution of COD, colour, turbidity and SS removal efficiency in different processes in the Mann–Whitney test. Figure 5(a) shows that the COD removal efficiency distribution of the DOF process is more concentrated, and the removal efficiency of the ECS process is scattered. This inconsistency can be attributed to variable influent quality and electrochemical oxidation characteristics in the ECS process. The removal efficiency has been reported to depend on the water quality of the treated wastewater when the current density is constant (Chen 2004; Anglada *et al.* 2009). Since equipment can only apply a constant current density during operation, the COD removal efficiency of the ECS process varies greatly. In addition, Figure 5(b) shows that the removal efficiency for colour in the ECS process is also unstable. Figure 5(c) and 5(d) show the distribution of turbidity and SS removal efficiency in the two processes. Figure 5(c) and 5(d) show that the turbidity and SS removal efficiency of the DOF process is above 90%, while the removal efficiency of these two parameters in the ECS process is distributed between 20 and 80%, showing that the DOF process exhibits a significant advantage in suspended pollutant removal performance. In the ECS process, due to the high conductivity of the influent water and the characteristics of the ECS process itself, the sacrificial electrodes made of iron and aluminium will quickly dissolve (Mollah *et al.* 2001, 2004). So, consumable metal plates such as iron or aluminium are not used as sacrificial electrodes. Therefore, the ECS process has hardly removal ability for SSs. Therefore, the DOF process replaced the ECS process, and the removal efficiency of all indicators exhibited a significant increase.

Economic evaluation

The present tests were designed to determine the difference in operating costs between the DOF and the ECS processes.

The operating cost of the DOF process applied to the DFWTP secondary effluent (without considering labour costs and equipment depreciation costs) includes the following main parts: flocculant agent (i.e., PAC) cost, polymer coagulant aid (i.e., PAM) cost and power consumption. The operating cost of the original ECS process includes the following main parts: reducing agent (i.e., sodium bisulfite) cost and power consumption. The unit price of PAC solid is 2,000 RMB/t, the unit price

Table 2 | Hypothesis testing summary of the Mann–Whitney test

Null hypothesis	Test	Significance	Decision
The distribution of COD removal efficiency is the same across categories of Groups.	Independent-samples Mann–Whitney *U* test	0.00	Reject the null hypothesis
The distribution of colour removal efficiency is the same across categories of Groups.	Independent-samples Mann–Whitney *U* test	0.00	Reject the null hypothesis
The distribution of turbidity removal efficiency is the same across categories of Groups.	Independent-samples Mann–Whitney *U* test	0.00	Reject the null hypothesis
The distribution of SS removal efficiency is the same across categories of Groups.	Independent-samples Mann–Whitney *U* test	0.00	Reject the null hypothesis

Notes: Asymptotic significance is displayed. The significance level is 0.050.

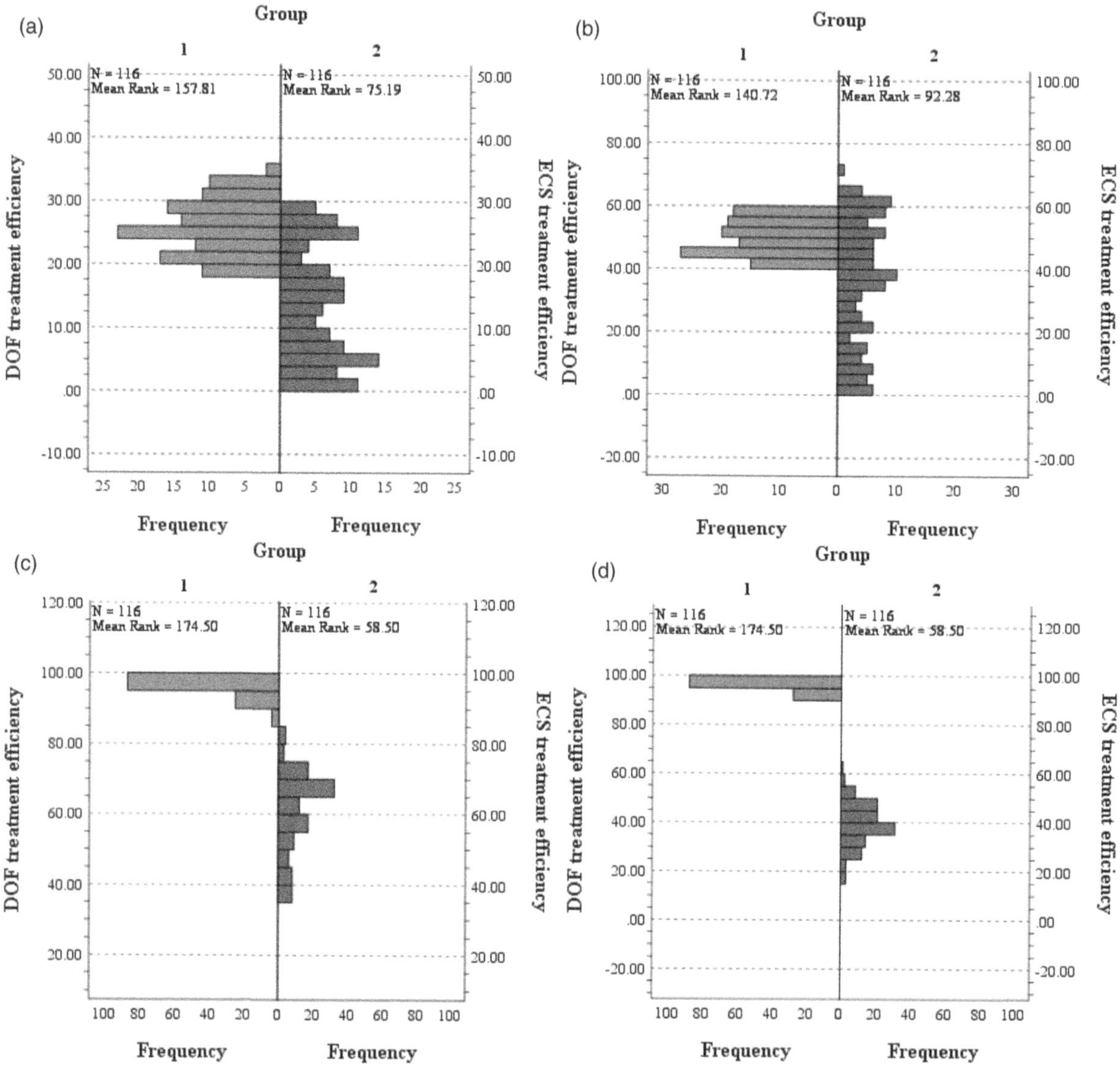

Figure 5 | Independent-samples, Mann–Whitney *U* test of DOF and ECS (a) COD, (b) colour, (c) turbidity and (d) SS removal efficiency.

of PAM solid is 7,900 RMB/t, the unit price of sodium bisulfite solid is 2,400 RMB/t, the electricity price is calculated as 0.45 RMB/kW·h, and the sludge treatment fee is 0.01 RMB/m^3. When the treatment capacity is 50 m^3/h, the corresponding treatment costs for ozone flotation and electrocatalysis are 1.77 and 3.34 RMB/m^3, respectively. The specific calculation results are shown in Table 3.

Table 3 | DOF and ECS process operating cost calculation

Project	PAC dosage (g/m^3)	PAM dosage (g/m^3)	$NaHSO_3$ dosage (g/m^3)	Pharmacy fee (RMB)	Electricity bill (RMB)	Sludge treatment fee (RMB)	Total price (RMB)
DOF	550	2.5	–	1.22	0.55	0.01	1.78
ECS	–	–	780	1.87	1.47	–	3.34

The reduced operating cost of the DOF process comes from the following two aspects. First, compared with the ECS process, the energy consumption of the DOF process exhibited a significant decrease. At the same time, because the DOF process does not produce Cl_2 and HClO, the DOF process will not cause equipment corrosion and personnel health threats. Second, since HClO is not produced, there is no need to add reducing agents before the membrane treatment process is entered. Improvements in the above two aspects led to an improvement in the working environment of workers, removal performance and treatment economy.

CONCLUSIONS

To solve the problems of unstable COD, turbidity and SS removal for electrocatalytic processes with unstable operation of the subsequent ultrafiltration membrane–reverse osmosis membrane in CFWTPs, the study indicated that the DOF process exhibited favourable COD, colour, turbidity and SS removal efficiency, and the removal efficiency reached 25.4, 49.9, 95 and 96%, respectively. Meanwhile, the DOF process effluent can meet the influent requirements of the subsequent dual membrane system. The treatment cost was apparently reduced by 47% (i.e., 1.77 RMB/m^3 for the DOF process) compared with the original ECS process. This work offers valuable insights into the application of the DOF process in the enhanced treatment of the CFWTP secondary effluent.

CONFLICT OF INTEREST

The authors declare that they have no known competing financial interests or personal relationships that could have appeared to influence the work reported in this paper.

ACKNOWLEDGEMENTS

This work was supported by the National Key R&D Programme of China (No. 2019YFB2103003), the National Natural Science Foundation of China (Nos 51708443 and 52070151), the Key Research and Development Project of Shaanxi Province (No. 2019ZDLSF05-03) and the New Style Think Tank of Shaanxi Universities.

DATA AVAILABILITY STATEMENT

All relevant data are included in the paper or its Supplementary Information.

REFERENCES

Ahmad, A., Puasa, S. & Abiding, S. 2006 Crossflow ultrafiltration for removing direct-15 dye from wastewater of textile industry. *ASEAN Journal on Science and Technology for Development* **23** (3), 207–216.

Anglada, A., Urtiaga, A. & Ortiz, I. 2009 Contributions of electrochemical oxidation to waste-water treatment: fundamentals and review of applications. *Journal of Chemical Technology & Biotechnology* **84** (12), 1747–1755.

Barati, R. & Liang, J.-T. 2014 A review of fracturing fluid systems used for hydraulic fracturing of oil and gas wells. *Journal of Applied Polymer Science* **131** (16), n/a.

Bogacki, J. P., Marcinowski, P., Naumczyk, J. & Wiliński, P. 2017 Cosmetic wastewater treatment using dissolved air flotation. *Archives of Environmental Protection* **43** (2), 65–73.

Chen, G. 2004 Electrochemical technologies in wastewater treatment. *Separation and Purification Technology* **38** (1), 11–41.

Clark, K. A., Chanda, D., Balte, P., Karmaus, W. J., Cai, B., Vena, J., Lawson, A. B., Mohr, L. C., Gibson, J. J. & Svendsen, E. R. 2013 Respiratory symptoms and lung function 8–10 months after community exposure to chlorine gas: a public health intervention and cross-sectional analysis. *BMC Public Health* **13** (1), 945.

Collivignarelli, M. C., Abbà, A., Miino, M. C. & Damiani, S. 2019 Treatments for color removal from wastewater: state of the art. *Journal of Environmental Management* **236**, 727–745.

Ding, P., Chu, L. & Wang, J. 2016 Biological treatment of actual petrochemical wastewater using anaerobic/anoxic/oxic process and the microbial diversity analysis. *Applied Microbiology and Biotechnology* **100** (23), 10193–10202.

Dixon, M., Staaks, C., Fabris, R., Vimonses, V., Chow, C. W. K., Panglisch, S., van Leeuwen, J. A. & Drikas, M. 2013 The impact of optimised coagulation on membrane fouling for coagulation/ultrafiltration process. *Desalination and Water Treatment* **51** (13), 2718–2725.

Erkanlı, M., Yilmaz, L., Culfaz-Emecen, P. Z. & Yetis, U. 2017 Brackish water recovery from reactive dyeing wastewater via ultrafiltration. *Journal of Cleaner Production* **165**, 1204–1214.

Estrada, J. M. & Bhamidimarri, R. 2016 A review of the issues and treatment options for wastewater from shale gas extraction by hydraulic fracturing. *Fuel* **182**, 292–303.

Fakhru'l-Razi, A., Pendashteh, A., Abdullah, L. C., Biak, D. R. A., Madaeni, S. S. & Abidin, Z. Z. 2009 Review of technologies for oil and gas produced water treatment. *Journal of Hazardous Materials* **170** (2), 530–551.

Gao, S., Du, M., Tian, J., Yang, J., Yang, J., Ma, F. & Nan, J. 2010 Effects of chloride ions on electro-coagulation-flotation process with aluminum electrodes for algae removal. *Journal of Hazardous Materials* **182** (1–3), 827–834.

Gordalla, B. C., Ewers, U. & Frimmel, F. H. 2013 Hydraulic fracturing: a toxicological threat for groundwater and drinking-water? *Environmental Earth Sciences* **70** (8), 3875–3893.

Hayes, T. D. 2009 *Sampling and Analysis of Water Streams Associated with the Development of Marcellus Shale Gas*. Gas Technology Institute, Des Plaines, IL.

He, Y.-M., Chen, K.-Y. & Zhang, T.-Y. 2016 Advanced treatment of heavy oil wastewater for reuse by the combination of microwave enhanced coagulation and iron/carbon micro-electrolysis. *Journal of Water Reuse and Desalination* **6** (1), 40–49.

Jin, P. K., Wang, X. C. & Hu, G. 2006 A dispersed-ozone flotation (DOF) separator for tertiary wastewater treatment. *Water Science and Technology* **53** (9), 151–157.

Jin, X., Jin, P. & Wang, X. 2015 A study on the effects of ozone dosage on dissolved-ozone flotation (DOF) process performance. *Water Science and Technology* **71** (9), 1423–1428.

Jin, P., Jin, X., Bjerkelund, V. A., Østerhus, S. W., Wang, X. C. & Yang, L. 2016 A study on the reactivity characteristics of dissolved effluent organic matter (EfOM) from municipal wastewater treatment plant during ozonation. *Water Research* **88**, 643–652.

Jin, X., Jin, P., Hou, R., Yang, L. & Wang, X. C. 2017 Enhanced WWTP effluent organic matter removal in hybrid ozonation-coagulation (HOC) process catalyzed by Al-based coagulant. *Journal of Hazardous Materials* **327**, 216–224.

Jin, P., Song, J., Wang, X. C. & Jin, X. 2018 Two-dimensional correlation spectroscopic analysis on the interaction between humic acids and aluminum coagulant. *Journal of Environmental Sciences* **64**, 181–189.

Jin, X., Wang, W., Wang, S., Jin, P., Wang, X. C., Zhang, W., An, W. & Wang, Y. 2019a Application of a hybrid gravity-driven membrane filtration and dissolved ozone flotation (MDOF) process for wastewater reclamation and membrane fouling mitigation. *Journal of Environmental Sciences (China)* **81**, 17–27.

Jin, X., Wang, Y., Zhang, W., Jin, P., Wang, X. C. & Wen, L. 2019b Mechanism of the hybrid ozonation-coagulation (HOC) process: comparison of preformed Al 13 polymer and in situ formed Al species. *Chemosphere* **229**, 262–272.

Komatsu, K., Imai, A., Matsushige, K., Nara, F. & Kawasaki, N. 2008 Characterization of dissolved organic matter (DOM) in Lake Kasumigaura and several DOM sources using method of three-dimensional excitation-emission matrix fluorescence spectra. *Mizu Kankyō Gakkaishi* **31** (5), 261–267.

Lee, B. H., Song, W. C., Manna, B. & Ha, J. K. 2008 Dissolved ozone flotation (DOF) – a promising technology in municipal wastewater treatment. *Desalination* **225** (1), 260–273.

Lester, Y., Yacob, T., Morrissey, I. & Linden, K. G. 2014 Can we treat hydraulic fracturing flowback with a conventional biological process? The case of guar gum. *Environmental Science & Technology Letters* **1** (1), 133–136.

Lester, Y., Ferrer, I., Thurman, E. M., Sitterley, K. A., Korak, J. A., Aiken, G. & Linden, K. G. 2015 Characterization of hydraulic fracturing flowback water in Colorado: implications for water treatment. *Science of The Total Environment* **512**, 637–644.

Lira-Barragán, L. F., Ponce-Ortega, J. M., Serna-González, M. & El-Halwagi, M. M. 2016 Optimal reuse of flowback wastewater in hydraulic fracturing including seasonal and environmental constraints. *AIChE Journal* **62** (5), 1634–1645.

Lucas, M. S., Peres, J. A. & Li Puma, G. 2010 Treatment of winery wastewater by ozone-based advanced oxidation processes (O_3, O_3/UV and O_3/UV/H_2O_2) in a pilot-scale bubble column reactor and process economics. *Separation and Purification Technology* **72** (3), 235–241.

Luek, J. L. & Gonsior, M. 2017 Organic compounds in hydraulic fracturing fluids and wastewaters: a review. *Water Research* **123**, 536–548.

Malik, P. 2004 Dye removal from wastewater using activated carbon developed from sawdust: adsorption equilibrium and kinetics. *Journal of Hazardous Materials* **113** (1–3), 81–88.

Mikocka-Walus, A. & Andrews, J. M. 2016 Experience with and attitudes toward psychotherapy and antidepressants among patients with inflammatory bowel disease and functional gastrointestinal disorders: an online patient survey to inform system design. *Gastroenterology Nursing* **39** (4), 278–286.

Mollah, M. Y. A., Schennach, R., Parga, J. R. & Cocke, D. L. 2001 Electrocoagulation (EC)–science and applications. *Journal of Hazardous Materials* **84** (1), 29–41.

Mollah, M. Y., Morkovsky, P., Gomes, J. A., Kesmez, M., Parga, J. & Cocke, D. L. 2004 Fundamentals, present and future perspectives of electrocoagulation. *Journal of Hazardous Materials* **114** (1–3), 199–210.

Olsson, O., Weichgrebe, D. & Rosenwinkel, K.-H. 2013 Hydraulic fracturing wastewater in Germany: composition, treatment, concerns. *Environmental Earth Sciences* **70** (8), 3895–3906.

Pereira, M. D. S., Borges, A. C., Muniz, G. L., Heleno, F. F. & Faroni, L. R. D. A. 2020 DOF/DAF comparison for the treatment of milk industry wastewater. *Revista Ambiente & água* **15** (5), 1.

Qurie, M., Abbadi, J., Scrano, L., Mecca, G., Bufo, S. A., Khamis, M. & Karaman, R. 2013 Inland treatment of the brine generated from reverse osmosis advanced membrane wastewater treatment plant using epuvalisation system. *International Journal of Molecular Sciences* **14** (7), 13808–13825.

Schultz, R., Skoumal, R. J., Brudzinski, M. R., Eaton, D., Baptie, B. & Ellsworth, W. 2020 Hydraulic fracturing-Induced seismicity. *Reviews of Geophysics* **58** (3), n/a.

Shih, J.-S., Saiers, J. E., Anisfeld, S. C., Chu, Z., Muehlenbachs, L. A. & Olmstead, S. M. 2015 Characterization and analysis of liquid waste from Marcellus Shale gas development. *Environmental Science & Technology* **49** (16), 9557–9565.

Slutz, J. A., Anderson, J. A., Broderick, R. & Horner, P. H. 2012 Key Shale Gas Water Management Strategies: An Economic Assessment Tool; SPE-157532, p. 15; In International Conference on Health, Safety and Environment in Oil and Gas Exploration and Production, 2012; SPE/APPEA International Conference on Health, Safety, and Environment in Oil and Gas Exploration and Production, Perth, Australia.

Song, J., Jin, X., Wang, X. C. & Jin, P. 2019 Preferential binding properties of carboxyl and hydroxyl groups with aluminium salts for humic acid removal. *Chemosphere* **234**, 478–487.

Stringfellow, W. T., Camarillo, M. K., Domen, J. K., Sandelin, W. L., Varadharajan, C., Jordan, P. D., Reagan, M. T., Cooley, H., Heberger, M. G. & Birkholzer, J. T. 2017 Identifying chemicals of concern in hydraulic fracturing fluids used for oil production. *Environmental Pollution (1987)* **220** (PA), 413–420.

Wang, Y.-p., Wang, L.-j., Peng, P.-y. & Lu, T.-h. 2006 Treatment of naphthalene derivatives with iron-carbon micro-electrolysis. *Transactions of Nonferrous Metals Society of China* **16** (6), 1442–1447.

Wang, L., Li, Y.-J., Xiong, Y., Tan, W.-B., Zhang, L.-Y., Li, X., Wang, X.-S., Xu, J.-F., Li, T.-T., Wang, J.-S., Cai, M.-X., Xi, B.-D. & Wang, D.-H. 2017 Spectroscopic characterization of DOM and the nitrogen removal mechanism during wastewater reclamation plant. *PLoS One* **12** (11), e0187355.

Wang, J., Liu, M., Bentley, Y., Feng, L. & Zhang, C. 2018a Water use for shale gas extraction in the Sichuan Basin, China. *Journal of Environmental Management* **226**, 13–21.

Wang, R., Jin, X., Wang, Z., Gu, W., Wei, Z., Huang, Y., Qiu, Z. & Jin, P. 2018b A multilevel reuse system with source separation process for printing and dyeing wastewater treatment: a case study. *Bioresource Technology* **247**, 1233–1241.

Williams, L., Macnaghten, P., Davies, R. & Curtis, S. 2017 Framing 'fracking': exploring public perceptions of hydraulic fracturing in the United Kingdom. *Public Understanding of Science* **26** (1), 89–104.

Ye, Q. & Ahammed, F. 2020 Quantification of relationship between annual daily maximum temperature and annual daily maximum rainfall in South Australia. *Atmospheric and Oceanic Science Letters* **13** (4) 286–293.

Yost, E. E., Stanek, J., DeWoskin, R. S. & Burgoon, L. D. 2016 Overview of chronic oral toxicity values for chemicals present in hydraulic fracturing fluids, flowback, and produced waters. *Environmental Science & Technology* **50** (9), 4788–4797.

Zhang, Z. 2017 Treatment of oilfield wastewater by combined process of micro-electrolysis, Fenton oxidation and coagulation. *Water Science and Technology* **76** (11–12), 3278–3288.

Zhang, M., Wang, C., Peng, Y., Wang, S., Jia, F. & Zeng, W. 2016 Organic substrate transformation and sludge characteristics in the integrated anaerobic anoxic oxic–biological contact oxidation (A2/O–BCO) system treating wastewater with low carbon/nitrogen ratio. *Chemical Engineering Journal* **283**, 47–57.

Zhou, X.-J., Guo, W.-Q., Yang, S.-S., Zheng, H.-S. & Ren, N.-Q. 2013 Ultrasonic-assisted ozone oxidation process of triphenylmethane dye degradation: evidence for the promotion effects of ultrasonic on malachite green decolorization and degradation mechanism. *Bioresource Technology* **128**, 827–830.

Zodi, S., Potier, O., Lapicque, F. & Leclerc, J.-P. 2009 Treatment of the textile wastewaters by electrocoagulation: effect of operating parameters on the sludge settling characteristics. *Separation and Purification Technology* **69** (1), 29–36.

Zonoozi, M. H., Moghaddam, M. R. A. & Arami, M. 2008 Removal of acid red 398 dye from aqueous solutions by coagulation/flocculation process. *Environmental Engineering and Management Journal* **7** (6), 695–699.

First received 29 October 2020; accepted in revised form 24 December 2020. Available online 15 March 2021

© 2021 The Authors

Adsorption of tylosin in wastewater by iron-rich farmland soil and the effect of iron reduction and common cations

Luoying Lin, Xingbao Hu, Jianming Liang, Zhujiang Huang, Guangwei Yu and Yunxiao Chong

ABSTRACT

Livestock wastewater reused in farmland may cause tylosin to stay in farmland soil. Under the influence of some factors, such as irrigation and fertilization, tylosin may desorb and diffuse into the water environment. Batch adsorption experiments and soil column flooding experiments were set up to investigate the effects of several cations and iron reduction on the adsorption, desorption and oxidation removal of tylosin in iron-rich farmland soils (red soil). The results showed that tylosin can be adsorbed by the red soil. The coexistence of these cations significantly reduced its adsorption capacity. The order of influence was as follows: $Ca^{2+} > Mg^{2+} > K^{+} > NH_4^{+} > Na^{+}$. This means that some agricultural farming measures, such as the application of chemical fertilizers, would release the adsorbed tylosin into the farmland. Anaerobic iron reduction and massive production of ferrous ions did not affect the adsorption and desorption of tylosin in the red soil column. Moreover, the ferrous iron could activate persulfate to generate hydroxyl radicals and sulfate radicals which oxidized and removed the tylosin adsorbed in the soil column. Therefore, the iron reduction that occurred during flooding was not a factor causing tylosin release, which provided a way for tylosin in iron-rich farmland soils to be oxidized and removed.

Key words | antibiotic, cation, iron reduction, iron-rich farmland soil, tylosin

Luoying Lin
Xingbao Hu
Zhujiang Huang
Guangwei Yu
Yunxiao Chong (corresponding author)
The College of Natural Resource and Environment,
South China Agricultural University,
Guangzhou 510642,
China
E-mail: *cyx04@scau.edu.cn*

Jianming Liang
Tech and Environmental Group Co., Ltd,
Guangzhou 510000,
China

HIGHLIGHTS

- All of Ca^{2+}, Mg^{2+}, K^{+}, NH_4^{+} and Na^{+} significantly reduced the tylosin adsorption capacity.
- The iron reduction did not affect the adsorption and desorption.
- Persulfate largely removed the tylosin adsorbed in red soil pretreated by iron reduction.
- The Fe^{2+} from iron reduction could activate persulfate to generate free radicals.
- Horizontal flow constructed wetland (HSCW) assisted by persulfate has the potential to remove antibiotics from livestock wastewater.

INTRODUCTION

As an antibiotic widely used in poultry and livestock, tylosin (TYL, $C_{46}H_{77}NO_{17}$, MW 916.10, water solubility: 5 g/L, log K_{ow}: 1.63 (McFarland *et al.* 1997), pK_a: 7.1 (Wollenberger *et al.* 2000)) exists largely in livestock manure in the form of maternal or metabolites (Halling-Sørensen *et al.* 2001; Aust *et al.* 2008; García-Sánchez *et al.* 2013). Hoese *et al.* (2009) found that tylosin could reach 0.3 mg/g in the swine manure. There was 1–30 mg/L tylosin in some manure storage tanks (Kolz *et al.* 2005). In many areas, livestock wastewater was reused for farmland

This is an Open Access article distributed under the terms of the Creative Commons Attribution Licence (CC BY 4.0), which permits copying, adaptation and redistribution, provided the original work is properly cited (http://creativecommons.org/licenses/by/4.0/).

doi: 10.2166/wrd.2021.106

irrigation and fertilization after anaerobic treatment. However, antibiotics, such as tylosin, were rarely eliminated during anaerobic treatment (Ben *et al.* 2009) and would diffuse to farmland soil with wastewater reuse. It is reported that the adsorption distribution coefficient (K_d) of tylosin for some types of farmland soil in the United States was 24–65 L/kg (Hu & Coats 2009) and 10.4–387 L/kg in the Netherlands (Ter Laak *et al.* 2006b), indicating that farmland soil has a strong adsorption capacity for tylosin. Part of the tylosin remaining in farmland soil might be degraded (Hu & Coats 2007; Sassman *et al.* 2007), whereas some factors might cause its further diffusion to natural water environments, endangering the safety of aquatic organisms (Wollenberger *et al.* 2000).

The mineral particles in the soil, especially those containing rich Fe(III), play an important role in the adsorption of tylosin (Guo *et al.* 2013, 2016; Call *et al.* 2019). However, when the soil is flooded (wetland soil or paddy farming) and forms an anaerobic environment, dissimilatory iron reduction (DIR) reactions are likely to occur. During DIR, some bacteria can use Fe(III) in the soil as an electron acceptor, coupled to oxidation of organic matter. The energy released in iron reduction is captured by the bacteria for growth (Lovley 1991). Studies found that DIR caused the release of phosphate and heavy metals absorbed on the surface of soil mineral particles containing Fe(III) (Wang *et al.* 2016; Upreti *et al.* 2019). It is unclear whether tylosin would also be released into the soil solution due to DIR and then spread to downstream rivers with the discharge of water.

Meanwhile, as ion exchange was the main mechanism of adsorption of tylosin on iron minerals, it is easily affected by the ionic strength (Ter Laak *et al.* 2006a; Essington *et al.* 2010). Call *et al.* (2019) revealed that both Na^+ and Ca^{2+} in solution could compete with tylosin for soil surface adsorption sites. Many cations may be brought to the farmland soil by irrigation or fertilization, such as ammonium ions in chemical nitrogen fertilizers, and calcium and magnesium ions in chemical phosphate fertilizers. These cations may also be a factor leading to the release and diffusion of tylosin in farmland.

To prevent tylosin from entering farmland soil and natural water environment, the tylosin in the wastewater needs to be treated before reuse. DIR in the soil may also provide a potential way for the oxidation and removal of tylosin. Ferrous iron generated by DIR could be adsorbed on the surface of the soil and form the Fe(II)/Fe(III) solid-phase interface. It is well known that Fe(II) and Fe(III) are the most efficient transition metals for activating hydrogen peroxide (Anipsitakis & Dionysiou 2004). Sulfate radicals ($SO_4^{\bullet-}$, $E^\circ = 2.5$–3.1 V) produced by the persulfate (PS) have strong oxidation properties. Zhong *et al.* (2015) revealed iron filings containing Fe_3O_4 combined with persulfate can immediately generate a large number of radicals for contaminant degradation. Therefore, the Fe(II)/Fe(III) solid-phase interface of the soil may have the potential to activate persulfate for sulfate radicals, leading to the removal of tylosin.

To verify the above speculation, red soils rich in iron oxides were used to conduct the following investigations: (1) the adsorption characteristics of tylosin on iron-rich red soil and the influence of main cations existing in chemical fertilizer; (2) the effect of DIR on the adsorption of tylosin; and (3) the removal of tylosin by persulfate in red soil pretreated by DIR.

MATERIALS AND METHODS

Tylosin and iron-rich red soil

Tylosin tartrate (TYL, potency: ≥800 units/mg tylosin, CAS No.: 74610-55-2) was purchased from Aladdin® (Shanghai, China) and stored at 4 °C. The red soil was collected from farmland in the Leizhou Peninsula, Guangdong Province. It was air-dried and sieved through 0.15 mm for use. The total iron content in red soil was 53.7 g/kg, and the main iron minerals were hematite (Fe_2O_3) and magnetite (Fe_3O_4).

Effects of different cations

The batch adsorption experiments using 2.0000 ± 0.0005 g red soil and 50 mL tylosin solution were conducted in Erlenmeyer flasks. The adsorption isotherm experiments had different initial concentrations ranging from 0.5 to 100 mg/L. Common cations (Na^+, K^+, NH_4^+, Ca^{2+} and Mg^{2+}) contained in fertilizers were selected as coexisting ions, and their effects on the adsorption of tylosin were investigated. The concentration of these ions was set at 0.01 mol/L and was, respectively, formed by adding NaCl, KCl, NH_4Cl, $CaCl_2$ and $MgCl_2$ to the experimental system. The initial concentration of tylosin was set at 5 mg/L.

According to the results of this experiment, the calcium ion having the greatest influence on the adsorption of tylosin was selected for the desorption experiment.

Effect of DIR

A 2.5 L glass bottle was used to build a sealed flooded soil system (Supplementary Figure S1). Because the red soil particles were tiny and the pore water in the soil column was limited, quartz sand was added to increase pore water. Preliminary experiments showed that tylosin was not adsorbed by quartz sand (Supplementary Figure S2). The glass bottle was filled with about 10 cm soil column, including 90 g red soil, 1,500 g quartz sand and 10 g paddy soil, providing iron-reducing bacteria. During the experiment, nitrogen was purged into the bottle to form the anaerobic environment. The adsorption efficiency of tylosin in the soil column system was verified by preliminary experiments (Supplementary Figure S3).

To explore the effect of DIR on the adsorption of tylosin, three experimental groups were set up. They were named as DIR pretreatment, Continuous DIR and No DIR. The soil columns were submerged with a solution with glucose concentration at 2,000 mg/L in both DIR pretreatment and Continuous DIR systems, and No DIR was treated using a solution without glucose. Except for glucose, the other components of the solutions remained the same (Supplementary Table S1). All three systems were kept in anaerobic incubation, and the ferrous ion in pore water was monitored. When ferrous ion concentration no longer increased, the incubation was terminated and the solution was discharged. Then, one solution containing 2 mg/L tylosin was added into the No DIR and the DIR pretreatment to flood the soil columns. Another solution added to the Continuous DIR included both 2 mg/L tylosin and 2,000 mg/L glucose. Subsequently, pore water in three systems was taken at 1 h, 2, 4, 6, 8 and 10 days to determine the concentrations of tylosin and ferrous ion.

Fe(II)-activated persulfate oxidation

The experimental device was the same as that described in the section 'Effect of DIR'. Four groups were set up this time. All groups were pretreated to produce ferrous ions and to allow tylosin adsorption in the soil column similar to that in DIR pretreatment in the section 'Effect of DIR'. Then, the soil column in four groups was flooded again by different concentrations (0, 3, 6 and 9 mmol/L) of sodium persulfate ($Na_2S_2O_8$) solution when the pretreatment solution was discharged. After 1 day of reaction, sodium persulfate solution was also discharged, and 0.02 mol/L calcium chloride solution was added for desorption. Before the sodium persulfate solution and calcium chloride solution were drained, pore water samples in the soil column of four groups were taken to determine the tylosin concentration.

Radical verification

To verify the types of radicals involved in the Fe(II)-activated persulfate oxidation reaction, methanol and *tert*-butyl alcohol (TBA) were used as the scavenger of $^{\bullet}OH$ and $SO_4^{\bullet -}$. It was reported that methanol can react with both $^{\bullet}OH$ and $SO_4^{\bullet -}$, while TBA reacts mainly with $^{\bullet}OH$ (Eibenberger *et al.* 1978; Buxton *et al.* 1988; Anipsitakis & Dionysiou 2004). When there was excessive methanol or TBA in the experiment, the radicals would first react with them, instead of the target pollutants, indicating that the radicals have participated in the reaction. The experiment was divided into three groups: the control check group, the methanol group and the TBA group. Similarly, all groups were pretreated to produce ferrous ions and to allow tylosin (with the initial concentration of 20 mg/L) to be adsorbed in the soil column. First, the soil columns in three groups after pretreatment were treated by 0.02 mol/L calcium chloride solution to desorb tylosin for 1 day. Then, 10 mL methanol and 10 mL TBA were slowly infiltrated into the soil column in the methanol group and the TBA group, respectively. After 6 h, 10 mL sodium persulfate solution with concentration at 50 g/L was infiltrated into the soil column in all groups again, reaching the concentration of persulfate in pore water of the soil column at 3 mmol/L. After 1 day of reaction, water samples were taken to determine the tylosin concentration.

Analytical methods

The concentrations of tylosin were detected by reverse-phase high-performance liquid chromatography (Agilent 1260 HPLC) with Agilent ZARBAX SB-C18 column (5 μm, 4.6×250 mm) and a UV detector (wavelength at 290 nm for tylosin). The mobile phase (at a flow rate of 0.5 mL/min) was a mixture of 0.01 mol/L KH_2PO_4 (65%) (pH = 2 adjusted by formic acid) and acetonitrile (35%). The injection volume was 100 μL. The column temperature was 30 °C.

The pH value was detected by a Shanghai Rex PHSJ-5 laboratory pH meter. The ferrous ion concentration in the reaction solution was detected by the phenanthroline method, and the soil samples were immediately transferred to 0.5 mol/L HCl (solid–liquid ratio of 1 : 50) to extract for 24 h (García-Balboa *et al.* 2011).

Data analysis

All experiments were in triplicate, and all data represented the average of three independent experiments with error bars representing standard errors. Microsoft® Excel was used for statistical data. One-way ANOVA in IBM SPSS 25 was used to compare the significant differences between treatments with *p*-value <0.05, suggesting a statistically significant difference. Origin 2018 was used to draw the chart.

RESULTS AND DISCUSSION

Adsorption isotherm of tylosin

The adsorption isotherm of tylosin could be well described by the Henry model and the Freundlich model (Figure 1). Compared with them, the correlation coefficient of 0.861 for Langmuir equation fitting was much lower. The Freundlich equation is an empirical formula suitable for the case where the surface of the adsorbent is uneven. The $1/n$ value characterized the degree of nonlinearity of the adsorption isotherm and the strength of the adsorption reaction. The $1/n$ value here was 0.661, indicating that the adsorption was nonlinear. It is suggested that there may be some specific interactions with functional groups of the soil in

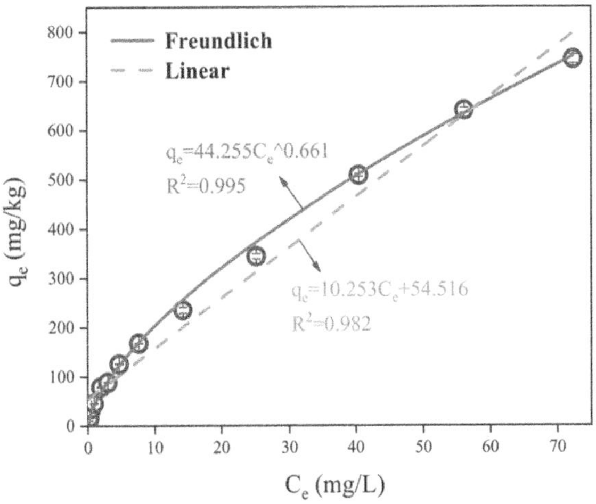

Figure 1 | Tylosin adsorption isotherm on the red soil (initial concentration was 0.5–100 mg/L; contact time was 24 h and temperature was 25 °C).

addition to surface adsorption (Pei *et al.* 2014). For example, polar groups (phenolic and carboxylic groups) in soil organic matter could be bonded to sulfonamides via hydrogen bonds and van der Waals forces (Thiele-Bruhn & Aust 2004; Kahle & Stamm 2007). Meanwhile, when the $1/n$ value was less than 1, the adsorption ability of soil on pollutants decreased with the increase in pollutant concentration, which may be due to the pollutants occupying the high energy sites of the soil first and then being followed by lower energy sites (Guo *et al.* 2013). This is a characteristic of heterogeneous media. The adsorption distribution coefficient (K_d) for tylosin on red soil was 10.253 L/kg, which was similar to previous studies. Zhang *et al.* (2011) had reported K_d for tylosin on agricultural soils ranging from 1.7 to 12 L/kg. Ter Laak *et al.* (2006a) found that K_d of tylosin on another agricultural soil loamy sand was 8.9 L/kg. The result indicated that tylosin could retain in the red soil.

Effects of different cations

Figure 2 shows the adsorption capacity of tylosin on red soil with the coexistence of some cations. It can be clearly seen that these cations significantly decreased the adsorption of tylosin. Among them, the adsorption capacity order was as follows: $Na^+ > NH_4^+ > K^+ > Mg^{2+} > Ca^{2+}$. Tylosin is an ionizable compound with a pK_a of 7.1. It was positively charged under acidic conditions and exists in the form of TYL^+, so adsorption sites of red soil particles were easily

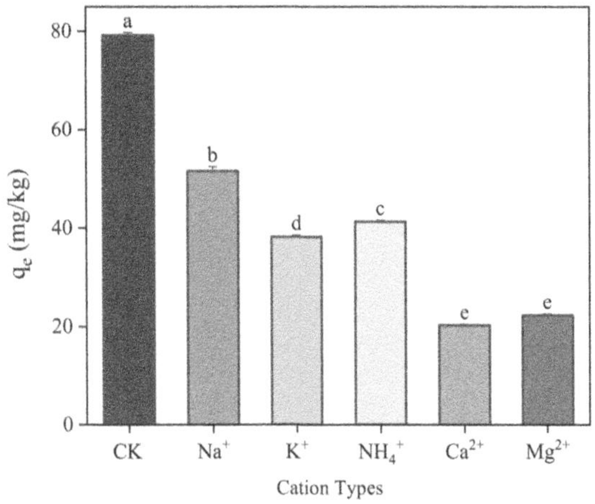

Figure 2 | Effects of different cations' coexistence on the adsorption capacity of tylosin (different letters indicate significant differences at $p < 0.05$).

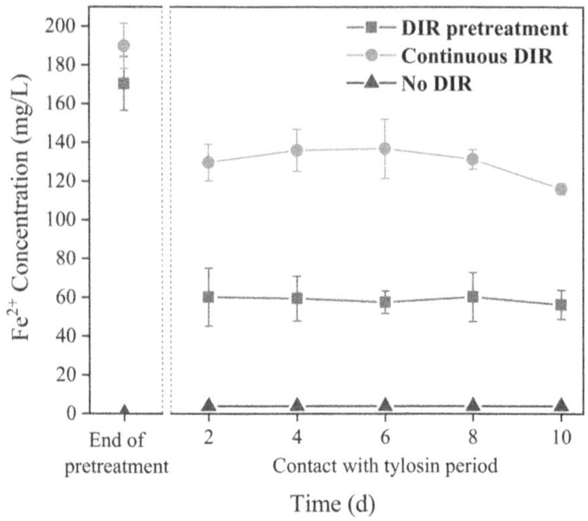

Figure 3 | Ferrous ion concentration in pore water of three groups of soil columns with different pretreatment.

competed by these cations. Moreover, the influence of divalent cations appeared greater than that of monovalent cations, which may be attributed to higher charge densities of divalent cations (Kutzner *et al.* 2014).

Calcium ion was the most competitive. When Ca^{2+} coexisted, the adsorption of tylosin was reduced by 74.4%. The result of desorption experiments showed that, compared with ultrapure water, calcium chloride could release more tylosin adsorbed on red soil (Supplementary Figure S4). This means that many conventional farming measures would cause the tylosin adsorbed on soil to be desorbed and released. The greatest impact may be the treatment of soil acidification (using lime largely) and the application of chemical phosphate fertilizers (calcium phosphate as the main ingredient). Both of them would cause a large quantity of calcium ions to enter the soil. In addition, the application of liquid fertilizers containing NH_4^+, Na^+, Mg^{2+} and Ca^{2+} together, such as integrated water and fertilizer irrigation or biogas slurry (Yu *et al.* 2010), will also be a risk factor leading to the release of tylosin.

Effect of DIR on tylosin adsorption

The glucose in the pretreatment solution caused the reduction of iron oxides in both DIR pretreatment and Continuous DIR systems and a large number of ferrous ions were released into the pore water, reaching 170 and 189.7 mg/L (Figure 3), respectively. In the No DIR system, since the pretreatment liquid did not contain glucose, there were very few ferrous ions in pore water after anaerobic incubation. After adding the solution containing tylosin, the ferrous ion in pore water of the soil column in three types of systems showed significant differences (Figure 3; $p < 0.05$). It was maintained at about 60 mg/L in the DIR pretreatment, which should be caused by the desorption of ferrous ions adsorbed on the red soil particles. The Continuous DIR system had a much higher concentration than the DIR pretreatment, reaching 130 mg/L. This was because the simultaneous addition of glucose with tylosin caused the iron oxides in the red soil to continuously reduce. Ferrous ion was still very low in No DIR (Figure 3).

However, the adsorption of tylosin in the soil columns of the three types of systems did not show significant differences ($p > 0.05$), and their concentrations in pore water and the variation tendency over time were also similar. In all of the systems, the rapid adsorption of tylosin on the red soil in the soil column made it decrease to about 0.1 mg/L in the initial 1 h (Figure 4), and then a small amount of desorption occurred over time, stabilizing between 0.2 and 0.4 mg/L. This means that, within 1 h of contact with tylosin solution, the red soil column in DIR pretreatment and Continuous DIR where iron reduction had occurred still maintained the same quick adsorption capacity for tylosin as that in No DIR without iron reduction. Due to the richness of iron oxides in red soil (Ma *et al.* 2010; Yu *et al.*

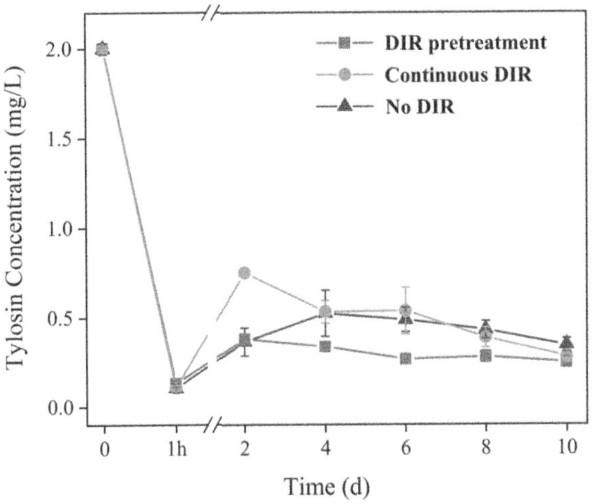

Figure 4 | Tylosin concentration changes in pore water of three groups of soil columns with different pretreatment.

2016), the proportion of reduced iron oxides was relatively small (Supplementary Table S3), so it could not affect the adsorption sites of tylosin on the red soil.

The continued reduction of iron oxides in the red soil column of Continuous DIR did not lead to an increase of tylosin in the pore water, which indicated that the adsorbed tylosin did not undergo a large amount of desorption. This further showed that the DIR, which can cause changes in the chemical form of ferric iron in red soil, had no effect on the adsorption and desorption of tylosin. It was speculated that the inhibitory effect of tylosin on iron-reducing bacteria makes iron oxides adsorbing tylosin difficult to reduce. Tylosin has a great influence on soil microbial activity and microbial community (Westergaard *et al.* 2001; Aldén Demoling & Bååth 2008). Therefore, the reduction of iron oxides may be part of the soil particles that did not adsorb tylosin. In addition, it seemed that ferrous ions did not compete for tylosin adsorption sites of red soil particles like divalent cations such as calcium and magnesium ions. The reason may be the low concentration of ferrous ion, which was lower than that of calcium and magnesium ions in the section 'Effects of different cations'. For the actual iron-rich farmland soil, although the types of organic matter that could induce iron reduction contained in most farmlands were diverse, the iron concentrations in pore water were far lower than that in this study (Garnier *et al.* 2017). This means that the iron reduction and ferrous ions caused by flooding in the soil would be lower than the level of this study. Therefore, it can be inferred that DIR in farmland soil would not affect the adsorption and desorption of tylosin.

Removal of tylosin by the reduced red soil and persulfate

In this study, four groups of red soil columns (pretreated with iron reduction) that adsorbed tylosin were contacted with persulfate solutions of different concentrations and contacted with calcium chloride solution after the persulfate solution was discharged. The tylosin concentration in pore water of each group of soil columns is shown in Figure 5. Among them, the tylosin in the soil column contacted with a solution containing no persulfate (with a concentration of 0) reached 0.126 mg/L, and those with a persulfate concentration of 3, 6 and 9 mmol/L, respectively, were 0.034, 0.029 and 0.021 mg/L. These were significantly lower than the one without persulfate. After contact with calcium chloride solution, tylosin in these three groups of soil columns was also much lower than the first one. This indicated that after the reaction with persulfate, the tylosin adsorbed on the red soil column was largely removed, with the result that the portion of the residual ones that could be released into the pore water was also greatly

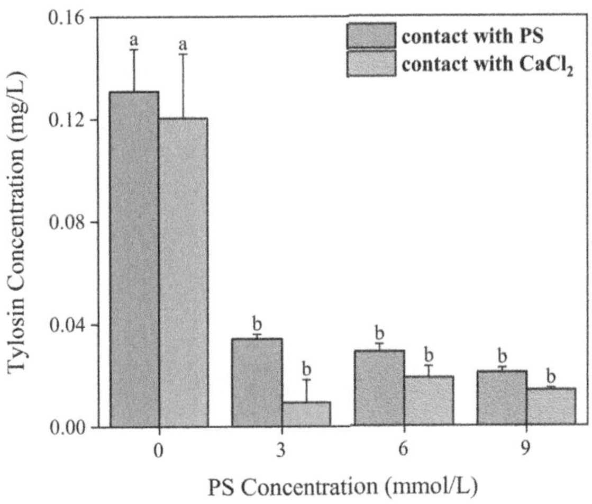

Figure 5 | The tylosin concentration in pore water of four groups of red soil columns after contacting with different persulfate (PS) solutions and calcium chloride solutions successively (different letters indicate significant differences between treatment processes at $p < 0.05$).

reduced. However, the removal of tylosin did not increase with increase of persulfate. Therefore, for the amount of tylosin adsorbed by the soil columns, 3 mmol/L persulfate could achieve the maximum removal. In addition, the variation of ferrous iron before and after reaction in the first group soil column was also much lower than those in the other three groups (Figure 6). Obviously, the ferrous iron in the soil column of the other three groups was involved in the reaction. It can be judged that the Fe(II)-activated persulfate oxidation reaction between ferrous iron and persulfate in these soil columns was likely to occur; the removal of tylosin should be the result of free radical oxidation produced by the reaction. Meanwhile, the variation of ferrous content in red soil columns increased with the increase in persulfate concentration, attributed to the reaction between excessive persulfate and ferrous iron, which was similar to a previous study (Xu & Li 2010).

Radical verification

Figure 7 shows the changes of tylosin in pore water of the soil column of different systems before and after persulfate exposure. After oxidation, the tylosin concentration in the methanol or TBA coexisting system was significantly higher than that in the control system. This means that they prevented the removal of tylosin. Methanol is usually

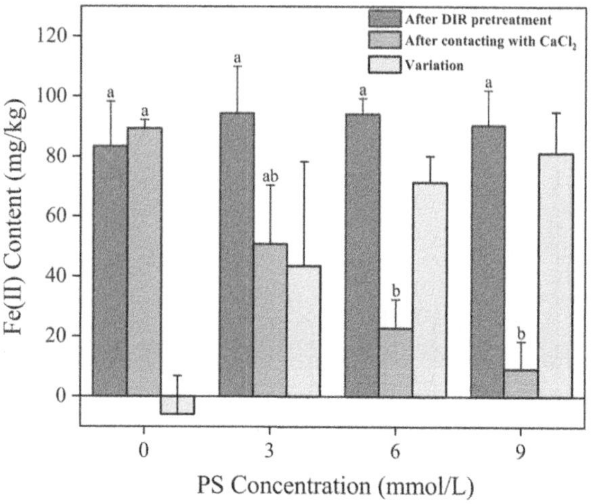

Figure 6 | The change of ferrous content of four groups of red soil columns contacting with different persulfate solutions (different letters indicate significant differences between treatment processes at $p < 0.05$).

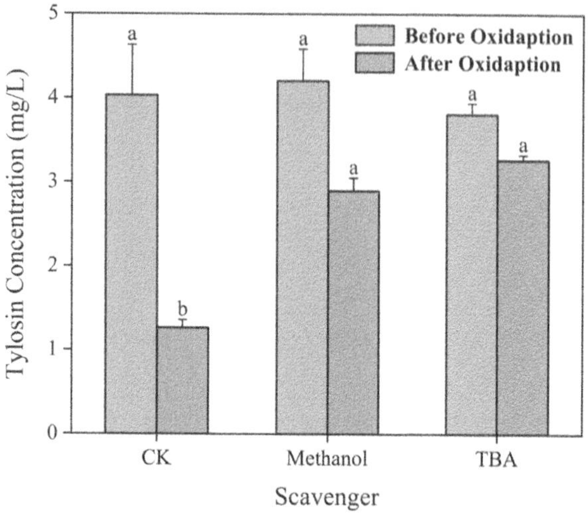

Figure 7 | The changes of tylosin in pore water of the soil column of different systems before and after persulfate exposure (different letters indicate significant differences between the treatment groups at $p < 0.05$).

used as a scavenger for $^{\cdot}OH$ and $SO_4^{\bullet-}$, and TBA was for $SO_4^{\bullet-}$. Therefore, the result confirmed that the two radicals, $^{\cdot}OH$ and $SO_4^{\bullet-}$, participated in the reaction. This is similar to the result of ferrous-activated persulfate oxidation of 2,4-dichlorophenol (Anipsitakis & Dionysiou 2004). $SO_4^{\bullet-}$ formed from Fe(II)-activated persulfate (Equation (1)), and $^{\cdot}OH$ may be from $SO_4^{\bullet-}$ initiated other reactions (Equation (2)) (Eibenberger *et al.* 1978):

$$Fe^{2+} + S_2O_8^{2-} \rightarrow Fe^{3+} + SO_4^{\bullet-} + SO_4^{2-} \quad (1)$$

$$SO_4^{\bullet-} + H_2O \rightarrow {}^{\bullet}OH + HSO_4^{-} \quad (2)$$

This further confirmed that the ferrous iron formed by the anaerobic reduction of the iron-rich soil could activate the persulfate to form an Fe(II)-activated persulfate oxidation system, which in turn caused the adsorbed tylosin to be oxidized and removed. Since free radical was not selective, other organic pollutants could also be removed in this reaction system. Therefore, it is also a potential way to remove other antibiotics in livestock wastewater. The removal was achieved in the following way. Before reusing, the wastewater first entered into a horizontal flow constructed wetland (HFCW) using iron-rich soil as substrate and periodically adding persulfate during operation.

CONCLUSION

Tylosin was retained in the iron-rich red soil by adsorption. The adsorption isotherm was well described by the Henry model and the Freundlich model. A variety of cations contained in chemical fertilizers had an impact on the adsorption of tylosin. Divalent calcium ions and magnesium ions had the greatest impact, which could cause a large amount of adsorbed tylosin to be released. Therefore, tylosin entering the farmland soil was unstable. Anaerobic iron reduction and ferrous iron production had no significant effect on the adsorption and desorption of tylosin in the iron-rich red soil column. The ferrous iron from iron reduction could activate the persulfate to form free radicals, causing the adsorbed tylosin in the soil column to be oxidized and removed. This result can be used as a way to remove tylosin.

ACKNOWLEDGEMENT

This work was supported by the National Natural Science Foundation of China (Nos. 51778245 and 51378226).

DECLARATION OF INTEREST

The authors declare that they have no known competing financial interests or personal relationships that could have appeared to influence the work reported in this paper.

DATA AVAILABILITY STATEMENT

All relevant data are included in the paper or its Supplementary Information.

REFERENCES

Aldén Demoling, L. & Bååth, E. 2008 No long-term persistence of bacterial pollution-induced community tolerance in tylosin-polluted soil. *Environmental Science & Technology* **42** (18), 6917–6921. doi:10.1021/es8004706.

Anipsitakis, G. P. & Dionysiou, D. D. 2004 Radical generation by the interaction of transition metals with common oxidants. *Environmental Science & Technology* **38** (13), 3705–3712. doi:10.1021/es035121o.

Aust, M. O., Godlinski, F., Travis, G. R., Hao, X., McAllister, T. A., Leinweber, P. & Thiele-Bruhn, S. 2008 Distribution of sulfamethazine, chlortetracycline and tylosin in manure and soil of Canadian feedlots after subtherapeutic use in cattle. *Environmental Pollution* **156** (3), 1243–1251. doi:10.1016/j.envpol.2008.03.011.

Ben, W., Qiang, Z., Pan, X. & Chen, M. 2009 Removal of veterinary antibiotics from sequencing batch reactor (SBR) pretreated swine wastewater by Fenton's reagent. *Water Research* **43** (17), 4392–4402. doi:10.1016/j.watres.2009.06.057.

Buxton, G. V., Greenstock, C. L., Helman, W. P. & Ross, A. B. 1988 Critical review of rate constants for reactions of hydrated electrons, hydrogen atoms and hydroxyl radicals ($^{\cdot}OH/^{\cdot}O^-$ in aqueous solution. *Journal of Physical and Chemical Reference Data* **17** (2), 513–886. doi:10.1063/1.555805.

Call, J. J., Essington, M. E. & Rakshit, S. 2019 The cation exchange behavior of tylosin in loess-derived soil. *Chemosphere* **233**, 615–624. doi:10.1016/j.chemosphere.2019.06.028.

Eibenberger, H., Steenken, S., O'Neill, P. & Schulte-Frohlinde, D. 1978 Pulse radiolysis and electron spin resonance studies concerning the reaction of $SO_4^{\cdot -}$ with alcohols and ethers in aqueous solution. *The Journal of Physical Chemistry* **82** (6), 749–750. doi:10.1021/j100495a028.

Essington, M. E., Lee, J. & Seo, Y. 2010 Adsorption of antibiotics by montmorillonite and kaolinite. *Soil Science Society of America Journal* **74** (5), 1577. doi:10.2136/sssaj2009.0283.

García-Balboa, C., Vicente, M. S., Blázquez, M. L., González, F., Muñoz, J. A. & Ballester, A. 2011 Iron speciation in dissimilatory Fe(III)-reducing cultures. *Geomicrobiology Journal* **28** (4), 371–379. doi:10.1080/01490451.2010.506100.

García-Sánchez, L., Garzón-Zúñiga, M. A., Buelna, G., Moeller-Chávez, G. E., Noyola, A., Avilez-Flores, M. & Estrada-Arriaga, E. B. 2013 Occurrence of tylosin in swine wastewater in Mexico. *Water Science and Technology* **68** (4), 894–900. doi:10.2166/wst.2013.323.

Garnier, J., Garnier, J.-M., Vieira, C. L., Akerman, A., Chmeleff, J., Ruiz, R. I. & Poitrasson, F. 2017 Iron isotope fingerprints of redox and biogeochemical cycling in the soil-water-rice plant system of a paddy field. *Science of The Total Environment* **574**, 1622–1632. doi:10.1016/j.scitotenv.2016.08.202.

Guo, X., Yang, C., Dang, Z., Zhang, Q., Li, Y. & Meng, Q. 2013 Sorption thermodynamics and kinetics properties of tylosin and sulfamethazine on goethite. *Chemical Engineering Journal* **223**, 59–67. doi:10.1016/j.cej.2013.02.115.

Guo, X., Miao, Y., Yang, C., Zhang, Q., Gao, L. & Hu, Y. 2016 Sorption of tylosin on black carbon from different sources. *Journal of Environmental Chemical Engineering* **4** (3), 3393–3400. doi:10.1016/j.jece.2016.07.012.

Halling-Sørensen, B., Jensen, J., Tjørnelund, J. & Montforts, M. H. M. M. 2001 Worst-case estimations of predicted environmental soil concentrations (PEC) of selected veterinary antibiotics and residues used in Danish agriculture. *Pharmaceuticals in the Environment* 143–157. doi:10.1007/978-3-662-04634-0_13.

Hoese, A., Clay, S. A., Clay, D. E., Oswald, J., Trooien, T., Thaler, R. & Carlson, C. G. 2009 Chlortetracycline and tylosin runoff from soils treated with antimicrobial containing manure. *Journal of Environmental Science and Health, Part B* **44** (4), 371–378. doi:10.1080/03601230902801075.

Hu, D. & Coats, J. R. 2007 Aerobic degradation and photolysis of tylosin in water and soil. *Environmental Toxicology and Chemistry* **26** (5), 884–889. doi:10.1897/06-197r.1.

Hu, D. & Coats, J. R. 2009 Laboratory evaluation of mobility and sorption for the veterinary antibiotic, tylosin, in agricultural soils. *Journal of Environmental Monitoring* **11** (9), 1634–1638. doi:10.1039/b900973f.

Kahle, M. & Stamm, C. 2007 Sorption of the veterinary antimicrobial sulfathiazole to organic materials of different origin. *Environmental Science & Technology* **41** (1), 132–138. doi:10.1021/es061198b.

Kolz, A. C., Moorman, T. B., Ong, S. K., Scoggin, K. D. & Douglass, E. A. 2005 Degradation and metabolite production of tylosin in anaerobic and aerobic swine-manure lagoons. *Water Environment Research* **77** (1), 49–56. doi:10.2175/106143005(41618.

Kutzner, S., Schaffer, M., Börnick, H., Licha, T. & Worch, E. 2014 Sorption of the organic cation metoprolol on silica gel from its aqueous solution considering the competition of inorganic cations. *Water Research* **54**, 273–283. doi:10.1016/j.watres.2014.01.042.

Lovley, D. R. 1991 Dissimilatory Fe(III) and Mn(IV) reduction. *Microbiological Reviews* **55** (2), 259–287. doi:10.1128/mmbr.55.2.259-287.1991.

Ma, L., Xu, R. & Jiang, J. 2010 Adsorption and desorption of Cu(II) and Pb(II) in paddy soils cultivated for various years in the subtropical China. *Journal of Environmental Sciences* **22** (5), 689–695. doi:10.1016/s1001-0742(09)60164-9.

McFarland, J. W., Berger, C. M., Froshauer, S. A., Hayashi, S. F., Hecker, S. J., Jaynes, B. H., Jefson, M. R., Kamicker, B. J., Lipinski, C. A., Lundy, K. M., Reese, C. P. & Vu, C. B. 1997 Quantitative structure–activity relationships among macrolide antibacterial agents: in vitro and in vivo potency against *Pasteurella multocida*. *Journal of Medicinal Chemistry* **40** (9), 1340–1346. doi:10.1021/jm960436i.

Pei, Z., Yang, S., Li, L., Li, C., Zhang, S., Shan, X., Wen, B. & Guo, B. 2014 Effects of copper and aluminum on the adsorption of sulfathiazole and tylosin on peat and soil. *Environmental Pollution* **184**, 579–585. doi:10.1016/j.envpol.2013.09.038.

Sassman, S. A., Sarmah, A. K. & Lee, L. S. 2007 Sorption of tylosin A, D, and A-aldol and degradation of tylosin A in soils. *Environmental Toxicology and Chemistry* **26** (8), 1629. doi:10.1897/07-007r.1.

Ter Laak, T. L., Gebbink, W. A. & Tolls, J. 2006a The effect of pH and ionic strength on the sorption of sulfachloropyridazine, tylosin, and oxytetracycline to soil. *Environmental Toxicology and Chemistry* **25** (4), 904. doi:10.1897/05-232r.1.

Ter Laak, T. L., Gebbink, W. A. & Tolls, J. 2006b Estimation of soil sorption coefficients of veterinary pharmaceuticals from soil properties. *Environmental Toxicology and Chemistry* **25** (4), 933. doi:10.1897/05-229r.1.

Thiele-Bruhn, S. & Aust, M.-O. 2004 Effects of pig slurry on the sorption of sulfonamide antibiotics in soil. *Archives of Environmental Contamination and Toxicology* **47** (1). doi:10.1007/s00244-003-3120-8.

Upreti, K., Maiti, K. & Rivera-Monroy, V. H. 2019 Microbial mediated sedimentary phosphorus mobilization in emerging and eroding wetlands of coastal Louisiana. *Science of The Total Environment* **651**, 122–133. doi:10.1016/j.scitotenv.2018.09.031.

Wang, Y., Liu, X., Si, Y. & Wang, R. 2016 Release and transformation of arsenic from As-bearing iron minerals by Fe-reducing bacteria. *Chemical Engineering Journal* **295**, 29–38. doi:10.1016/j.cej.2016.03.027.

Westergaard, K., Müller, A., Christensen, S., Bloem, J. & Sørensen, S. 2001 Effects of tylosin as a disturbance on the soil microbial community. *Soil Biology and Biochemistry* **33** (15), 2061–2071. doi:10.1016/s0038-0717(01)00134-1.

Wollenberger, L., Halling-Sørensen, B. & Kusk, K. 2000 Acute and chronic toxicity of veterinary antibiotics to *Daphnia magna*. *Chemosphere* **40** (7), 723–730. doi:10.1016/s0045-6535(99)00443-9.

Xu, X.-R. & Li, X.-Z. 2010 Degradation of azo dye Orange G in aqueous solutions by persulfate with ferrous ion. *Separation and Purification Technology* **72** (1), 105–111. doi:10.1016/j.seppur.2010.01.012.

Yu, F.-B., Luo, X.-P., Song, C.-F., Zhang, M.-X. & Shan, S.-D. 2010 Concentrated biogas slurry enhanced soil fertility and tomato quality. *Acta Agriculturae Scandinavica, Section B – Plant Soil Science* **60** (3), 262–268. doi:10.1080/09064710902893385.

Yu, H.-Y., Li, F.-B., Liu, C.-S., Huang, W., Liu, T.-X. & Yu, W.-M. 2016 Iron redox cycling coupled to transformation and immobilization of heavy metals: implications for paddy rice safety in the red soil of South China. *Advances in Agronomy* **137**, 279–317. doi:10.1016/bs.agron.2015.12.006.

Zhang, Q., Yang, C., Dang, Z. & Huang, W. 2011 Sorption of tylosin on agricultural soils. *Soil Science* **176** (8), 407–412. doi:10.1097/ss.0b013e3182247420.

Zhong, H., Brusseau, M. L., Wang, Y., Yan, N., Quig, L. & Johnson, G. R. 2015 In-situ activation of persulfate by iron filings and degradation of 1,4-dioxane. *Water Research* **83**, 104–111. doi:10.1016/j.watres.2015.06.025.

First received 29 October 2020; accepted in revised form 18 January 2021. Available online 16 February 2021

 © 2021 The Authors

Feeding preparation strategy for the supercritical water oxidation (SCWO) system for disposing of liquid hazardous waste

Bo Wang, Meng Lu, Lixin Zhao, Zhezhou Zhang, Yihua Zhao and Songyan Qin

ABSTRACT

Supercritical water oxidation (SCWO) technology has a great potential application for the disposal of liquid hazardous wastes. The feeding process is vital for the safety and normal operation of SCWO. This paper describes the feeding preparation strategy of SCWO with multilevel grouping and programming. Based on the physicochemical properties, the liquid hazardous wastes from various industrial manufacturers were first grouped into acid, neutral and alkaline groups. By C_n^2 mixing between samples, the primary grouping tanks were determined to be acid, alkali, amphoteric reacting, organic and organic high-chlorine liquid wastes. By distributing acid, alkali, amphoteric reacting liquid wastes into organic wastes, three parallel feeding routes are regulated for the homogenizing tank phase, which avoids the reaction between wastes. By calculating with the linear programming optimization model of MATLAB, the waste compatibility ratio of each feeding route was determined to meet the feeding requirements of SCWO. The feeding preparation strategy of this paper provides a practical instruction for the SCWO design and operation.

Key words | feeding progress, feeding strategy, liquid hazardous waste, supercritical water oxidation

Bo Wang
Meng Lu
Lixin Zhao (corresponding author)
Zhezhou Zhang
Songyan Qin
School of Environmental Science and Safety Engineering,
Tianjin University of Technology,
Tianjin 300384,
China
E-mail: *qinsongyan@vip.126.com*

Bo Wang
Yihua Zhao
Tianjin Eco-City Water Investment and Construction Co. Ltd,
Tianjin 300467,
China

HIGHLIGHTS

- Supercritical water oxidation (SCWO) technology has great potential application for the disposal of liquid hazardous wastes.
- This paper describes the feeding preparation strategy of SCWO with multilevel grouping and programming.
- The feeding preparation strategy of this paper provides a practical instruction for the SCWO design and operation.

This is an Open Access article distributed under the terms of the Creative Commons Attribution Licence (CC BY 4.0), which permits copying, adaptation and redistribution, provided the original work is properly cited (http://creativecommons.org/licenses/by/4.0/).

doi: 10.2166/wrd.2021.089

GRAPHICAL ABSTRACT

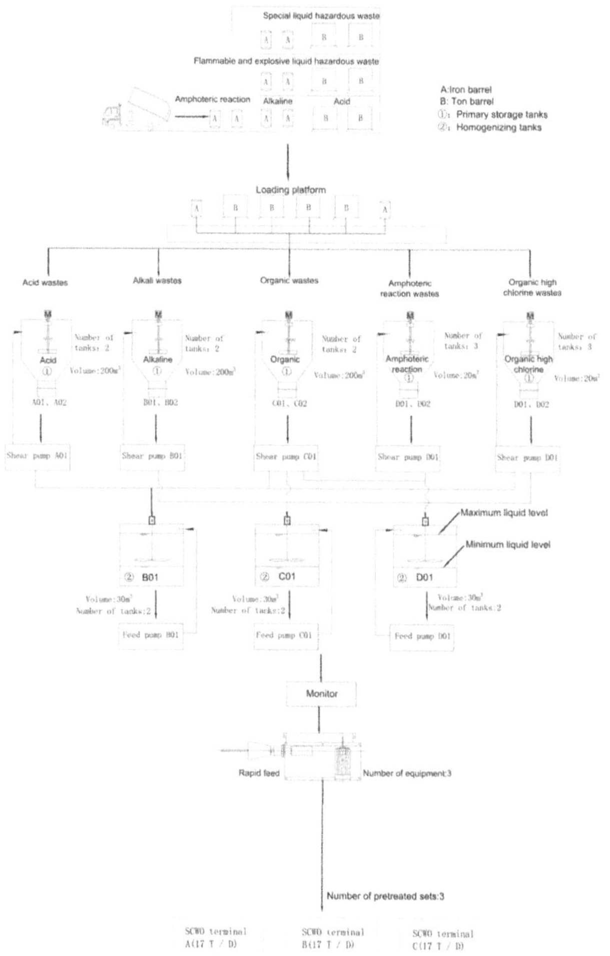

INTRODUCTION

Hazardous waste liquor, which is identified based on the National Catalogue of Hazardous Wastes (NCHWs, hazardous waste list in China) (Ministry of Environmental Protection of the People's Republic of China 2016) or standards (Ministry of Ecology and Environment of the People's Republic of China 2019), usually has one or more hazardous characteristics, such as corrosiveness, toxicity, flammability, reactivity and infectivity. These characteristics would cause environmental pollution incidents or casualties, if the hazardous waste is inappropriately disposed of (Duan *et al.* 2008; Quah & Cockerham 2017; Yang *et al.* 2020). Therefore, the safe disposal and management of hazardous wastes are essential for humans' wellbeing worldwide (El-Fadel *et al.* 2001). Traditional ways to dispose of hazardous waste are landfill and incineration, but the secondary pollution from landfill leachate (Verma & Kumar 2017) or incineration discharged air still pose health threats to humans.

Supercritical water oxidation (SCWO) is gradually becoming an attractive technique for hazardous waste disposal (Harradine *et al.* 1993; Yang *et al.* 2019) because of its high efficiency, stable operation, wide application and lack of secondary pollution. The mechanism of SCWO is to use supercritical state water as a reaction medium, which exploits the unique solvating properties (TC = 374.15°C, PC = 22.12 MPa), to provide enhanced solubility of organic reactants, so that the organic matter is completely dissolved in water to convert into small molecular compounds and then eventually turned to CO_2, N_2, water, etc. The composed elements such as Cl, S and P are converted into chloride, sulfate and phosphate, and the metal is converted into a non-hazardous substance such as steady-state oxides (Franck 1984). In recent years, China has encouraged SCWO's application to hazardous waste disposal. The first demonstration plant for SCWO was begun in 2015, aiming at harmless disposal of organic waste collected from various industrial factories in Tianjin. The feeding requirements for the demonstrated SCWO process are listed in Table 1. To get the qualified feeding, the grouping strategy is vital because it ensures the stable operating status, maximizes the economic cost and improves the safety management of the SCWO process. However, current research literature on SCWO in currently operational plants is still scarce, especially for the SCWO's feeding preparation process. Besides, the different type of SCWO reactors have their own specific feed requirements, with the feeding source and nature also varying in different regions. Therefore, the operating parameters of feeding preparation units such as material grouping and mixing ratio should be obtained through laboratory studies based on the characters of feeding hazardous liquor.

The feeding process of liquid hazardous waste for SCWO studied in this paper is shown in Figure 1. The liquid hazardous waste enters the primary storage tanks through the raw material collecting platform, and the waste liquid in different primary storage tanks pumps into the corresponding homogeneous tanks based on sample compatibility; meanwhile, the calorific value and pH value are adjusted in homogeneous tanks. The qualified material is fed through the spray or injection pump into the SCWO terminal. Although the feeding process of liquid hazardous waste is relatively conventional, operating parameters such as raw material grouping and mixing ratio need to be determined through laboratory studies.

This paper will propose a feeding preparation strategy for an upcoming demonstrated SCWO process in China. Physicochemical properties of the samples from industrial factories were analyzed first, and then compatibility tests of samples were carried out to determine the group of primary tanks. An algorithm model of compatible feeding was constructed for the homogenizing tank. This paper provides a guiding framework for the design and operation of the feeding treatment process of the liquid hazardous wastes disposal based on the SCWO system.

Table 1 | Feeding requirements for the SCWO system in a demonstration plant in Tianjin, China

Index	Range
pH	6–10
Calorific value	5.10 ± 5% MJ/kg
Total chlorine	<10,000 mg/L
Total phosphorus	<20 mg/L
Total nitrogen	<10,000 mg/L
Total salinity	<80 g/L
Particle size	<500 μm

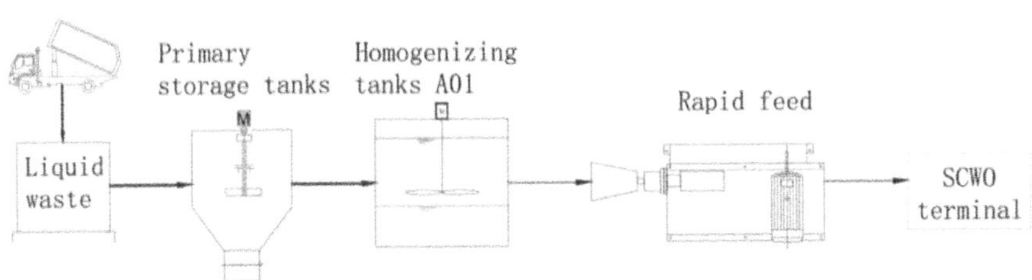

Figure 1 | Liquid hazardous waste feeding process of SCWO terminal.

METHODOLOGY

Hazardous waste properties analysis

Liquid hazardous waste samples were taken from 14 industrial plants (Table 2). The physicochemical properties and the corresponding analysis method are listed in Supplementary Table S1.

Table 2 | Industry liquid hazardous waste samples

Number	Liquid hazardous waste information
1	Machinery factory washing waste liquid (1)
2	Water-based paint plant degreaser (2)
3	Resources recycling company metal waste acid (3)
4	Water-based paint factory waste liquid one (4)
5	Water-based paint factory water-containing solvent waste liquid (5)
6	Machinery plant sodium carbonate waste liquid (6)
7	Machinery factory cleaning agent waste liquid (7)
8	Machinery factory cutting waste liquid (8)
9	PPG paint factory water-based cleaning liquid (9)
10	Drug laboratory centrifuge mother liquor (10)
11	Water-based paint factory waste liquid two (11)
12	Food factory containing benzene residue (12)
13	Pharmaceutical factory waste organic solvent (13)
14	Polyurethane factory adhesive waste (14)

Grouping liquid hazardous waste for the primary storage tanks

In the primary storage tank stage, raw liquid hazardous wastes were grouped into acid, alkaline, organics and specific wastes based on their physicochemical properties. Then, the compatibility among the same group wastes was tested by analyzing the composition and carrying out the intermixing experiment. The composition of liquid hazardous wastes was analyzed by GC-MS (ISO 2016). The intermixing experiment of liquid hazardous wastes was carried out to judge the compatibility between the same group samples to see if they could be put into the same storage tanks. The same group samples were paired at C_n^2 combination, the pair was intermixed at equal proportions to observe if exothermic, precipitation and gas production phenomenon emerged. If there was any reaction, the reacted pair would be picked out for further re-pair until all stable coexisting materials were determined. The mixing experiments were conducted in the Erlenmeyer flasks, the pair samples were added in a volume of 50 mL, and then the flask was plugged by silica gel plug. The plug was perforated with an electronic thermometer and a 10-inch balloon. The flasks were oscillated continuously at room temperature for 30 min at 150 r/min to observe if the heat released or balloon bulged.

The coexisting materials were mixed at equal ratios, and then the physicochemical properties of the mixed liquids were tested as 2.1 described to check if any physicochemical properties of the mixed liquids changed. A total of seven important parameters of pH, particle size, salt content, P, Cl, N and calorific value were analyzed based on their physicochemical properties, then the same group hazardous wastes will be mixed for further storage.

Mixing scheme for the homogenizing tank

The homogenizing tank is the core step to provide the qualified feeding waste for the SCWO reactor. The quality standard of the feeding waste is listed in Table 1. To meet the feeding requirement meanwhile maximum consuming low calorific value wastes during the homogenizing phase. The mixing scheme of liquid wastes from primary tanks should be designed based on the compatible mixing regulations, and the feasibility of the process route must be verified by experiments. Then, the theoretical prediction algorithm model is established to obtain the optimal scheme for the stable operation of the system.

Compatible mixing regulations

The safety and stability during mixing wastes from primary tanks should be guaranteed. Any potential reaction should be avoided after the liquid wastes from different primary tanks are pumped into the homogenizing tanks.

The mixing proportion of high phosphorus-containing material from the primary storage tank into homogenizing tanks should be strictly controlled.

Organic high-chlorine liquid waste should be dosed as combustion improver upon reaching the upper limit of Cl because of its high calorific value and Cl content.

Compatible feeding test experiment

After determining the feeding route based on compatible mixing regulations, the mixing experiment was carried out to test the compatibility of materials from different primary tanks by flask test. Because of the stratification characteristics of oil–water mixture in a static state, it is necessary to verify if oil–water mixture can be maintained at the uniform state by agitation shaking.

pH adjustment

The pH of the SCWO reactor should be in the range of 6–9, 30% sodium hydroxide is used in the experiment of the acid group and alkali adjustment, and the proportion is determined.

Compatibility feeding algorithm model

Based on the results of the compatible feeding test, MATLAB software was used to build the batching model. Algorithms and procedures are attached. The linear programming algorithm was applied to determine the mixing portion of wastes from various primary tanks by calculating the ingredients of wastes.

RESULTS AND DISCUSSION

Physicochemical properties of hazardous waste

Table 3 shows that among 14 samples, three samples are water-insoluble organic, one sample stratified into an organic upper layer and water-soluble under layer. The rest of the samples are water-soluble. Therefore, the samples have been classified into acid, neutral and alkaline groups based on the pH value for the primary storage tanks.

The calorific value of hazardous waste from different types of industries varied significantly, and there are huge differences between organic liquids and water-soluble liquids. The organic liquid samples such as samples (12), (2up), (13) and (14) have calorific values more than 12.80 MJ/kg, which is considered as flammable. Meanwhile, the water-soluble liquid sample has a relatively low calorific value, which can be defined as a non-combustible material. The calorific value of samples (5) and (10) is obviously higher than that of other water-soluble liquid samples, and the calorific value of other samples is generally low or almost zero.

As to the Cl and P percentage, benzene residue organic sample (12) contains a much higher Cl, mechanical plant cleaning agent waste (7) and waterborne paint factory waste sample (11) contain much higher phosphorus, which are all more than SCWO system feed requirement. Other elements are in an acceptable percentage. Therefore, it is necessary to focus on these three types of materials in the subsequent compatibility feeding process.

As to the insoluble particulate matter, only cleaning agent waste liquor (7) from the machinery plant has a tiny amount of particulate matter larger than 500 μm. The solid content of this sample was about 0.68%, of which only 1.75% was particulate matter larger than 500 μm. This tiny amount could be separated by the fine sieve pre-filtration process before the material enters the storage tank.

Analysis of organic components

The main soluble and volatile organic components of liquid waste are shown in Supplementary Table S2. Waterborne coating plant deoiling agent organic phase (2up), pharmaceutical plant waste organic solvent (13) and polyurethane plant adhesive waste (14) also contain a certain amount of benzene series, such as toluene, ethylbenzene, and a large number of volatile ketones and esters, such as acetone and ethyl acetate, which are harmful to human blood, nervous systems and reproductive systems. Compared with other water-soluble liquid samples, water-based paint factory aqueous solvent waste liquid (5) and drug laboratory centrifugal mother liquid (10) contain a higher concentration of organic matter, mainly alcohols with a low boiling point, so their flash point is low and can easily cause fires. There are nine kinds of organic components in benzene residue (12) from the food factory, almost all of them aromatic hydrocarbons, and more than half of them containing the Cl group, which is toxic to the organism and can cause direct harm to human health. In addition to the drug laboratory centrifugal mother liquid (10), the organic components in other water-soluble liquid samples are low and mainly aliphatic hydrocarbons, mostly alkanes, ketones, ethers, alcohols and esters.

Primary storage tank grouping

The compatibility of samples in the acid and alkaline group

As shown in Figure 1, primary storage tanks are extremely important units for the long-term stable and safe storage of the raw liquid hazardous wastes, and also act as the ingredient supplying cabinets for the next homogenizing steps. Raw

Table 3 | Physicochemical properties of samples

Sample code \ Phase		pH	Calorific value (MJ/kg)	Cl (mg/L)	P (mg/L)	N (mg/L)	Salt content(g/L)	Particle size (μm)	S (mg/L)	I (μg/L)	F (mg/L)	COD (mg/L)
1	WS	2.31	0.00	1,221.30	0.08	341.41	3.00	<500	0.00	341.67	NO	142.00
2down	WS	4.24	0.00	26.10	0.14	94.74	0.00	NO	0.00	698.83	NO	5,700.00
3	WS	1.39	0.01	924.50	0.70	480.22	3.18	<500	3,590.03	1,742.11	197.83	13,333.00
4	WS	6.19	2.54	409.70	3.39	93.86	2.07	<500	136.38	795.25	NO	237,383.00
5	WS	5.90	7.23	32.18	1.03	94.36	4.62	<500	699.71	1,676.83	NO	628,696.00
6	WS	10.32	0.00	211.80	0.69	96.08	16.81	<500	0.00	394.25	NO	61.00
7	WS	9.58	0.57	32.80	865	94.96	2.10	>500	134.82	0.00	NO	55,340.00
8	WS	7.10	0.62	12,187.50	5.49	94.96	0.60	<500	911.82	237.45	NO	47,692.00
9	WS	8.30	0.81	35.20	9.06	96.13	3.29	<500	222.74	0.00	NO	57,857.00
10	WS	7.46	4.21	14.71	0.88	361.28	0.34	<500	849.10	177.25	NO	359,829.00
11	WS	7.45	0.77	27.82	491.10	95.69	9.77	<500	661.93	922.67	NO	69,868.00
12	Organic	NO	21.97	147,264.88	0.63	7,873.20	NO	NO	1,258.20	0.00	NO	NO
2up	Organic	NO	34.77	1,081.40	0.88	345.11	NO	<500	414.50	1,097.50	NO	NO
13	Organic	NO	36.50	1,233.80	1.30	234.90	NO	NO	317.20	604.92	NO	NO
14	Organic	NO	24.40	2,706.10	5.45	290.10	NO	NO	352.40	0.00	NO	NO

WS, water solubility; NO, none.

liquid hazardous wastes were transported in individual barrels to the primary storage tank stage, so they should first be divided into different groups. Reasonable grouping and mixing the raw liquid hazardous wastes could reduce the number of primary tanks, which would reduce investment and facilitate safety management. More importantly, when the same grouping wastes are mixed into the same storage tank, they must stay stable and compatible. That is, overheating, sedimentation and excessive pressure should be strictly prevented in the primary tank after the same grouping wastes are mixed.

The C_n^2 reaction between samples in acid and alkaline liquid hazardous waste is shown in Table 4. The water-based paint factory waste liquid (4) could result in the milky white turbidity and yellow particle sticking on the wall. After adding water-based paint factory aqueous solvent waste liquid (5), the phenomenon is more obvious, indicating that the water-based paint factory waste liquid (4) and corresponding aqueous solvent waste liquid (5) are easy to react with other materials in the acid group. This water-soluble colloidal material will quickly precipitate after mixing with the strong acid waste liquid. The alkaline group showed the water-based paint factory waste liquid (11) reacted with the rest of the alkaline samples, forming black particles sticky on the tube wall. This may be ascribed to neutral (pH = 7.45) of water-based paint factory waste liquid two (11), the water-soluble colloidal substance is quickly precipitated after being mixed with the higher alkali waste. Therefore, the samples from water-based paint factory waste liquid (4, 5 and 11) should be picked out and defined as reactive materials.

The compatibility of samples in organic groups

There was no obvious reaction phenomenon in the samples of the organic group, and it was preliminarily concluded that the organic group samples could be mixed completely. After mixing, the organic components solved mutually well and formed a stable solution. However, the benzene residue from the food factory (12) has an extremely high Cl content; considering its large calorific value, the benzene residue can be used as a liquid combustion improver meanwhile diluting its total Cl content. Therefore, an isolated tank for holding the benzene residue is proposed in the primary stage.

Physicochemical properties of the grouped mixture

Based on the physicochemical properties and equal portion mixing experiment, the liquid hazardous wastes from 14 industries can be initially grouped into five categories, namely acid, alkali, amphoteric reactions, organic and organic high chlorine in the primary storage tanks phase (Table 5).

The physicochemical properties of the mixture of each group are shown in Table 6. The actual measured values are close to the calculated values obtained by summing

Table 4 | Material compatibility reaction in the acid group and the alkaline group

Acid group

Samples	1	2down	3	4	5
1	–				
2down	–	–			
3	–	–	–		
4	–	–	White bonding, sticky bottom precipitation	–	
5	White and orange granules are bonded and wall-precipitated	–	Wall precipitation of white insoluble particulate matter	–	–

Alkaline group

Samples	6	7	8	9	10	11
6	–					
7	–	–				
8	–	–	–			
9	–	–	–	–		
10	–	–	–	–	–	
11	Black insoluble particles	Black insoluble particles	–	Black insoluble particles	–	–

Note: '–' indicates no phenomenon.

Table 5 | Primary storage tanks grouping

Category	Liquid hazardous waste
Acidic	Mechanical washing waste liquid (1), water-based paint factory deoiling agent aqueous phase waste liquid (2down), resource recycling company metal waste acid (3)
Alkaline	Machinery plant sodium carbonate waste liquid (6), mechanical cleaning agent waste liquid (7), mechanical cutting waste liquid (8), PPG paint factory water-based cleaning liquid (9), drug laboratory centrifugal mother liquid (10)
Amphoteric reaction	Water-based paint factory waste liquid (4), water-based paint factory waste liquid (11), water-based paint factory aqueous solvent waste liquid (5)
Organic	Waterborne coating plant deoiling agent organic phase (2up), pharmaceutical plant waste organic solvent (13), polyurethane plant adhesive waste (14)
Organic high chlorine	Food factory containing benzene residue (12)

the individual portion, which illustrated no reaction among the same group wastes.

Homogenizing tank grouping

Three feeding routes are regulated for the homogenizing tank phase, that is (1) the mixing of the acid, the organic high-chlorine and the organic group wastes, (2) the mixing of the alkali, organic high-chlorine and organic group wastes and (3) the mixing of the amphoteric reaction and the organic wastes. There is no reaction observed in each route after three group wastes are mixed, and the mixture can be uniformly mixed by shaking well. After standing, the mixture gradually shows a stratification phenomenon and quickly returns to the turbid state if stirred and shaken again. The physicochemical properties of the mixture in each route do not change significantly after mixing (Table 7).

The feeding plan

The liquid hazardous wastes of various industries are transported to the plant area by hazardous chemical tank truck, and then enter into each temporary storage area according to the group. The materials in each storage area enter the corresponding primary storage tank through the loading platform and mix evenly. Then, the materials in each primary storage tank are divided into three routes and transported to the homogenizing tank according to the compatibility ratio. The material after uniform mixing meets the feeding requirements through monitoring, and then it is fed quickly through the high-speed feed pump in Figure 2.

Table 6 | Physicochemical properties of materials in various storage tanks after equal mixing

Category	pH	Calorific value (MJ/kg)	Cl (mg/L)	P (mg/L)	N (mg/L)	Salt content (g/L)	Particle size (μm)
Acidic	1.84	0.00	735.60	0.31	312.80	2.16	<500
Alkaline	9.92	1.12	2,485.60	176.20	146.20	4.89	<500
Amphoteric reaction	6.36	3.83	162.40	165.20	95.80	5.72	<500
Organic	NO	32.23	1,685.20	2.50	284.80	NO	NO
Organic high chlorine	NO	21.97	147,264.88	0.63	7,873.20	NO	NO

NO, none.

Table 7 | The physicochemical properties of the mixture in each feeding route

Storage tank ratio	Calorific value (MJ/kg)	Cl (mg/L)	P (mg/L)	N (mg/L)	Salt content (g/L)
Acid:Organic high chlorine:Organic (1:1:1)	16.72	48,427.86	1.21	3,015.60	0.74
Alkali:Organic high chlorine:Organic (1:1:1)	17.14	51,434.42	64.82	2,546.70	1.58
Amphoteric reaction:Organic (1:1)	17.85	902.81	85.56	185.19	1.91

Table 8 | Application results of the algorithm model

	Compatibility ratio									
Route	**Acidic**	**Alkaline**	**Amphoteric reacting**	**Organic**	**Organic perchlorides**	**Calorific value (MJ/kg)**	**Cl (mg/L)**	**P (mg/L)**	**N (mg/L)**	**Salt content (g/L)**
1	13.13			1.87	1.00	5.10	9,993.60	0.587	776.60	1.69
2		17.60		1.50	1.00	5.10	9,998.90	17.59	550.20	4.49
3			73.07	1.00		5.10	233.60	19.86	96.80	3.53

Compatibility feeding algorithm model

(1) Objective function

The ingredients based on the SCWO system should consider the largest possible consumption of materials in acid storage tanks, alkali storage tanks and amphoteric reacting waste storage tanks, and then preferentially select materials in organic high-chlorine storage tanks. Adjust to the upper limit of Cl and increase the calorific value, and finally adjust the calorific value with a small amount of material in the organic storage tank. The objective function is

$$\text{Max} f(x) = x_1 + x_2$$

where $f(x)$ represents the total proportion of liquid hazardous waste in the batch, %; x_1 represents the proportion of the material in the acid storage tank (or the material in the alkali storage tank or the material in the amphoteric tank), %; x_2 represents the proportion of material in the organic high-chlorine storage tank in the batch, %.

(2) Constraints

The constraints of the ingredients in the model are the feed requirements of the SCWO system:

$$\begin{cases} A_1x_1 + A_2x_2 + A_3x_3 \leq 10,000 \\ B_1x_1 + B_2x_2 + B_3x_3 \leq 10,000 \\ C_1x_1 + C_2x_2 + C_3x_3 \leq 20 \\ D_1x_1 + D_2x_2 + D_3x_3 = 5.1 \\ E_1x_1 + E_2x_2 + E_3x_3 \leq 80 \\ x_1 + x_2 + x_3 = 1 \\ 0 \leq x_1,\ x_2,\ x_3 \leq 1 \end{cases}$$

where x_1, x_2, x_3 respectively represent the materials in the acid storage tank in the batch (the material in the alkali storage tank or the material in the double reaction storage tank), the material in the organic storage tank, and the material in the organic high-chlorine storage tank, %; A_1, A_2, A_3 respectively represent the Cl content of the material in the corresponding storage tank, mg/L; B_1, B_2, B_3 respectively represent the N content of the corresponding material in the storage tank, mg/L; C_1, C_2, C_3 respectively represent the P content of the material in the corresponding tank, mg/L; D_1, D_2, D_3 respectively represent the calorific value of the material in the corresponding tank, MJ/kg; E_1, E_2, E_3 respectively represent the salt content of the material in the corresponding tank, g/L.

(3) MATLAB software editing

Linprog command is applied to find the optimal solution to the linear programming problem, the parameter form is [x, fval] = linprog (f, A, b, Aeq, beq, lb, ub).

The programming language of this study can be edited as:

```
f = [−1; −1];
A = [ A1 A2 A3; B1 B2 B3; C1 C2 C3; E1 E2 E3];
b = [10000;10000;20;80];
Aeq = [D1 D2 D3;1 1 1];
beq = [5.1;1];
lb = [0;0;0];
ub = [1;1;1];
[x, fval] = linprog (f, A, b, Aeq, beq, lb, ub)
```

[x, fval]: indicates that x is the optimal solution and fval is the optimal value.

f: represents the coefficient vector in front of each variable in the objective function. If it is the minimum value problem, then f is the coefficient of each variable. If it is the maximum value problem, then f is the opposite of the coefficient of each variable;

A and b: represent the matrix A and the vector b in the inequality constraint $A \times x \leq b$;

Aeq and beq: represent the matrix Aeq and the vector beq in the equality constraint $Aeq \times x = beq$;

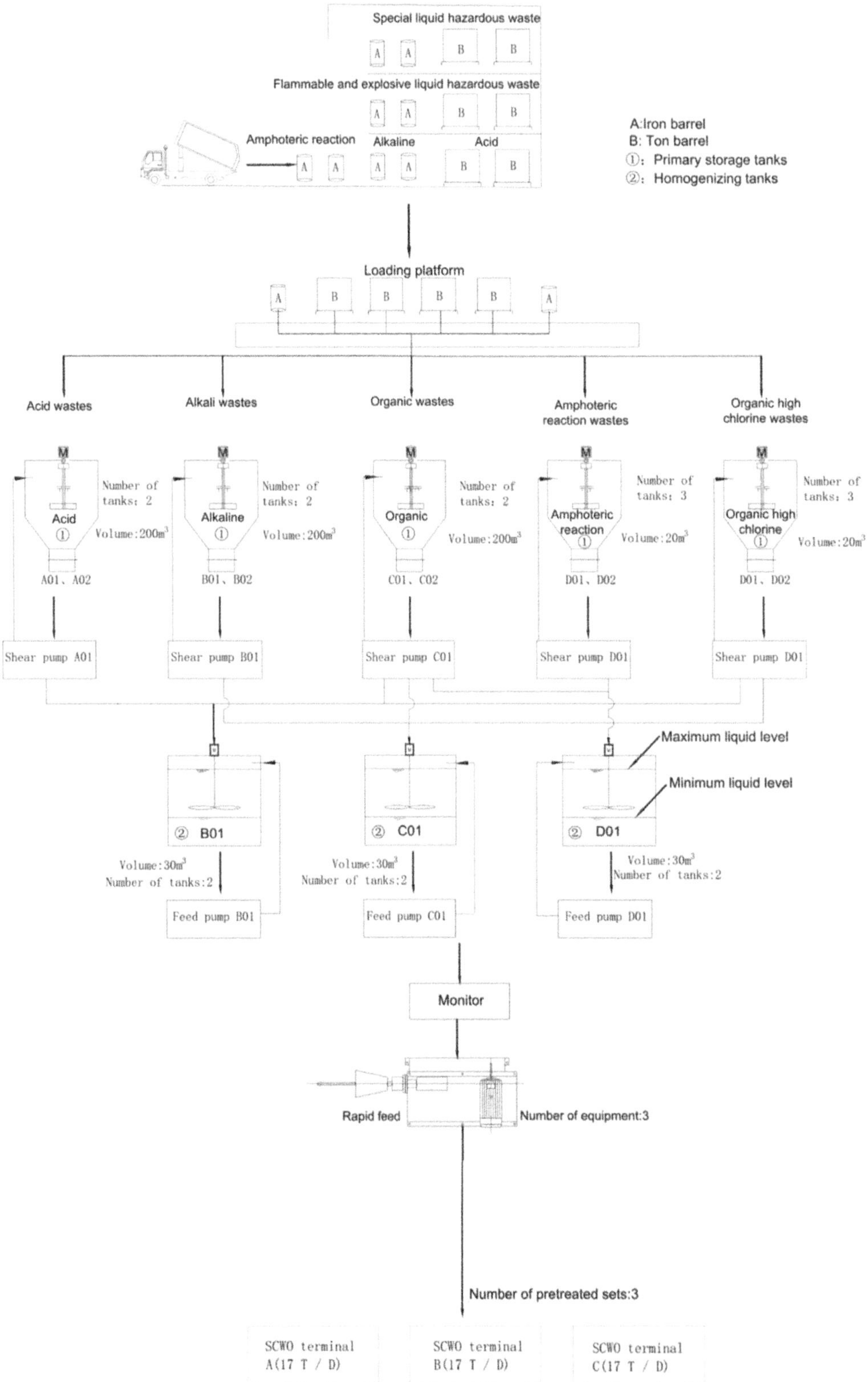

Figure 2 | General process of the feed pretreatment process for liquid hazardous waste.

Lb and ub: vectors representing the upper and lower bounds of the independent variables, respectively.

By using the linear programming optimization model of MATLAB, the optimal compatibility schemes of three feeding routes are simulated and calculated, respectively, in Table 8. The linear programming optimization algorithm model can be applied to three compatible feeding routes, and the materials after the compatibility meet the feeding index of the SCWO system. Therefore, it is certain to be applied in practical industrial operations.

CONCLUSIONS

This paper proposes a systematic strategy on how to design the feeding process of SCWO to not only safely meet the feed requirement but also maximally dispose of liquid hazardous waste. The feeding step of the storage tank was proposed as primary and homogenizing tanks. Based on the composition identification of the unknown hazard wastes, the hazards wastes were grouped to acid, alkaline, amphoteric reacted, organic and organic perchlorides in the primary tank. Then, three mixing routes for the homogenizing tank from the hazard wastes of the primary tank were proposed, that is, the acid, alkali or amphoteric reacted individually mixed with organic and organic perchlorides. Primary storage tanks help optimize process costs and safety management while stabilizing the feed. With the MATLAB programming, the material compatibility optimization model was established based on the linear programming algorithm. The model showed that the first, second and third feeding route compatibility ratio was 13.13 : 1.87 : 1.00, 17.60 : 1.50 : 1.00 and 73.07 : 1.00, respectively. With this strategy, the largest amount of liquid hazardous wastes could be disposed of without adding external fuel, which maximizes economic benefits while guaranteeing the safety.

ACKNOWLEDGEMENT

This work was funded by the Hebei key research and development project (20373604D).

DATA AVAILABILITY STATEMENT

All relevant data are included in the paper or its Supplementary Information.

REFERENCES

Duan, H. B., Huang, Q. F., Wang, Q., Zhou, B. Y. & Li, J. H. 2008 Hazardous waste generation and management in China: a review. *Journal of Hazardous Materials* **158**, 221–227.

El-Fadel, M., Zeinati, M., El-Jisr, K. & Jamali, D. 2001 Industrial-waste management in developing countries: the case of Lebanon. *Journal of Environmental Management* **61**, 281–300.

Franck, E. U. 1984 Physicochemical properties of supercritical solvents (Invited lecture). *Berichte der Bunsengesellschaft für Physikalische Chemie* **88**, 820–825.

Harradine, D. M., Buelow, S. J., Dell'Orco, P. C., Dyer, R. B., Foy, B. R., Robinson, J. M., Sanchez, J. A., Spontarelli, T. & Wander, J. D. 1993 Oxidation chemistry of energetic materials in supercritical water. *Hazardous Waste & Hazardous Materials* **10**, 233–246.

ISO 2016 *Water Quality – Determination of Volatile Organic Compounds in Water – Method Using Headspace Solid-Phase Micro-Extraction (HS-SPME) Followed by Gas Chromatography-Mass Spectrometry (GC-MS)*. Washington, DC, USA.

Ministry of Ecology and Environment of the People's Republic of China 2019 *Identification Standards for Hazardous Waste General Rules*. Beijing, China.

Ministry of Environmental Protection of the People's Republic of China 2016 *Directory of National Hazardous Wastes*. Beijing, China.

Quah, S. R. & Cockerham, W. C. 2017 *International Encyclopedia of Public Health*, 2nd edn. Academic Press, Elsevier, New York, USA.

Verma, M. & Kumar, R. N. 2017 Coagulation and electrocoagulation for co-treatment of stabilized landfill leachate and municipal wastewater. *Journal of Water Reuse and Desalination* **8**, 234–243.

Yang, J., Wang, S., Li, Y., Zhang, Y. & Xu, D. 2019 Novel design concept for a commercial-scale plant for supercritical water oxidation of industrial and sewage sludge. *Journal of Environmental Management* **233**, 131–140.

Yang, P., Zhang, L. J., Wang, X. J. & Wang, Z. L. 2020 Exploring the management of industrial hazardous waste based on recent accidents. *Journal of Loss Prevention in the Process Industries* **67**, 104224.

First received 22 September 2020; accepted in revised form 19 January 2021. Available online 9 February 2021

© 2021 The Authors

1,4-dioxane degradation using a pulsed switching peroxi-coagulation process

Yaobin Lu, Hualei Shi, Jialiang Yao, Guangli Liu, Haiping Luo and Renduo Zhang

ABSTRACT

Widely used in chemical product manufacture, 1,4-dioxane is one of the emerging contaminants, and it poses great risk to human health and the ecosystem. The aim of this study was to degrade 1,4-dioxiane using a pulsed switching peroxi-coagulation (PSPC) process. The electrosynthesis of H_2O_2 on cathode and Fe^{2+} production on iron sacrifice anode were optimized to enhance the 1,4-dioxane degradation. Under current densities of 5 mA/cm^2 (H_2O_2) and 1 mA/cm^2 (Fe^{2+}), 95.3 ± 2.2% of 200 mg/L 1,4-dioxane was removed at the end of 120 min operation with the optimal pulsed switching frequency of 1.43 Hz and pH of 5.0. The low residual H_2O_2 and Fe^{2+} concentrations were attributed to the high pulsed switching frequency in the PSPC process, resulting in effectively inhibiting the side reaction during the ·OH production and improving the 1,4-dioxane removal with low energy consumption. At 120 min, the minimum energy consumption in the PSPC process was less than 20% of that in the conventional electro-Fenton process (7.8 ± 0.1 vs. 47.0 ± 0.6 kWh/kg). The PSPC should be a promising alternative for enhancing 1,4-dioxane removal in the real wastewater treatment.

Key words | 1,4-dioxane degradation, electro-Fenton, pulsed switching circuit, sacrifice iron anode

Yaobin Lu
Guangdong Provincial Key Laboratory of Water Quality Improvement and Ecological Restoration for Watersheds, Institute of Environmental and Ecological Engineering,
Guangdong University of Technology,
Guangzhou 510006,
China

Yaobin Lu
Hualei Shi
Guangli Liu (corresponding author)
Haiping Luo
Renduo Zhang
Guangdong Provincial Key Laboratory of Environmental Pollution Control and Remediation Technology, School of Environmental Science and Engineering,
Sun Yat-sen University,
Guangzhou 510006,
China
E-mail: *liugl@mail.sysu.edu.cn*

Jialiang Yao
The Affiliated High School of South China Normal University,
Guangzhou 510630,
China

HIGHLIGHTS

- 1,4-dioxiane was efficiently removed in the PSPC process.
- Residual H_2O_2 and Fe^{2+} were minimized by optimizing pulsed switching circuits.
- The energy consumption in the PSPC reduced to 20% of that in the EF.

GRAPHICAL ABSTRACT

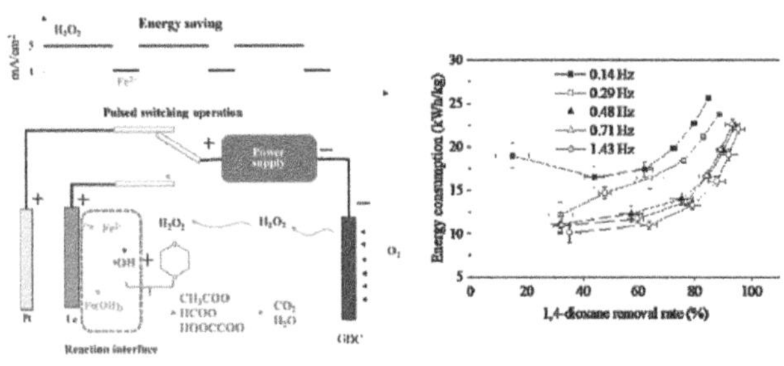

This is an Open Access article distributed under the terms of the Creative Commons Attribution Licence (CC BY 4.0), which permits copying, adaptation and redistribution, provided the original work is properly cited (http://creativecommons.org/licenses/by/4.0/).

doi: 10.2166/wrd.2021.092

INTRODUCTION

As an important solvent stabilizer, reaction agent, and reaction media, 1,4-dioxane ($C_4H_8O_2$) has been widely used in the manufacturing processes of chemical products such as paints, varnishes, lacquers, cosmetics, resins, and deodorants (Clercq *et al.* 2010; Barndõk *et al.* 2016a). Being possibly carcinogenic to humans and chemically stable, 1,4-dioxane can pose a great risk to human health and the ecosystem (Clercq *et al.* 2010; Takeuchi & Tanaka 2020; Zhao *et al.* 2020). Since 1,4-dioxane is bio-refractory, indicated by the low ratio of biochemical oxygen demand (BOD) to chemical oxygen demand (COD) (i.e. 0.06), non-volatility, and its miscibility with water (Nakagawa *et al.* 2016; Radcliffe & Page 2020; Rossum 2020; Somda *et al.* 2020), it is difficult to effectively remove it in typical biological wastewater treatments (Mahendra *et al.* 2013; Huang *et al.* 2018; Xu *et al.* 2020). For example, complete decomposition of 100 mg/L 1,4-dioxane in the activated sludge processes required 7 days (Sei *et al.* 2010). Even with pure culture, more than 25 h was needed for the 1,4-dioxane removal (Sun *et al.* 2010; Sei *et al.* 2013). The biological co-metabolism can be used to enhance the 1,4-dioxane degradation. However, additional nutrients, such as tetrahydrofuran and lactate (Sekar & DiChristina 2014), may increase the treatment cost (Hand *et al.* 2015; Zhang *et al.* 2016; Chen *et al.* 2020; Fan *et al.* 2020; Lin *et al.* 2020). Thus, it is necessary to develop an efficient method for 1,4-dioxane degradation.

The advanced oxidation processes (AOPs) have been developed for 1,4-dioxane removal in recent years. Compared with the biological treatment process, AOPs can produce high reactive and non-selective oxidant (i.e. ·OH) (Elkacmi & Bennajah 2019), and thus are a fast and efficient alternative to 1,4-dioxane degradation (Clercq *et al.* 2010; Barndõk *et al.* 2016a). Because of its good performance, easy control, and environmental friendliness (Wang *et al.* 2016), the Fenton process is one of the powerful AOPs for 1,4-dioxane removal. The Fenton process could reduce 1,4-dioxane from 100 to 0.5 mg/L within 6 h (Nakagawa *et al.* 2016). However, the high risk of H_2O_2 transport limits application of the Fenton process (Brillas & Martínez-Huitle 2015; Gao *et al.* 2015; Wang *et al.* 2016). The risks of H_2O_2 transport can be avoided by in-situ H_2O_2 production using the electro-Fenton (EF) process (Brillas & Martínez-Huitle 2015). As a combined EF process, the peroxi-coagulation (PC) process utilizes a sacrificial iron anode and a gas diffusion cathode (GDC) for H_2O_2 generation (Brillas *et al.* 2009). Fe^{2+} ions can release from the sacrificial iron anode to catalyze H_2O_2 for hydroxyl radicals (·OH) production. Excess Fe^{2+} dissolution from the iron anode can form $Fe(OH)_3$ precipitation as a coagulant. However, excess Fe^{2+} may also result in severe OH scavenger and high iron sludge production (Sun & Pignatello 1993; Benhadji *et al.* 2016).

Recently, a pulsed switching peroxi-coagulation (PSPC) process was developed for effectively controlling Fe^{2+} consumption (Lu *et al.* 2018). The controllable Fe^{2+} release in the auxiliary anode and H_2O_2 electrosynthesis in the cathode of PSPC was useful for maximum ·OH production and minimum iron sludge (Lu *et al.* 2018). Under current densities of 5.0 mA/cm^2 (H_2O_2) and 0.5 mA/cm^2 (Fe^{2+}), and the pulsed switching frequency of 1.0 s (H_2O_2): 0.3 s (Fe^{2+}), 500 mg/L 2,4-dichlorophenoxyacetic acid was completely removed in the PSPC within 240 min (Lu *et al.* 2018). The energy consumption for 500 mg/L 2,4-dichlorophenoxyacetic acid removal in the PSPC was only 50% of that in the conventional EF process (68 ± 6 vs. 136 ± 10 kWh/kg TOC) (Lu *et al.* 2018). The iron consumption in the PSPC was only ~5% of that in the conventional PC process. Although PSPC showed great potential in the 2,4-dichlorophenoxyacetic acid removal, the application of PSPC is still in its infancy and needs further testing for other refractory organics. Moreover, the physicochemical characteristics of different refractory organics can affect its removal in the AOPs (Clercq *et al.* 2010; Lu *et al.* 2018). For example, 1,4-dioxane has higher solubility and toxicity than 2,4-dichlorophenoxyacetic acid (Clercq *et al.* 2010; Lu *et al.* 2018). The optimal operational conditions for 1,4-dioxane removal in the PSPC may be different from those for 2,4-dichlorophenoxyacetic acid removal. The experiments on 1,4-dioxane removal in the PSPC should be not only important for enhancing 1,4-dioxane degradation in the wastewater treatment but also for expanding the potential application of PSPC. Thus, the objective of this study was to investigate the feasibility of the PSPC process for 1,4-dioxane degradation. The effect of H_2O_2 and Fe^{2+} production, pulsed switching frequency and current density were tested on the 1,4-dioxane removal. Five control experiments were carried out to distinguish the PSPC process from the EF and PC processes. The main intermediates of 1,4-dioxane degradation and the residual Fe^{2+} and H_2O_2 were identified to discuss the mechanism of efficient 1,4-dioxane degradation in the PSPC process.

MATERIALS AND METHODS

PSPC setup

The PSPC process was investigated in an undivided electrochemical cell with a cylindrical chamber (diameter × length = 3 × 4.5 cm). A platinum wire electrode (CHI115, CH Instrument, Inc., Shanghai, China) and an iron mesh (diameter of 3 cm, thickness of 0.3 mm, 90 meshes) were used as main and auxiliary anodes, respectively. A gas diffusion cathode (GDC) was constructed with a conductive gas diffusion layer, a catalyst layer, and a supporting layer of stainless steel mesh (90 meshes) (Wang *et al.* 2017). Carbon black powder (EC-300 J, Hesen, Shanghai, China) was used as a catalyst for H_2O_2 production. The effective surface area of the GDC was 7 cm^2. The distance between the GDC and Pt anode was 3.5 cm. The distance between the GDC and Pt anode was 2 cm. In-situ H_2O_2 and Fe^{2+} production was driven by the power supply (IT6700, ITECH Electronic Co., Ltd, Nanjing, China) with constant current. Two time relays with an accuracy of 0.1 s (ZYS48-S, Zhuoyi Electronic Co., Ltd, Shanghai, China) were used to control the pulsed switching circuits and the running time of H_2O_2 and Fe^{2+} productions in PSPC. A solution of 0.1 M Na_2SO_4 and 200 mg/L 1,4-dioxane (99%, Merck) was used as the electrolyte. The pH of the electrolyte was adjusted to 5.0 with H_2SO_4 or NaOH. The electrolyte was recycled in the electrochemical cell using a peristaltic pump (BT-100, Qite, China) with a flow rate of 36 mL/min.

Experimental procedure

The current density for H_2O_2 was set at 5.0 mA/cm^2 to produce H_2O_2 efficiently with low energy consumption based on the results of our preliminary tests. In order to avoid excess iron oxidization into Fe^{3+}, the current density for Fe^{2+} production was controlled at 1 mA/cm^2. The molar concentration ratio of Fe^{2+} to total Fe in the solution was >95% under the iron anode operation with 1 mA/cm^2. Thus, the pulsed switching ratio of H_2O_2 and Fe^{2+} productions were tested under the current densities of 5 mA/cm^2 for H_2O_2 and 1 mA/cm^2 for Fe^{2+}, including 4 s (H_2O_2): 4 s (Fe^{2+}) (i.e. 1.00), 4:3 s (i.e. 1.33), 4:2 s (i.e. 2.00), 4:1 s (i.e. 4.00), and 4:0.5 s (i.e. 8.00). The pulsed switching frequency, including 4:3 s (0.14 Hz), 2:1.5 s (0.29 Hz), 1.2:0.9 s (0.48 Hz), 0.8:0.6 s (0.71 Hz), and 0.4:0.3 s (1.43 Hz), was tested under the fixed pulsed switching ratio between H_2O_2 and Fe^{2+} productions of 1.33 (4/3 s), respectively. Under the pulsed switching frequency of 1.43 Hz, the 1,4-dioxane removal was tested under different current densities, including 15 mA/cm^2 (H_2O_2) +3 mA/cm^2 (Fe^{2+}), 10 mA/cm^2 (H_2O_2) +2 mA/cm^2 (Fe^{2+}), 5 mA/cm^2 (H_2O_2) +1 mA/cm^2 (Fe^{2+}), and 2.0 mA/cm^2 (H_2O_2) +0.4 mA/cm^2 (Fe^{2+}), respectively.

Five controls were used to distinguish the PSPC process from others, such as EF and PC processes (Brillas *et al.* 2009; Ahangarnokolaei *et al.* 2017). Control 0 was to test the 1,4-dioxane adsorption in the cell without H_2O_2 and Fe^{2+} production (Table 1). Control 1 was a conventional PC process to keep a continuous production of H_2O_2 and Fe^{2+} without the pulsed switching frequency in the cell under the current densities of 5 mA/cm^2 (H_2O_2) and 1 mA/cm^2 (Fe^{2+}). Control 2 was a conventional EF process to keep a continuous H_2O_2 production of 5 mA/cm^2, and add 7.9 mM Fe^{2+} into the cell without iron anode operation. Control 3 was to keep a pulsed H_2O_2 production at 5 mA/cm^2 with the pulsed switching frequency of 0.4:0.3 s (1.43 Hz), and 7.9 mM Fe^{2+} added into the cell without iron anode operation. Control 4 was to

Table 1 | Operational conditions in five controls

Control experiment	Current density for H_2O_2 production (mA/cm^2)	Current density for Fe^{2+} production (mA/cm^2)
Control 0	0	0
Control 1	5	1
Control 2	5	0[b]
Control 3	5[a]	0[b]
Control 4	5	0

[a]Pulsed H_2O_2 production at 5 mA/cm^2 and the pulsed switching frequency of 0.4:0.3 s (1.43 Hz).

[b]7.9 mM Fe^{2+} addition into cell without iron anode operation.

keep a continuous H_2O_2 production of 5 mA/cm^2, and add 7.9 mM Fe^{2+} into the cell without iron anode operation. The addition of 7.9 mM Fe^{2+} was equal to the total Fe^{2+} production in the PSPC process under the pulsed switching frequency ratio of 0.4:0.3 s, and current densities of 5 mA/cm^2 (H_2O_2) and 1 mA/cm^2 (Fe^{2+}) within 120 min.

Analysis and calculation

The H_2O_2 concentration was measured using a spectrophotometer (T6, Persee, Beijing, China) according to the titanium (IV) sulfate method (Barazesh *et al.* 2015). Fe^{2+} and total iron concentrations were determined with the phenanthroline spectrophotometric method (Xu *et al.* 2013). The total organic carbon (TOC) was measured using a Shimadzu TOC-L CPH analyzer (Shimadzu Co., Japan). The pH was measured by a pH meter (FE20, Mettler-Toledo, Switzerland). The total COD (TCOD) and soluble COD (SCOD) was determined using the dichromate standard method according to the samples filtrated without or with a 0.22 μm filter, respectively (Liu *et al.* 2015; Ye *et al.* 2017). A high performance liquid chromatograph (HPLC, P230II, Dalian Yilite Analytic Instrument Co. Ltd, China) was used to determine 1,4-dioxane concentration. A 10:90 (v/v) acetonitrile/water (phosphate buffer of pH 3) solution at 1.0 mL/min was used as the mobile phase. The chromatograph was equipped with a SinoChrom ODS-BP column (5 μm, 4.6 mm × 25 cm, Dalian Yilite Analytic Instrument Co. Ltd, China). The UV detector was set at 190 nm and the temperature was maintained at 30°. The intermediates of 1,4-dioxane degradation, including oxalic, acetic and formic acids, were quantified by an ion chromatograph (IC, CIC-D100, SHINE IC Solution Experts, China).

The energy consumption in the PSPC process was calculated based on the electricity consumption for H_2O_2 and Fe^{2+} production, except for the energy for electrolyte recirculation in the cell. The energy input (P, W) and energy consumption per kg 1,4-dioxane removal (EC, kWh/kg) were calculated as follows:

$$P = U_1 I_1 a + U_2 I_2 b \quad (1)$$

$$EC = \frac{1000Pt}{V_s \Delta C} \quad (2)$$

where V_s is the solution volume (L); I_1 and I_2 are the currents for H_2O_2 and Fe^{2+} production, respectively (A); U_1 and U_2 are the voltages for H_2O_2 and Fe^{2+} production, respectively (V); a is the ratio between the running time for H_2O_2 production and the whole operation time (h); b is the ratio between the running time for Fe^{2+} production and the whole operation time (h); t is the operation time (h); and ΔC is the decrement of 1,4-dioxane concentration in the PSPC process during the experiment (mg/L). The current efficiency of H_2O_2 production was estimated as previously described (Luo *et al.* 2015).

RESULTS

Effect of the pulsed switching ratio of H_2O_2 and Fe^{2+} productions on 1,4-dioxane degradation

In the PSPC process, the molar ratio of H_2O_2 and Fe^{2+} concentrations were determined by the pulsed switching ratio of H_2O_2 and Fe^{2+} production. For example, when the pulsed switching ratio of H_2O_2 and Fe^{2+} production increased from 1.00 to 8.00, the molar ratio of H_2O_2 and Fe^{2+} increased from 1.49 to 12.0. As shown in Figure 1(a), more than 80% of 1,4-dioxane was removed under different pulsed switching ratios of H_2O_2 and Fe^{2+} production within 120 min. The 1,4-dioxane removal slightly increased from 80.1 ± 1.8 to 89.6 ± 1.1% with the pulsed switching ratio increasing from 1.00 to 8.00. The low pulsed switching ratio resulted in low energy consumption (Figure 1(b)). The energy consumption under the pulsed switching ratio of 1.33 was 25.6 kWh/kg, which was only 76% of that under the pulsed switching ratio of 8.00. The energy consumption was almost kept stable at the pulsed switching ratio of 1.33 and 1.00. Therefore, the pulsed switching ratio of 1.33 could be suitable for 1,4-dioxane removal in the PSPC process. Correspondingly, the molar ratio of H_2O_2 and Fe^{2+} was 1.99, and the 1,4-dioxane removal efficiency was 84 ± 2%.

Effect of pulsed switching frequency of H_2O_2 and Fe^{2+} production on 1,4-dioxane degradation

Under the fixed pulsed switching ratio of 1.33, different pulsed switching frequency, including 0.14, 0.29, 0.48,

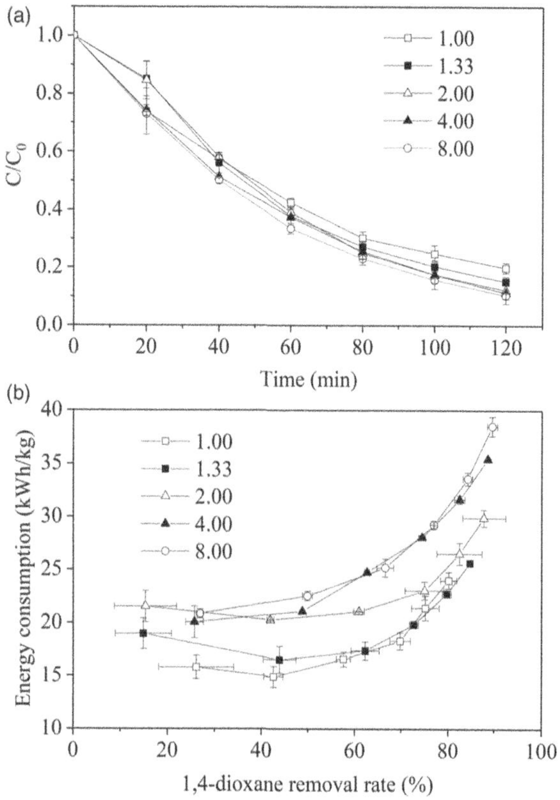

Figure 1 | (a) 1,4-dioxane removal and (b) energy consumption under different pulsed switching ratios of H_2O_2 and Fe^{2+} within 120 min (current densities of 5 mA/cm^2 for H_2O_2 production and 1 mA/cm^2 for Fe^{2+} production).

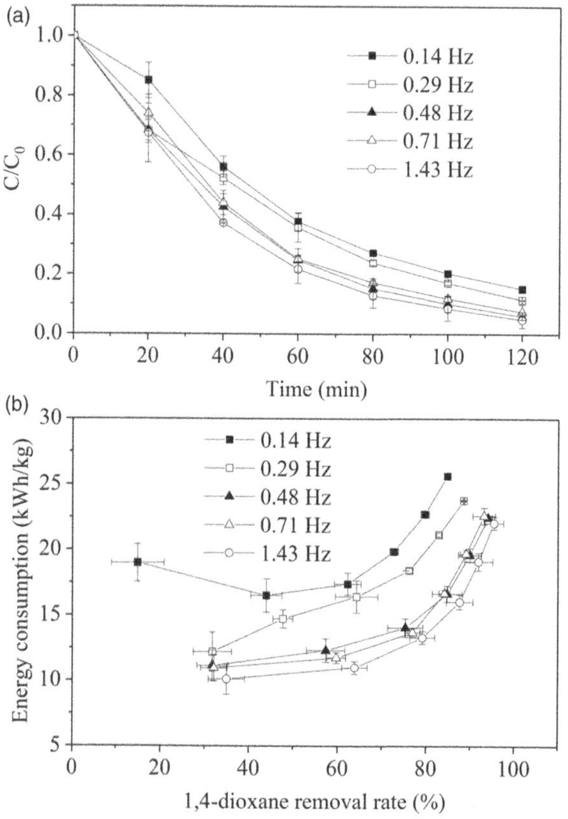

Figure 2 | Effect of pulsed switching frequency on (a) 1,4-dioxane removal and (b) energy consumption in the PSPC process (pulsed switching ratio of 1.33, current densities of 5 mA/cm^2 for H_2O_2 production and 1 mA/cm^2 for Fe^{2+} production).

0.71, and 1.43 Hz, was tested as shown in Figure 2. High pulsed switching frequency resulted in high 1,4-dioxane removal and low energy consumption. The 1,4-dioxane removal gradually increased from 84.8 ± 0.4 to 95.3 ± 2.2% with the pulsed switching frequency increasing from 0.14 to 1.43 Hz within 120 min. Based on Equation (1), the energy input for H_2O_2 and Fe^{2+} production was determined by the pulsed switching ratio of H_2O_2 and Fe^{2+} production. Under the pulsed switching ratio of 1.33, the energy input of the PSPC process was about 0.0718 W with 80% 1,4-dioxane removal, 96% of which was consumed by H_2O_2 generation. With the pulsed switching frequency increasing from 0.14 to 1.43 Hz, the energy consumption of the PSPC process with 80% 1,4-dioxane removal was decreased from 23 to 14 kWh/kg (Figure 2(b)). Therefore, the pulsed switching frequency of H_2O_2 and Fe^{2+} production was optimized to 0.4:0.3 s in the PSPC process. Correspondingly, 95.3 ± 2.2% of 1,4-dioxane was removed with the energy consumption of 22.1 ± 0.5 kWh/kg within 120 min.

Effect of current densities in the PSPC process on 1,4-dioxane degradation

Under the optimally pulsed switching frequency of 1.43 Hz, the effect of current densities in the PSPC process was tested as shown in Table 2. To keep the optimal molar ratio of H_2O_2 and Fe^{2+} production as described above, the current densities of H_2O_2 and Fe^{2+} production were changed proportionally, including 15 mA/cm^2 (H_2O_2) +3 mA/cm^2 (Fe^{2+}), 10 mA/cm^2 (H_2O_2) +2 mA/cm^2 (Fe^{2+}), 5 mA/cm^2 (H_2O_2) +1 mA/cm^2 (Fe^{2+}), and 2.0 mA/cm^2 (H_2O_2) +0.4 mA/cm^2 (Fe^{2+}). High current density of H_2O_2 and Fe^{2+} resulted in high 1,4-dioxane removal under the optimally pulsed switching frequency in the PSPC process (Table 2). However, even with low current densities such as 2.0 mA/cm^2 (H_2O_2) and 0.4 mA/cm^2 (Fe^{2+}), 89.8 ± 2.2% of 1,4-dioxane was still removed within 120 min.

Table 2 | The 1,4-dioxane removal in the PSPC processes and different controls

Time (min)	1,4-Dioxane removal (%)								
	Current densities (mA/cm²) in the PSPC process (pulsed switching frequency = 1.43 Hz)								
	15 (H_2O_2) + 3 (Fe^{2+})	10 (H_2O_2) + 2 (Fe^{2+})	5 (H_2O_2) + 1 (Fe^{2+})	2.0 (H_2O_2) + 0.4 (Fe^{2+})	Control 0	Control 1	Control 2	Control 3	Control 4
20	52.9 ± 1.7	50.4 ± 5.8	32.6 ± 9.9	28.2 ± 3.1	3.4 ± 3.1	33.0 ± 5.2	59.0 ± 0.2	57.3 ± 6.4	6.4 ± 3.8
40	78.7 ± 1.1	77.4 ± 4.7	62.8 ± 0.2	49.2 ± 6.5	4.8 ± 2.5	66.0 ± 3.6	65.0 ± 2.4	63.2 ± 1.5	11.3 ± 2.7
60	90.5 ± 1.7	88.7 ± 3.9	78.3 ± 4.8	64.2 ± 3.3	6.9 ± 3.3	82.4 ± 3.3	66.6 ± 2.1	65.4 ± 2.0	13.9 ± 3.9
80	95.2 ± 1.4	94.1 ± 3.1	87.1 ± 4.2	77.4 ± 5.4	10.2 ± 1.4	92.0 ± 2.7	68.8 ± 2.2	67.9 ± 1.2	18.0 ± 3.1
100	98.2 ± 0.1	96.6 ± 2.3	91.5 ± 4.0	83.9 ± 1.4	10.7 ± 1.4	96.5 ± 1.7	70.9 ± 1.7	69.8 ± 0.8	18.6 ± 2.3
120	98.8 ± 0.5	97.7 ± 1.7	95.3 ± 2.2	89.8 ± 2.2	11.3 ± 2.2	97.2 ± 2.0	74.2 ± 0.5	71.7 ± 0.4	19.0 ± 1.7

In terms of 1,4-dioxane removal within 120 min, the different processes were in the order as follows: PSPC ≈ Control 1 > Control 2 > Control 3≫Control 4 ≈ Control 0. Control 0 showed that the adsorption was not significant for 1,4-dioxane removal in the PSPC process. Without Fe^{2+} catalyst addition, 1,4-dioxane could be hardly removed by H_2O_2 oxidation (see Control 4). Control 1 (i.e. the conventional PC process) removed 1,4-dioxane efficiently but consumed ~18.2 mM Fe^{2+} from the iron electrode within 120 min. Control 2 (i.e. the EF process) removed less 1,4-dioxane than PC and PSPC processes. Control 3 showed that only pulsed H_2O_2 production with 7.9 mM Fe^{2+} addition did not remove 1,4-dioxane as efficiently as that in the PSPC process within 120 min (71.7 ± 0.4% vs. 95.3 ± 2.2%).

High current densities of H_2O_2 and Fe^{2+} production resulted in high energy consumption in the PSPC process (Table 3). The energy consumption under 15 mA/cm² (H_2O_2) +3 mA/cm² (Fe^{2+}) was 15 times higher than that under 2.0 mA/cm² (H_2O_2) +0.4 mA/cm² (Fe^{2+}) within 120 min (121.0 ± 0.6 vs. 7.8 ± 0.1 kWh/kg). Nevertheless, the minimum energy consumption in the PSPC process under 2.0 mA/cm² (H_2O_2) +0.4 mA/cm² (Fe^{2+}) was less than 20% of that in Control 1 and 2 within 120 min, respectively (Table 3).

Residual Fe^{2+} and H_2O_2 concentrations during 1,4-dioxane degradation in the PSPC process

Under the pulsed switching frequency of 1.43 Hz and current densities of 5 mA/cm² (H_2O_2) and 1 mA/cm² (Fe^{2+}), the residual Fe^{2+} and H_2O_2 concentrations during 1,4-dioxane degradation was determined in the PSPC process within

Table 3 | Energy consumption in the PSPC processes and different controls

Time (min)	Energy consumption (kWh/kg)						
	Current density (mA/cm²) in the PSPC process (pulsed switching frequency = 1.43 Hz)						
	15 (H_2O_2) + 3 (Fe^{2+})	10 (H_2O_2) + 2 (Fe^{2+})	5 (H_2O_2) + 1 (Fe^{2+})	2.0 (H_2O_2) + 0.4 (Fe^{2+})	Control 1	Control 2	Control 3
20	36.7 ± 1.2	20.4 ± 1.8	10.1 ± 1.2	4.5 ± 0.5	21.9 ± 2.9	9.8 ± 0.0	5.8 ± 0.6
40	49.0 ± 0.7	26.6 ± 0.8	11.0 ± 0.5	5.0 ± 0.6	21.6 ± 1.3	17.9 ± 0.5	10.4 ± 0.2
60	64.2 ± 1.2	34.8 ± 0.5	13.3 ± 0.5	5.6 ± 0.2	25.7 ± 1.2	26.2 ± 0.7	15.1 ± 0.4
80	81.9 ± 1.2	44.2 ± 0.2	16.0 ± 0.5	6.1 ± 0.4	30.4 ± 0.8	33.8 ± 0.9	19.3 ± 0.3
100	101.0 ± 0.0	52.5 ± 0.3	19.1 ± 0.7	7.0 ± 0.1	36.5 ± 0.8	41.0 ± 0.7	23.5 ± 0.2
120	121.0 ± 0.6	62.9 ± 0.8	22.1 ± 0.5	7.8 ± 0.1	43.5 ± 1.1	47.0 ± 0.6	27.5 ± 0.3

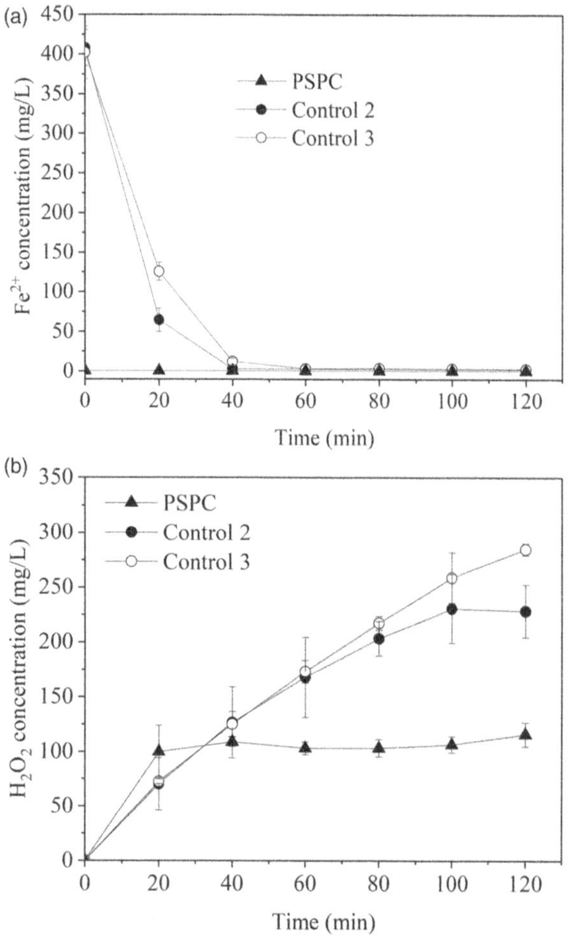

Figure 3 | Residual (a) Fe^{2+} and (b) H_2O_2 concentrations in PSPC process (pulsed switching frequency of 1.43 Hz, current densities of 5 mA/cm^2 for H_2O_2 production and 1 mA/cm^2 for Fe^{2+} production) and controls within 120 min.

120 min (Figure 3). The residual Fe^{2+} concentration was kept as low as 0.6 ± 0.5 mg/L within 120 min. The H_2O_2 concentration reached 100 mg/L and remained stable at 100 mg/L within 120 min. The residual Fe^{2+} and H_2O_2 concentrations in the PSPC process were different from that in Controls 2 and 3. With initial 7.9 mM Fe^{2+} addition, the residual Fe^{2+} concentration was decreased to <0.5 mg/L within 40 min in Controls 1 and 2. However, the residual H_2O_2 concentration was gradually increased to 280 and 250 mg/L in Controls 1 and 2, respectively. High residual Fe^{2+} concentration within 40 min and high residual H_2O_2 concentration within 40–120 min in Controls 1 and 2 may account for the lower 1,4-dioxane removal compared with that in the PSPC process. The residually soluble Fe^{3+} was not detected in all the tests, indicating that $Fe(OH)_3$ precipitated in the cell (Brillas *et al.* 2009).

Intermediates in the 1,4-dioxane degradation

Under the pulsed switching frequency of 1.43 Hz and current densities of 5 mA/cm^2 (H_2O_2) and 1 mA/cm^2 (Fe^{2+}), TCOD and SCOD in the cell were measured within 120 min as shown in Figure 4. The TCOD and SCOD concentrations decreased from 356 ± 28 to 119 ± 27 mg/L and from 356 ± 24 to 99 ± 10 mg/L within 120 min, respectively. Almost the same removals of TCOD and SCOD (67 vs. 72%) indicated that hydroxyl radical oxidation played the key role in the removal of 1,4-dioxane and its intermediates. Only 26.0% of TOC was removed in the PSPC system within 120 min, indicating that 1,4-dioxane was incompletely mineralized. High concentrations of the residual intermediates (e.g. small molecular organic acids) should account for low TOC removal in the 1,4-dioxane degradation in the PSPC process according to previous reports (Barndõk *et al.* 2016a). Acetic acid, formic acid, and oxalic acid were identified in the PSPC process (Figure 5). Under the pulsed switching frequency of 1.43 Hz and current densities of 5 mA/cm^2 (H_2O_2) and 1 mA/cm^2 (Fe^{2+}), the maximum formic acid concentration of 51.2 ± 0.3 mg/L was determined within 80 min. The concentrations of acetic acid, formic acid and oxalic acid were 10.0 ± 4.6, 31.1 ± 9.1, and 2.7 ± 0.8 mg/L at 120 min, respectively.

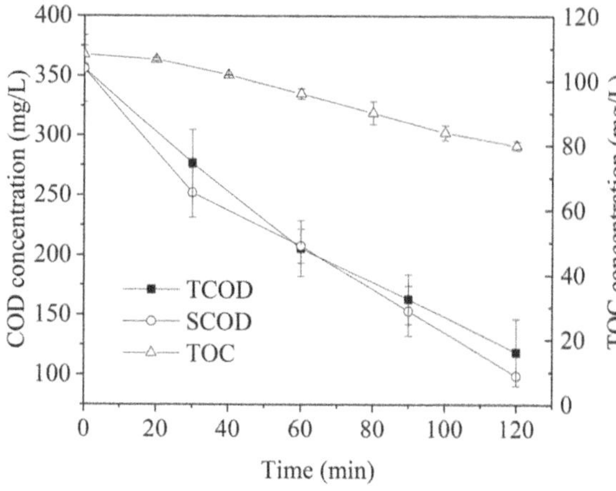

Figure 4 | Removal of TCOD, SCOD and TOC in the PSPC process within 120 min (pulsed switching frequency of 1.43 Hz, current densities of 5 mA/cm^2 for H_2O_2 production and 1 mA/cm^2 for Fe^{2+} production).

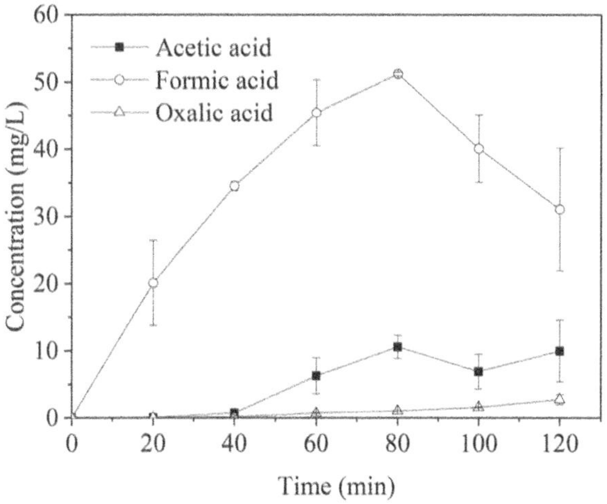

Figure 5 | Intermediate concentrations in the 1,4-dioxane degradation in the PSPC process (pulsed switching frequency of 1.43 Hz, current densities of 5 mA/cm^2 for H_2O_2 production and 1 mA/cm^2 for Fe^{2+} production).

DISCUSSION

During the hydroxyl radical production, many secondary reactions can occur in the Fenton process as follows:

$$Fe^{2+} + H_2O_2 + H^+ \rightarrow Fe^{3+} + H_2O + \cdot OH \quad K = 63M^{-1} \cdot s^{-1} \tag{3}$$

$$H_2O_2 + \cdot OH \rightarrow H_2O + HO_2 \cdot \quad k_2 = 2.7 \times 10^7 M^{-1} \cdot s^{-1} \tag{4}$$

$$Fe^{2+} + \cdot OH \rightarrow Fe^{3+} + OH^- \quad k_2 = 3.2 \times 10^8 M^{-1} \cdot s^{-1} \tag{5}$$

$$Fe^{2+} + HO_2 \cdot + H^+ \rightarrow Fe^{3+} + H_2O_2 \quad k_2 = 1.2 \times 10^6 M^{-1} \cdot s^{-1} \tag{6}$$

$$HO_2 \cdot + HO_2 \cdot \rightarrow H_2O_2 + O_2 \quad k_2 = 8.3 \times 10^5 M^{-1} \cdot s^{-1} \tag{7}$$

$$Fe^{3+} + H_2O_2 \rightarrow Fe^{2+} + HO_2 \cdot + H^+ \quad k_2 = 0.01 M^{-1} \cdot s^{-1} \tag{8}$$

The excess of Fe^{2+} or H_2O_2 may give rise to the side reactions (4), (5), (6), and (7). As previously reported, the ·OH attacks the organics to form hydroxylated derivatives with a reaction rate (k_2) of about 10^7–10^9 M^{-1} s^{-1} (Brillas & Martínez-Huitle 2015; Gao *et al.* 2015), which is comparable to that in the side reactions (4), (5), (6), and (7). The reaction rate of 63 $M^{-1} \cdot s^{-1}$ in the ·OH production (Equation (3)) was much lower than those in Equations (4)–(7). Excess of Fe^{2+} or H_2O_2 can act as powerful competitors to the organics such as 1,4-dioxane to react with the ·OH (Sun & Pignatello 1993; Muruganandham & Swaminathan 2004). Therefore, excess of Fe^{2+} or H_2O_2 can result in low 1,4-dioxane removal and high energy consumption in the EF and PC processes. However, with the pulsed switching circuits, the excess of Fe^{2+} and H_2O_2 production was minimized in the PSPC process (Figure 3, Tables 1 and 2). The low residual Fe^{2+} concentration in the PSPC process within 120 min could be attributed to the high reduction activity of Fe^{2+} released from the iron sacrifice electrode, which can be rapidly consumed as described in Equation (3). However, low residual Fe^{2+} concentration in Controls 2 and 3 within 40–120 min indicated that not enough Fe^{2+} catalyzed H_2O_2 to produce ·OH. The Fe^{2+} in Controls 2 and 3 may be regenerated via the side reaction such as Equation (8) with low reaction rate, inhibiting efficient ·OH production. Therefore, the PSPC could have higher 1,4-dioxane removal with lower energy consumption compared with other processes.

The optimal molar ratio of H_2O_2 to Fe^{2+} (summarized as $[H_2O_2]/[Fe^{2+}]$) was greatly changed depending on the organics, solutions, etc. The optimal $[H_2O_2]/[Fe^{2+}]$ was in the range of 5–11 for the chlorinated aliphatic organic removal in the Fenton process (Tang & Huang 1997). $[H_2O_2]/[Fe^{2+}] = 1000{:}1$ had been reported to produce the highest ·OH concentration in the Fenton process among different molar ratios (Fischbacher *et al.* 2017). The theoretically optimal $[H_2O_2]/[Fe^{2+}]$ was 2.0 based on the pulsed switching ratio of 1.33 in this study, which was different from $[H_2O_2]/[Fe^{2+}] = 6$ for the 2,4-dichlorophenoxyacetic acid degradation in the PSPC process (Lu *et al.* 2018). It indicated that the pulsed switching ratio of H_2O_2 to Fe^{2+} should be further investigated to optimize $[H_2O_2]/[Fe^{2+}]$ during the PSPC application into the real wastewater treatment.

With a continuous consumption of H_2O_2 and Fe^{2+} as shown in Equation (3), lower H_2O_2 and Fe^{2+} concentrations were accumulated under higher pulsed switching frequency in the PSPC process. Similar results had been reported in the Fenton process with the stepwise addition of H_2O_2 to avoid excess H_2O_2 accumulation and decrease H_2O_2 scavenging ·OH (Zhang *et al.* 2012; Gan & Li 2013). With the pulsed switching frequency increasing from 0.71 to 1.43 Hz, the 1,4-dioxane removal was not significantly improved (93.1 ± 2.6 vs. 95.4 ± 2.2%), indicating the limited effect of high pulsed switching frequency on the 1,4-dioxane removal. In addition, higher pulsed switching frequency will require higher precision of time relays and higher investment. Although the highest pulsed switching frequency of 1.43 Hz had the highest 1,4-dioxane removal with the lowest energy consumption in our PSPC process, the optimal pulsed switching frequency still needs further study in the PSPC process in future.

Because the residual H_2O_2 concentration was much higher than that of the residual Fe^{2+} concentration in the PSPC process, the ·OH production and 1,4-dioxane removal should occur close to the sacrifice iron electrode. However, with 7.9 mM Fe^{2+} addition in Control 2 (i.e. the EF process), the ·OH production and 1,4-dioxane removal could occur close to the cathode. Compared with the PSPC, the EF process may suffer from $Fe(OH)_3$ precipitation on the cathode surface, which can decrease the reaction area of H_2O_2 electro-generation in the cathode. Because of high solubility of 1,4-dioxane, the $Fe(OH)_3$ coagulation had no apparent effect on the 1,4-dioxane removal. The 1,4-dioxane degradation in the PSPC process was proposed as shown in Figure 6, based on the quantified intermediates. Various short chain organic acids have been identified in the intermediates of 1,4-dioxane degradation in AOPs (Barndõk *et al.* 2016a). In the UV photo-Fenton process, the intermediates such as acetic, oxalic, methoxyacetic and glycolic acids were identified during 7.3 g/L 1,4-dioxane degradation (Barndõk *et al.* 2016a). The maximum acetic acid concentration reached ~200 mg/L within 120 min (Barndõk *et al.* 2016a). In the solar photocatalysis using an NF-TiO_2 composite with TiO_2 nanoparticles, 140 mg/L 1,4-dioxane degradation resulted in the maximum formic acid of ~60 mg/L within 7 h (Barndõk *et al.* 2016b). Different AOPs including the PSPC showed that complete mineralization of 1,4-dioxane was greatly postponed due to high concentrations of short chain organic acids as intermediates (Barndõk *et al.* 2016a, 2016b). The short chain organic acids can be easily degraded by microbiology (Huang *et al.* 2017; Wu *et al.* 2020), thus the PSPC combined with a biological treatment process may be a potential way to mineralize 1,4-dioxane efficiently in future.

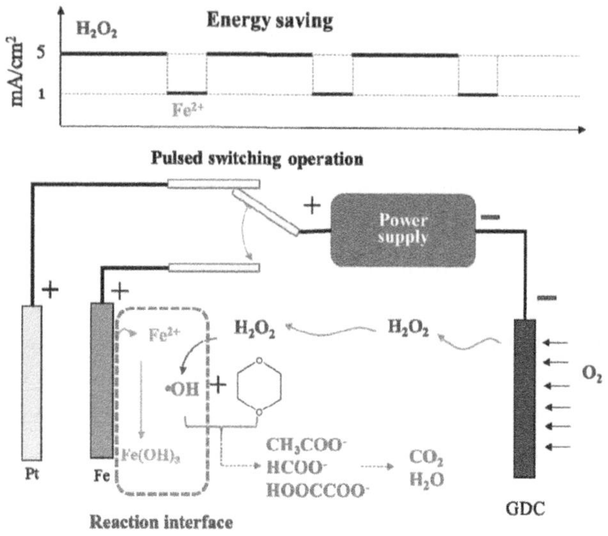

Figure 6 | Proposed degradation process of 1,4-dioxane in the PSPC.

CONCLUSIONS

The pulsed switching peroxi-coagulation (PSPC) process was used to remove 1,4-dioxane in this study. The pulsed switching ratio of H_2O_2 to Fe^{2+} production was optimized at 1.33. Under the optimized pulsed switching frequency of 1.43 Hz, the maximum 1,4-dioxane removal reached 98.8 ± 0.5% with the current densities of 15 mA/cm^2 (H_2O_2) +3 mA/cm^2 (Fe^{2+}) within 120 min. The minimum energy consumption reached 7.8 ± 0.1 kWh/kg with the current densities of 2.0 mA/cm^2 (H_2O_2) +0.4 mA/cm^2 (Fe^{2+}). At the end of 120 min operation, the residual Fe^{2+} concentration was kept as low as 0.6 ± 0.5 mg/L. Under the current densities of 5.0 mA/cm^2 (H_2O_2) +1 mA/cm^2 (Fe^{2+}), the residual H_2O_2 concentration reached 100 mg/L and remained stable at 100 mg/L. Low residual H_2O_2 and Fe^{2+} concentrations were attributed to high pulsed switching frequency in the PSPC process, resulting in effectively inhibiting the side reaction during the ·OH production and improving the 1,4-dioxane removal with low energy consumption. The quantified intermediates showed that the formic acid reached the maximum concentration of 51.2 ± 0.3 mg/L within 80 min. High concentrations of intermediates significantly hindered the mineralization of 1,4-dioxane.

ACKNOWLEDGEMENTS

This work was partly supported by grants from the Science and Technology Program of Guangzhou, China (No.

201804010450), the National Natural Science Foundation of China (Nos. 51608547, 51978676 and 51308557), the Guangdong Provincial Key Laboratory Project (2019B121203011), and the Fundamental Research Funds for the Central Universities (No. 19lgjc08).

DATA AVAILABILITY STATEMENT

All relevant data are included in the paper or its Supplementary Information.

REFERENCES

Ahangarnokolaei, M. A., Ganjidoust, H. & Ayati, B. 2017 Optimization of parameters of electrocoagulation/flotation process for removal of Acid Red 14 with mesh stainless steel electrodes. *Journal of Water Reuse and Desalination* **8** (2), 278–292.

Barazesh, J. M., Hennebel, T., Jasper, J. T. & Sedlak, D. L. 2015 Modular advanced oxidation process enabled by cathodic hydrogen peroxide production. *Environmental Science & Technology* **49** (12), 7391–7399.

Barndõk, H., Blanco, L., Hermosilla, D. & Blanco, Á. 2016a Heterogeneous photo-Fenton processes using zero valent iron microspheres for the treatment of wastewaters contaminated with 1,4-dioxane. *Chemical Engineering Journal* **284**, 112–121.

Barndõk, H., Hermosilla, D., Han, C., Dionysiou, D. D., Negro, C. & Blanco, Á. 2016b Degradation of 1,4-dioxane from industrial wastewater by solar photocatalysis using immobilized NF-TiO2 composite with monodisperse TiO2 nanoparticles. *Applied Catalysis B Environmental* **180**, 44–52.

Benhadji, A., Taleb Ahmed, M., Djelal, H. & Maachi, R. 2016 Electrochemical treatment of spent tan bath solution for reuse. *Journal of Water Reuse and Desalination* **8** (1), 123–134.

Brillas, E. & Martínez-Huitle, C. A. 2015 Decontamination of wastewaters containing synthetic organic dyes by electrochemical methods. An updated review. *Applied Catalysis B: Environmental* **166–167**, 603–643.

Brillas, E., Sirés, I. & Oturan, M. A. 2009 Electro-Fenton process and related electrochemical technologies based on Fenton's reaction chemistry. *Chemical Reviews* **109** (12), 6570–6631.

Chen, Z., Wu, G., Wu, Y., Wu, Q. & Hu, H. Y. 2020 Water Eco-Nexus Cycle System (WaterEcoNet) as a key solution for water shortage and water environment problems in urban areas. *Water Cycle* **1**, 71–77.

Clercq, J., Van de Steene, E., Verbeken, K. & Verhaege, M. 2010 Electrochemical oxidation of 1,4-dioxane at boron-doped diamond electrode. *Journal of Chemical Technology and Biotechnology* **85**, 1162–1167.

Elkacmi, R. & Bennajah, M. 2019 Advanced oxidation technologies for the treatment and detoxification of olive mill wastewater: a general review. *Journal of Water Reuse and Desalination* **9** (4), 463–505.

Fan, W., Yang, X., Wang, Y. & Huo, M. 2020 Loopholes in the current reclaimed water quality standards for clogging control during aquifer storage and recovery in China. *Water Cycle* **1**, 13–18.

Fischbacher, A., von Sonntag, C. & Schmidt, T. C. 2017 Hydroxyl radical yields in the Fenton process under various pH, ligand concentrations and hydrogen peroxide/Fe(II) ratios. *Chemosphere* **182**, 738–744.

Gan, P. P. & Li, S. F. Y. 2013 Efficient removal of Rhodamine B using a rice hull-based silica supported iron catalyst by Fenton-like process. *Chemical Engineering Journal* **229** (4), 351–363.

Gao, G., Zhang, Q., Hao, Z. & Vecitis, C. D. 2015 Carbon nanotube membrane stack for flow-through sequential regenerative electro-Fenton. *Environmental Science and Technology* **49** (4), 2375–2383.

Hand, S., Wang, B. & Chu, K. H. 2015 Biodegradation of 1,4-dioxane: effects of enzyme inducers and trichloroethylene. *Science of Total Environment* **520**, 154–159.

Huang, H., Chen, B.-Y. & Zhu, Z.-R. 2017 Formation and speciation of haloacetamides and haloacetonitriles for chlorination, chloramination, and chlorination followed by chloramination. *Chemosphere* **166**, 126–134.

Huang, H., Wu, J., Ye, J., Ye, T., Deng, J., Liang, Y. & Liu, W. 2018 Occurrence, removal, and environmental risks of pharmaceuticals in wastewater treatment plants in south China. *Frontiers of Environmental Science & Engineering* **12** (6), 69–79.

Lin, S., Xiao, L., Hu, Z. & Zhan, X. 2020 Nutrient recovery from animal manure using bipolar membrane electrodialysis: study on product purity and energy efficiency. *Water Cycle* **1**, 54–62.

Liu, G., Zhou, Y., Luo, H., Cheng, X., Zhang, R. & Teng, W. 2015 A comparative evaluation of different types of microbial electrolysis desalination cells for malic acid production. *Bioresoure Technology* **198**, 87–93.

Lu, Y., He, S., Wang, D., Luo, S., Liu, A., Luo, H., Liu, G. & Zhang, R. 2018 A pulsed switching peroxi-coagulation process to control hydroxyl radical production and to enhance 2,4-Dichlorophenoxyacetic acid degradation. *Frontiers of Environmental Science & Engineering* **12** (5), 9.

Luo, H., Li, C., Wu, C. & Dong, X. 2015 In situ electrosynthesis of hydrogen peroxide with an improved gas diffusion cathode by rolling carbon black and PTFE. *RSC Advances* **5** (80), 65227–65235.

Mahendra, S., Grostern, A. & Alvarez-Cohen, L. 2013 The impact of chlorinated solvent co-contaminants on the biodegradation kinetics of 1,4-dioxane. *Chemosphere* **91** (1), 88–92.

Muruganandham, M. & Swaminathan, M. 2004 Decolourisation of Reactive Orange 4 by Fenton and photo-Fenton oxidation technology. *Dyes & Pigments* **63** (3), 315–321.

Nakagawa, H., Takagi, S. & Maekawa, J. 2016 Fered-Fenton process for the degradation of 1,4-dioxane with an activated carbon electrode: a kinetic model including active radicals. *Chemical Engineering Journal* **296**, 398–405.

Radcliffe, J. C. & Page, D. 2020 Water reuse and recycling in Australia- history, current situation and future perspectives. *Water Cycle* **1**, 19–40.

Rossum, T. V. 2020 Water reuse and recycling in Canada-history, current situation and future perspectives. *Water Cycle* **1**, 98–103.

Sei, K., Kakinoki, T., Inoue, D., Soda, S., Fujita, M. & Ike, M. 2010 Evaluation of the biodegradation potential of 1,4-dioxane in river, soil and activated sludge samples. *Biodegradation* **21** (4), 585–591.

Sei, K., Miyagaki, K., Kakinoki, T., Fukugasako, K., Inoue, D. & Ike, M. 2013 Isolation and characterization of bacterial strains that have high ability to degrade 1,4-dioxane as a sole carbon and energy source. *Biodegradation* **24** (5), 665–674.

Sekar, R. & DiChristina, T. J. 2014 Microbially driven Fenton reaction for degradation of the widespread environmental contaminant 1,4-dioxane. *Environmental Science & Technology* **48** (21), 12858–128567.

Somda, W., Tischbein, B. & Bogardi, J. 2020 Water use inside inland valleys agro-systems in the Dano Basin, Burkina Faso. *Water Cycle* **1**, 88–97.

Sun, Y. & Pignatello, J. J. 1993 Photochemical reactions involved in the total mineralization of 2,4-D by iron(3+)/hydrogen peroxide/UV. *Environmental Science & Technology* **27** (2), 304–310.

Sun, B., Ko, K. & Ramsay, J. A. 2010 Biodegradation of 1,4-dioxane by a Flavobacterium. *Biodegradation* **22** (3), 651–659.

Takeuchi, H. & Tanaka, H. 2020 Water reuse and recycling in Japan – history, current situation, and future perspectives. *Water Cycle* **1**, 1–12.

Tang, W. Z. & Huang, C. P. 1997 Stochiometry of Fenton's reagent in the oxidation of chlorinated aliphatic organic pollutants. *Environmental Technology* **18** (1), 13–23.

Wang, N., Zheng, T., Zhang, G. & Wang, P. 2016 A review on Fenton-like processes for organic wastewater treatment. *Journal of Environmental Chemical Engineering* **4** (1), 762–787.

Wang, W., Lu, Y., Luo, H., Liu, G. & Zhang, R. 2017 Effect of an improved gas diffusion cathode on carbamazepine removal using the electro-Fenton process. *RSC Advances* **7** (41), 25627–25633.

Wu, G., Yin, Q., Fang, G. & Hu, Z. 2020 Slow growers possess a high pollutant removal potential through granule formation for wastewater treatment. *Water Cycle* **1**, 63–69.

Xu, H. Y., Shi, T. N., Wu, L. C. & Qi, S. Y. 2013 Discoloration of methyl orange in the presence of Schorl and HO: kinetics and mechanism. *Water Air & Soil Pollution* **224** (10), 1–11.

Xu, A., Wu, Y. H., Chen, Z., Wu, G. & Hu, H. Y. 2020 Towards the new era of wastewater treatment of China: development history, current status, and future directions. *Water Cycle* **1**, 80–87.

Ye, B., Luo, H., Lu, Y., Liu, G., Zhang, R. & Li, X. 2017 Improved performance of the microbial electrolysis desalination and chemical-production cell with enlarged anode and high applied voltages. *Bioresource Technology* **244** (Pt 1), 913–919.

Zhang, H., Wu, X. & Li, X. 2012 Oxidation and coagulation removal of COD from landfill leachate by Fered–Fenton process. *Chemical Engineering Journal* **210** (4), 188–194.

Zhang, S., Gedalanga, P. B. & Mahendra, S. 2016 Biodegradation kinetics of 1,4-dioxane in chlorinated solvent mixtures. *Environmental Science & Technology* **50** (17), 9599–9607.

Zhao, Y., Ji, B., Liu, R., Ren, B. & Wei, T. 2020 Constructed treatment wetland: Glance of development and future perspectives. *Water Cycle* **1**, 104–112.

First received 16 October 2020; accepted in revised form 25 November 2020. Available online 17 February 2021

 © 2021 The Authors

Recycling of rural abandoned constructed wetlands: mariculture wastewater treatment

Yu Xin, Lin Liu, Lili Wei, Xu Huang and Chaoxiang Liu

ABSTRACT

This study aimed to investigate the behavioral shifts of constructed wetland (CW) when the treated water was changed from domestic wastewater to mariculture wastewater. The results showed that the average removal efficiencies of ammonium nitrogen (NH_4^+-N), total nitrogen and chemical oxygen demand (COD) were 29.54, 46.07 and 57.15% in mariculture wastewater, respectively, which were significantly lower than those in domestic wastewater (71.35, 66.34 and 74.98%, respectively). While there was no significant difference in the removal efficiency of nitrate and phosphate ($P > 0.05$) between the two systems. Based on the analysis of bacterial community and adsorption properties, the results further indicated that the removal mechanism of NH_4^+-N between both systems was mainly due to substrate adsorption: the maximum adsorption capacity of NH_4^+-N on the substrate in mariculture wastewater was 5,432 mg kg^{-1}, whereas that in domestic wastewater was 18,033 mg kg^{-1}. In terms of bacterial communities, the dominant bacteria at the family level were *Victivallaceae* (18.63%) in domestic wastewater and *Porphyromonadaceae* (18.37%) in mariculture wastewater, which showed the significant alteration to the bacterial community. In conclusion, this study showed that conventional CW could be used for treating wastewater from land-based marine aquaculture, while the operating conditions needed to be optimized in the process of application.

Key words | bacteria community, metabolism pathways, removal efficiency, substrate adsorption

Yu Xin
Lin Liu (corresponding author)
Lili Wei
Xu Huang
Chaoxiang Liu
Key Laboratory of Urban Pollutant Conversion, Institute of Urban Environment, Chinese Academy of Sciences,
Xiamen 361021,
China
E-mail: *lliu@iue.ac.cn*

Yu Xin
Lin Liu
University of Chinese Academy of Sciences,
Beijing 100049,
China

Lin Liu
Fujian Institute of Innovation, Chinese Academy of Sciences,
Fuzhou 350002,
China

HIGHLIGHTS

- We addressed changes in the reuse of constructed wetlands for treating mariculture wastewater.
- Significant inhibitive effects on NH_4^+-N, total nitrogen and COD removal occurred during the wastewater transformation.
- Substrate adsorption negatively affected NH_4^+-N removal efficiency in constructed wetland-treatment of mariculture wastewater.

INTRODUCTION

Recently, there has been growing interest in the development of intensive land-based marine aquaculture that could bring economic benefits to coastal areas with environmental sustainability (Martins *et al.* 2010). In a study on land-based Atlantic salmonin farming in China, Sun *et al.* (2016) reported that only 36.5–47.8% nitrogen (N) and 20.4–38.6% phosphorus (P) in the feed were digested by fish and the residual nutrients were dissolved in drainage.

This is an Open Access article distributed under the terms of the Creative Commons Attribution Licence (CC BY 4.0), which permits copying, adaptation and redistribution, provided the original work is properly cited (http://creativecommons.org/licenses/by/4.0/).

doi: 10.2166/wrd.2021.107

Without proper treatment, nutrient enrichment or eutrophication in aquatic environments can cause an increase in algae growth and a decrease in dissolved oxygen, which further destroys the ecological balance and structure (Liang *et al.* 2017). Therefore, the treatment of wastewater from marine aquaculture in an environmentally friendly way is crucial for sustainable intensification of aquaculture.

Many technologies have been widely employed for purification of mariculture wastewater including biological technologies and physicochemical equipment (Munavalli & Pise 2012; Zhao *et al.* 2016; Cao *et al.* 2019), while these approaches lead to some problems such as capital investment, energy consumption, maintenance requirements, secondary pollution and low removal efficiency (He *et al.* 2017). Apart from these methods, constructed wetland (CW) as a lower cost, environment friendly, work efficient and energy-saving technology with no production of secondary pollutants, have shown potential for the mariculture wastewater treatment (Teresa 2018; Wang *et al.* 2020a). In addition, with the development of the urban ecological civilization construction and the progress of rural domestic wastewater interception, many CWs that were originally used for domestic wastewater treatment in rural areas are now idled. Therefore, these CWs had the feasibility of being reused to treat mariculture wastewater. However, during the transformation of the wastewater, understanding of the key purification mechanisms of CWs for specific contaminants in mariculture wastewater remains unclear.

Therefore, the current study was conducted to (1) investigate and compare the removal efficiency of pollutants and substrate bacterial community of CW in treating domestic and mariculture wastewater and (2) clarify the influence mechanisms that cause and influence the changes in performance of pollutant removal during the transformation of the wastewater.

METHODOLOGY

Experiment design and operation

Experimental wetland system

In this study, two horizontal flow structure CW systems were constructed to treat simulated domestic wastewater (R1) and mariculture wastewater (R2). The volume of each CW was 100 L, and the inlet and outlet were arranged at the bottom and upper parts of the device, respectively. Each CW tank was filled with zeolite ($\varphi = 4$–14 mm; porosity $= 0.43$), which was taken from the CW used to treat domestic wastewater in our research group. During the experimental period, the hydraulic loading rate of each system was set to 20 cm d^{-1}.

Artificial wastewater was used in this study: mariculture wastewater was made of chemical oxygen demand (COD) (glucose) 100 mg L^{-1}, NH_4^+-N 25 mg L^{-1}, phosphate 10 mg L^{-1}, nitrate 30 mg L^{-1} and nitrite 5 mg L^{-1}, which were mixed with seawater; domestic wastewater used tap water and pollutants were consistent with mariculture wastewater. Water samples were collected at a fixed time from the inlet and outlet of each CW system. At the end of the experiment, substrate samples were collected from each CW system to analyze the bacterial community.

Adsorption properties of the substrate

A quantitative sterilized substrate of CW was placed in 250 mL conical flasks. 200 mL 30 mg L^{-1} ammonia solution and 200 mL 30 mg L^{-1} ammonia solution with 3% NaCl were added to two separate conical flasks, which were placed in a thermostatic oscillator at 50 r min^{-1}, 25 °C. Samples were taken at 0, 10, 30, 60, 120, 240, 360 and 480 min. Finally, the adsorption capacity of the substrate to NH_4^+-N was calculated according to the residual concentration of the solution in each conical flask. Based on the experimental results, the quasi first-order model, quasi second-order model and intramolecular diffusion model were used to fit the adsorption process of the substrate for ammonium nitrogen. The three models are shown as follows:

$$\text{Quasi first-order model: } \ln(C_e - C_t) = -K_1 t + \ln C_e \tag{1}$$

$$\text{Quasi second-order model: } \frac{t}{C_t} = \frac{1}{k_2 C_e^2} + \frac{1}{C_e} t \tag{2}$$

$$\text{Intramolecular diffusion model: } C_t = K_p t^{0.5} \tag{3}$$

where C_e is the amount of pollutants adsorbed by the substrate at adsorption equilibrium (mg kg^{-1}); C_t is the amount of pollutants adsorbed by the substrate at t (mg kg^{-1}); K_1 is

Quasi first-order model adsorption kinetic constant (min^{-1}); K_2 is Quasi second-order model adsorption kinetic constant ($kg\ mg^{-1}\ min^{-1}$) and K_p is the diffusion rate constant ($mg\ kg^{-1}\ min^{1/2}$).

A quantitative sterilized substrate of CW was added to 250 mL conical flasks and ammonia solution at 10, 15, 25, 30 and 35 $mg\ L^{-1}$ with and without 3% NaCl was added to two equal sets of these flasks. Then, the conical flasks were placed in a thermostatic oscillator at 50 $r\ min^{-1}$, 25 °C and equilibrated 10 h. Finally, the adsorption capacity of the substrate to ammonium nitrogen was calculated according to the residual concentration of the solution in each conical flask. Combined with the experimental results, the Langmuir, Freundlich equation and Dubinin–Radushkevich (D-R) equation can be used to describe the relationship between the adsorption quantities of the substrate material and the equilibrium concentration of the solution. The three models are shown as follows:

$$\text{Langmuir equation: } \frac{C_w}{C_s} = \frac{C_w}{Q_{max}} + \frac{1}{K_L} \tag{4}$$

$$\text{Freundlich equation: } \lg C_s = \lg K_f + \frac{1}{n} \lg C_w \tag{5}$$

$$\text{Dubinin–Radushkevich equation: } \ln C_s = \ln Q_{max} - \beta\varepsilon^2$$

$$\varepsilon = RT \ln\left(1 + \frac{1}{C_w}\right)$$

$$E = \frac{1}{(2\beta)^{1/2}} \tag{6}$$

where C_w is the concentration of pollutants in the solution at adsorption equilibrium ($mg\ L^{-1}$); C_s is the adsorption capacity per unit substrate material ($mg\ kg^{-1}$); Q_{max} is the maximum adsorption capacity per unit substrate material ($mg\ kg^{-1}$); K_L is the Langmuir equilibrium constant, the larger the value, the stronger the adsorption capacity ($L\ kg^{-1}$); K_f is the Freundlich coefficient, which represents the adsorptive capacity of the substrate ($L\ kg^{-1}$) and n is a constant of the equation, which can represent the adsorptive strength of the substrate. β is the D-R adsorption constant; ε is the Polanyi potential; R is the ideal gas constant, 8.314 $J\ mol^{-1}\ K^{-1}$; T is the absolute temperature (K) and E is the average adsorption energy ($kJ\ mol^{-1}$).

Sample analysis

Samples were analyzed for NH_4^+-N, nitrate, nitrite, total nitrogen, COD and phosphate. In R1, all pollutants were measured based on the national standard method (China E.P.A. 2002). In R2, considering the character of mariculture wastewater, the NH_4^+-N was measured by indophenol blue spectrophotometry, COD was measured by the potassium permanganate method and other pollutants were measured based on the national standard method. In addition, as nitrite can interfere with the use of COD as an indicator of organic matter content; thus, in this study, COD correction data were applied to characterize the removal of organic matter.

Substrate samples were extracted by shaking each sample in 100 mL of sterile physiological saline with 100 μL of Tween 80 detergent solution using a vortex mixer for 10 min. Then, the solution was filtered through a 0.22 μm polycarbonate filter (Millipore, MA, USA) to collect the microorganisms. All the processed samples were stored at −80 °C until microbial DNA extraction. The total DNA on the filter paper was extracted with E.Z.N.A.® Water DNA Kit (Omega Bio-Tek, Norcross, GA, USA) according to the manufacturer's protocol. Thereafter, the extracted DNA was separated using agarose gel (1.0%) at 150 V for 20 min to examine the quality of DNA. DNA purity and quantity were determined using a NanoDrop spectrophotometer (NanoDrop Technologies Inc., Wilmington, DE, USA). Extracted DNAs was diluted to 1 $ng{\cdot}(l\ \mu L)^{-1}$ with sterile water in a centrifuge tube and stored at −80 °C prior to high-throughput sequencing. Deep sequencing of the 16S rRNA gene amplicons from the R1 and R2 samples was performed using the Illumina HiSeq 2000. Polymerase chain reaction (PCR) was carried out using Phusion® High-Fidelity PCR Master Mix with GC Buffer (New England Biolabs), the 1 $ng{\cdot}(l\ \mu L)^{-1}$ of genomic DNA, the specific primer with Barcode and high-fidelity enzyme. The primer pair used for PCR was 341F (5′-CCTAYGGGRBGCASCAG-3′) and 806R (5′-GGACTACNNGGGTAT CTAAT-3′), targeting the V3-V4 hypervariable region of the bacterial 16S rRNA gene. The PCR products were separated by electrophoresis on a 2% agarose gel. Then, Uparse software was used to cluster all effective tags of all samples, obtain operational taxonomic units (OTUs) and conduct annotated analysis of species with the GreenGene database. Finally, PICRUSt function prediction analysis was used to clarify the variations in metabolic pathways at R1 and R2 at the gene level. After obtaining KEGG Orthology (KO) information according to OTU, the abundances of relevant functional enzyme-encoding genes involved in the metabolism of nitrogen and carbon were calculated based on OTU abundance and KO database.

Statistical analysis

In our experiment, the statistical analysis was performed using SPSS. One-way analysis of variance (ANOVA) was used to compare mean values between different treatments with the least significant difference (LSD) test at the significance level of 0.05.

RESULTS

Pollutant removal

The pollutant removal efficiencies in R1 and R2 are shown in Figure 1. During the transformation of the wastewater, there were significant differences ($P < 0.05$) in the removal efficiencies of NH_4^+-N, total nitrogen and COD in CW. In R1, the average removal efficiencies of NH_4^+-N, total nitrogen and COD were 71.35, 66.34 and 74.98%, respectively. In R2, the average removal efficiency of NH_4^+-N, total nitrogen and COD were all significantly lower than R1 ($P < 0.05$) at 29.54, 46.07 and 57.15%, respectively. However, the removal efficiency of phosphate and nitrate was not statistically significant between the two systems ($P > 0.05$). In R1, the average removal efficiencies of these two indices were 77.78 and 16.44%, and in R2, they were 75.31 and 16.78%, respectively.

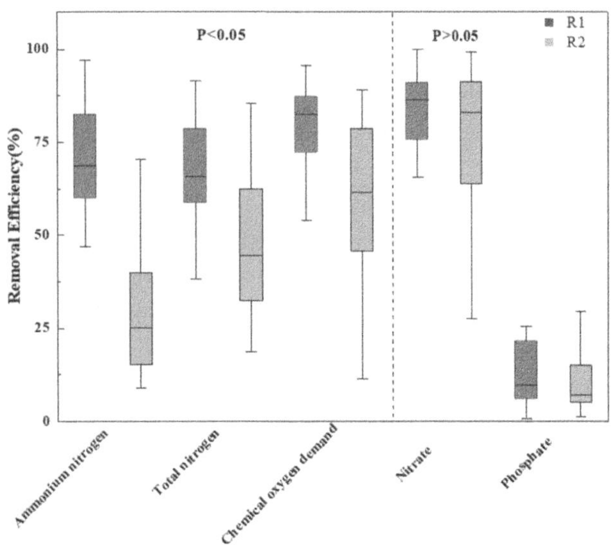

Figure 1 | Pollutant removal efficiency in both systems.

Bacterial community

The relative abundances of the dominant bacterial community (>1%) at phylum, class, order and family levels in substrate samples of each respective CW system are presented in Figure 2. During the transformation of the CW treatment process, the bacterial community retained high similarity at the phylum level. The dominant phylum in R1 and R2 was *Proteobacteria*, with relative abundances of 26.35 and 48.35%, respectively. However, there were several differences between these two systems at other taxonomic levels during the treatment of mariculture wastewater in CW. For example, in R1, *Lentisphaeria*, *Victivallales* and *Victivallaceae* were the dominant bacteria at class, order and family levels, with relative abundances of 18.66, 18.63 and 18.63%, respectively, whereas these bacteria were no longer predominant in R2, and their relative abundances were all reduced to 0.07%. In R2, *Bacteroidia*, *Bacteroidales* and *Porphyromonadaceae* had become the dominant bacteria at class, order and family levels and their relative abundances were 27.76, 27.76 and 18.37%, respectively. In addition, we found that the microorganisms in CW were not uniformly distributed on the surface of zeolite, and they mainly existed in the cracks of zeolite.

Carbon and nitrogen metabolism pathways in CW

To reveal the changes in the degradation pathways of nitrogen and carbon in different CW systems, relevant functional enzyme-encoding genes from R1 and R2 were analyzed based on function prediction analysis (Figures 3 and 4). The nitrogen metabolism pathways in R1 and R2 were the same and included four reduction pathways of nitrification, assimilation nitrate reduction, dissimilation nitrate

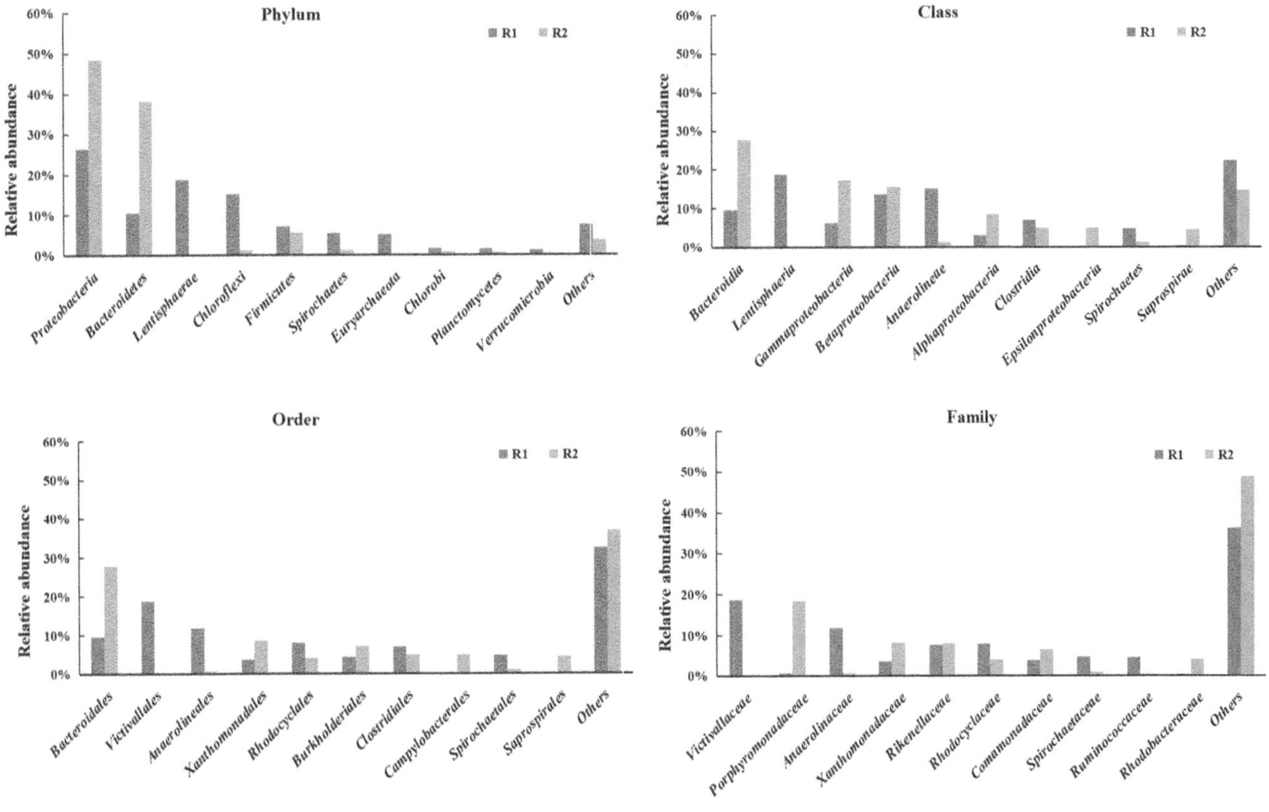

Figure 2 | Taxonomic affiliation of dominant bacterial diversity at phylum, class, order and family levels in different substrates.

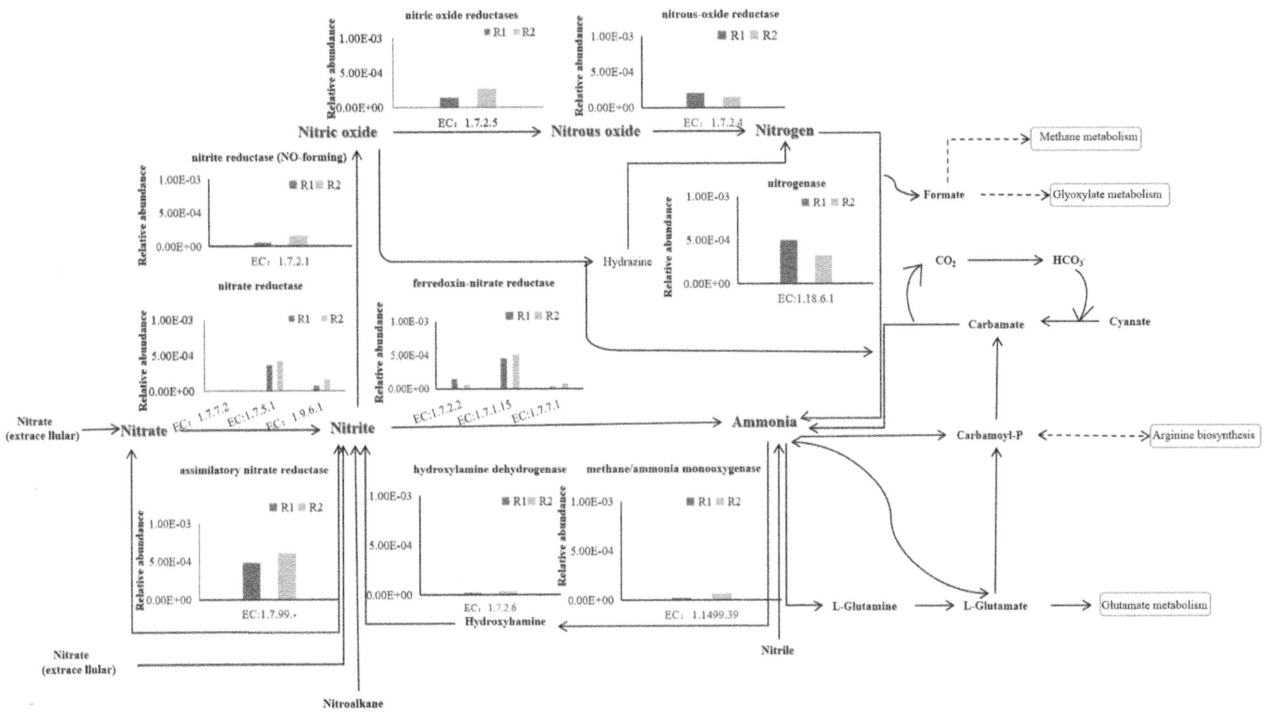

Figure 3 | Nitrogen metabolic pathways and relative abundance of enzyme-encoding genes in two CWs.

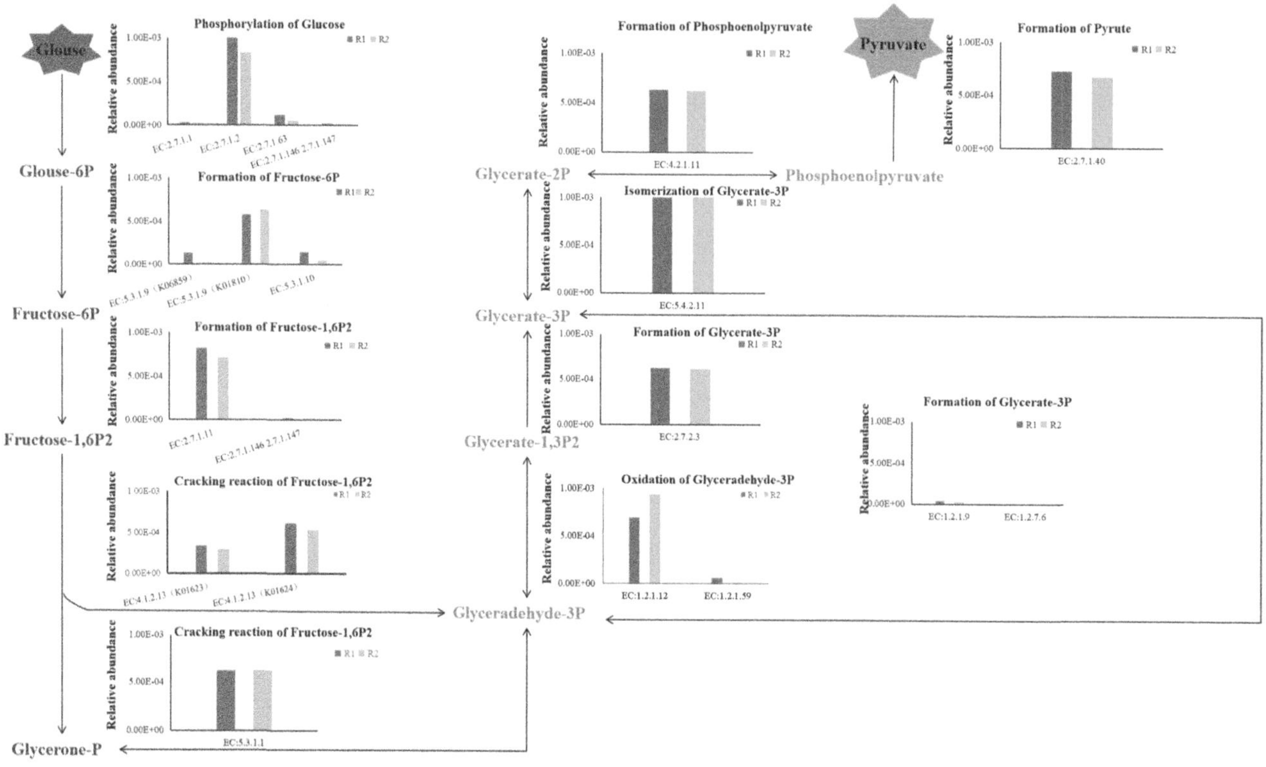

Figure 4 | Carbon metabolic pathways and relative abundance of enzyme-encoding genes in two CWs.

reduction and denitrification, and two oxidation pathways of nitrification and complete nitrification comammox, while the relative abundance of functional enzyme-encoding genes in different pathways were different. The relative abundance of encoding genes of nitrogenase (EC: 1.18.6.1) participating in nitrification was slightly higher in R1 (4.93×10^{-4}) than that in R2 (3.21×10^{-4}). During nitrification and complete nitrification comammox, the relative abundance of encoding genes of methane/ammonia monooxygenase, hydroxylamine dehydrogenase and assimilatory nitrate reductase (EC: 1.1499.39, EC: 1.7.2.6 and EC: 1.7.99, respectively) in R1 (2.27×10^{-5}, 2.16×10^{-5} and 4.78×10^{-4}, respectively) were slightly lower than in R2 (6.22×10^{-5}, 3.46×10^{-5} and 5.97×10^{-4}, respectively), and this phenomenon also existed in the process of assimilation nitrate reduction. At dissimilation nitrate reduction, the relative abundance of encoding genes of functional enzymes in R1 and R2 varied in different reaction stages. For example, at the nitrate to nitrite stage, the proportion of encoding genes of nitrate reductase/nitrite oxidoreductase (EC: 1.7.5.1) in R2 (4.09×10^{-4}) was higher than that in R1 (3.61×10^{-4}); however, at the nitrite to ammonium stage, the proportion of encoding genes of ferredoxin-nitrate reductase (EC: 1.7.2.2) in R2 (5.84×10^{-5}) was obviously lower than that in R1 (1.4×10^{-4}). For carbon metabolism pathways, we only focused on the Embden–Meyerhof pathway of glycolysis which was closely related to carbon source. The Embden–Meyerhof pathway begins with phosphorylation of glucose and ends in pyruvate formation. In this stage, the encoding genes of major functional enzymes were only marginally more abundant in R1 (9.20×10^{-3}) than in R2 (8.02×10^{-3}). For example, the relative abundance of encoding genes of glucokinase was obviously higher in R1 (1.05×10^{-3}) than in R2 (8.32×10^{-4}). Notably, this phenomenon implied that the removal efficiency of COD in R1 was higher than in R2.

Adsorption properties of NH_4^+-N

Since there was a greater significant difference in the NH_4^+-N removal efficiencies compared with that of the other pollutant removal efficiencies between the different

CW systems, the removal mechanism of NH_4^+-N was further analyzed by substrate the adsorption properties (Table 1). Adsorption kinetics are primarily used to describe the rate at which adsorbents adsorb solutes. The results shown in Table 1 indicated that in simulated wastewater, three models all could fit the adsorption process of NH_4^+-N on zeolite ($R^2 > 0.9$, $P < 0.05$). However, in wastewater with 3% NaCl, the quasi second-order model could fit the adsorption process of NH_4^+-N on zeolite ($R^2 = 0.91$) better than the intramolecular diffusion model ($R^2 = 0.81$) and the quasi first-order model ($R^2 = 0.84$). Therefore, we selected the quasi second-order model to analyze the difference in adsorption between the different systems. Based on these parameters, it could find that the adsorption rate of zeolite to NH_4^+-N in the wastewater (7.04×10^{-5} kg mg^{-1} min^{-1}) was lower than that in the wastewater with 3% NaCl (1.90×10^{-4} kg mg^{-1} min^{-1}). In addition, the quasi second-order model assumed that the adsorption rate was controlled by chemisorption (Liu *et al.* 2014), thereby the result also indirectly explained that the adsorption process of NH_4^+-N on zeolite in the different wastewater was chemisorption.

Table 1 | Parameters of adsorption properties for NH_4^+-N

	NH_4^+-N solution	NH_4^+-N solution with 3% NaCl
Quasi first-order model		
K_1 (min^{-1})	1.17×10^{-2}	9.2×10^{-3}
R^2	0.91	0.81
Quasi second-order model		
K_2 (kg mg^{-1} min^{-1})	7.04×10^{-5}	1.90×10^{-4}
R^2	0.95	0.91
Intramolecular diffusion model		
K_p (mg kg^{-1} min$^{1/2}$)	10.69	2.52
R^2	0.93	0.84
Freundlich		
K_f (L kg^{-1})	23.59	0.037
n	1.05	0.45
R^2	0.95	0.83
D-R		
Q_{max} (mg kg^{-1})	18,033	5,432
E (kJ mol^{-1})	10	8.3
R^2	0.80	0.94

Adsorption isotherm describes the solute molecule concentration between the liquid phase and the solid phase when the adsorption processes of the molecules are in equilibrium at a given temperature. In this experiment, the Langmuir equation was unable to fit the data ($R^2 < 0.3$, $P > 0.05$), so we selected the Freundlich equation and D-R equation to analyze the difference in adsorption among the different systems ($R^2 > 0.8$, $P < 0.05$). As shown in Table 1, the Freundlich equation and D-R equation fitted the NH_4^+-N data in simulated wastewater and wastewater with 3% NaCl ($R^2 > 0.9$, $P < 0.05$), respectively, and the respective K_f values, n values and Q_{max} values were ranked as follows: wastewater with 3% NaCl ($K_f = 0.037$ L kg^{-1}, $n = 0.45$, $Q_{max} = 5{,}432$ mg kg^{-1}) < wastewater ($K_f = 23.59$ L kg^{-1}, $n = 1.05$, $Q_{max} = 18{,}033$ mg kg^{-1}). The n values of the Freundlich model were used to describe the adsorption strength for NH_4^+-N were 1.05 in wastewater and 0.45 in wastewater with 3% NaCl. Based on these results, NH_4^+-N had a higher theoretical saturation capacity and adsorption strength in wastewater than in wastewater containing 3% NaCl. In addition, the E values of the D-R model indirectly explained the adsorption mechanism and the values lie within the range expected for chemical adsorption and physical adsorption (8–16 and <8 kJ mol^{-1}, respectively) (Fan *et al.* 2016). From Table 1, the values of E were 10 kJ mol^{-1} in wastewater and 8.3 kJ mol^{-1} in simulated wastewater with 3% NaCl, respectively. Thus, in our experiment, the adsorption mechanism of the zeolite for NH_4^+-N was chemisorption, which was consistent with the results of the adsorption kinetics simulation.

DISCUSSION

In this study, the inhibitive effects on NH_4^+-N and total nitrogen removal were observed during treatment transformation with inhibition ratios of 41.81 and 20.27%, respectively, which are consistent with the previous studies (Liang *et al.* 2017; Fu *et al.* 2019). However, there was no significant removal difference in phosphate and nitrate which was different from other studies. For example, Gao *et al.* (2015) found that the salt in mariculture wastewater could negatively influence phosphate removal in CW. Moreover, there are limited literature reviews on COD during CW for treating mariculture wastewater. While in our experiment, COD removal was also inhibited, and the inhibition ratio was 21.67%.

Generally, microbes play a key role in the nitrification and denitrification processes, which are widely considered as the major N removal mechanism in CWs (Vymazal 2007). Besides, substrate adsorption was another major way for N removal in CWs due to zeolite as the substrate in this experiment. From Figure 3, in nitrification, the relative abundance encoding genes of enzymes involved in mariculture wastewater (9.7×10^{-5}) were more abundant than those in domestic wastewater (4.5×10^{-5}). In contrast, the removal efficiency of NH_4^+-N in mariculture wastewater was significantly lower than that of domestic wastewater. Combining the results of adsorption properties, we believed the reason for this phenomenon was the adsorption property of NH_4^+-N on the substrate. Natural zeolite is a kind of silicate with silica-oxygen tetrahedron and alumina-oxygen tetrahedron as its skeleton and it has large specific surface area (SSA), strong ion exchange capacity and high static adsorption capacity for NH_4^+-N (Inglezakis *et al.* 2018; Wang *et al.* 2018, 2020b). However, the adsorption process is affected by environmental factors, such as organic matter, PH and salinity. Mariculture wastewater is a form of saline wastewater, and when sodium ions are present in the water, they can displace the cations on the zeolite surface (Behin *et al.* 2019), thereby reducing the steric hindrance and increasing the adsorption rate of zeolite for ammonium ions. Simultaneously, sodium ions would compete with the ammonium ions for the adsorption sites on the zeolite surface, so that the adsorption capacity of NH_4^+-N on zeolite would decrease during the transformation of wastewater as shown in Table 1.

Compared with NH_4^+-N, nitrate removal efficiency did not significantly differ between the two systems. In our research, we thought that there were two main reasons for this phenomenon: (1) low substrate adsorption capacity and (2) insufficient carbon compounds in the denitrification process. A previous study concluded that the affinity for adsorption of nitrate onto Al oxides was weak (Zhou & Haynes 2010). Although there was a salinity difference between the two systems, this did not affect the adsorption for nitrate on the substrate. In denitrification, theoretically, the reduction 1.0 g nitrate to nitrogen requires approximately 4.4 g of organic matter (COD). However, the organic content in both systems could not meet the requirements of denitrification. Therefore, the carbon source became a limiting factor in the denitrification process which limited the significant difference in nitrate removal efficiency between different systems. Combining these two factors, there was no difference in the nitrate removal efficiency during the transformation of the wastewater. This conclusion also could be supported by the analysis results of the bacterial community and nitrogen metabolism pathway. As shown in Figures 2 and 3, in denitrification, the relative abundance of encoding genes of enzymes in mariculture wastewater was not inhibited; several genes even had elevated expression. For example, the relative abundance of *nir*K, which is a key functional gene in denitrification (Jones & Hallin 2010) in R2 (1.4×10^{-4}) was greater than that in R1 (5.2×10^{-5}). This result indicated that there was no effect on denitrification after the transformation of the wastewater objective. In terms of total nitrogen, we thought that the NH_4^+-N removal efficiency in mariculture wastewater was significantly lower than that in domestic wastewater, which directly led to lower total nitrogen removal efficiency in mariculture wastewater.

The removal of COD in CW mainly depends on the metabolic activities of microorganisms. As shown in Figure 2, the relative abundances of the bacterial community at the family level in each CW were significantly different. The dominant bacteria changed from *Victivallaceae* to *Porphyromonadaceae* during the transformation of the wastewater. Previous studies have shown that the growth of *Victivallis* is chemo-organotrophic and restricted to a variety of sugars (Plugge & Zoetendal 2014). The CW microbes were originally used for domestic wastewater treatment, *Victivallis* was abundant in domestic wastewater, and significantly improved the removal efficiency of COD in domestic wastewater compared with that in mariculture wastewater. In addition, as shown in Figure 4, relevant functional enzyme-encoding genes involved in glycolysis were more abundant in domestic wastewater than that in mariculture wastewater, which also implied that the removal efficiency of COD may be higher in domestic wastewater than that in mariculture wastewater, and this result was consistent with COD removal.

With regards to phosphate, CWs can effectively remove phosphorus through filtration, adsorption, co-precipitation,

ion exchange, plant absorption and microbial decomposition by taking advantage of the physical, chemical and biological coordination (Vymazal 2007). However, in our experiment, we did not add the plants to the CW, and the most important point was that there was no biofilm-enhanced phosphorus removal mechanism in our systems because the environmental conditions were not suitable for the growth of phosphorus accumulating organisms. Thus, in our study, the phosphate removal mechanism mainly depended on the biosynthetic utilization and substrate adsorption. In general, the maximum phosphate uptake rate of bacteria was achieved under a higher concentration ratio of organic carbon to phosphate in the wastewater. However, the organic content in both systems was all insufficient. In addition, Zhou & Haynes (2010) found that the relative affinities for adsorption of anions onto Al oxides often follow the order: phosphate > chloride ion. Therefore, chlorine ions in mariculture wastewater did not affect the adsorption of phosphate onto the substrate. Considering these two factors, there was no significant difference in the phosphate removal efficiency between the two systems. Besides, zeolites are a class of crystalline aluminosilicates that have a net negative charge due to the substitution of Si (IV) by Al (III) in the Si tetrahedral (Nikolakis 2005). Therefore, adsorption of anions, such as phosphate, by zeolites is generally lower than that of cations (Haynes 2015).

CONCLUSION

Conventional CW can be applied in mariculture wastewater treatment, while the process was affected by wastewater characteristic, and it had negative effects on NH_4^+-N, total nitrogen and COD removal efficiency. These negative effects on COD were mainly caused by salts inhibiting the function of microorganisms in the substrate which will further change the community structures; meanwhile, salts would weaken NH_4^+-N removal efficiency by the substrate due to the competition for uptake into ion. This study provides both a new strategy for reuse of abandoned CW in rural areas and a new understanding of the removal mechanism of mariculture wastewater in CW.

ACKNOWLEDGEMENTS

This work was supported by Strategic Priority Research Program of the Chinese Academy of Sciences, China [XDA23030203]; Youth Innovation Promotion Association of Chinese Academy of Sciences, China [2020308] and Science and Technology Project of Fujian Province, China [2018Y0083].

DATA AVAILABILITY STATEMENT

All relevant data are included in the paper or its Supplementary Information.

REFERENCES

Behin, J., Ghadamnan, E. & Kazemian, H. 2019 Recent advances in the science and technology of natural zeolites in Iran. *Clay Minerals* **54** (2), 131–144.

Cao, S., Jing, Z., Yuan, P., Wang, Y. & Wang, Y. 2019 Performance of constructed wetlands with different substrates for the treated effluent from municipal sewage plants. *Journal of Water Reuse and Desalination* **9** (4), 452–462.

China, E. P. A. 2002 *The Standard Methods of Water and Wastewater Monitoring and Analysis of China*, 4th edn. China Environmental Science Press, Beijing.

Fan, H., Shi, L., Shen, H., Chen, X. & Xie, K. 2016 Equilibrium, isotherm, kinetic and thermodynamic studies for removal of tetracycline antibiotics by adsorption onto hazelnut shell derived activated carbons from aqueous media. *RSC Advances* **6** (111), 109983–109991.

Fu, G., Han, J., Yu, T., Huangshen, L. & Zhao, L. 2019 The structure of denitrifying microbial communities in constructed mangrove wetlands in response to fluctuating salinities. *Journal of Environment Management* **238**, 1–9.

Gao, F., Yang, Z., Li, C. & Jin, W. 2015 Saline domestic sewage treatment in constructed wetlands: study of plant selection and treatment characteristics. *Desalination Water Treatment* **53** (3), 593–602.

Haynes, R. J. 2015 Use of industrial wastes as media in constructed wetlands and filter beds-prospects for removal of phosphate and metals from wastewater streams. *Critical Reviews in Environmental Science and Technology* **45** (10), 1041–1103.

He, H., Chen, Y., Li, X., Cheng, Y., Yang, C. & Zeng, G. 2017 Influence of salinity on microorganisms in activated sludge processes: a review. *International Biodeterioration & Biodegradation* **119**, 520–527.

Inglezakis, V. J., Fyrillas, M. M. & Stylianou, M. A. 2018 Two-phase homogeneous diffusion model for the fixed bed sorption of heavy metals on natural zeolites. *Microporous Mesoporous Materials* **266**, 164–176.

Jones, C. M. & Hallin, S. 2010 Ecological and evolutionary factors underlying global and local assembly of denitrifier communities. *ISME Journal* **4** (5), 633–641.

Liang, Y., Zhu, H., Banuelos, G., Yan, B., Shutes, B., Cheng, X. W. & Chen, X. 2017 Removal of nutrients in saline wastewater using constructed wetlands: plant species, influent loads and salinity levels as influencing factors. *Chemosphere* **187**, 52–61.

Liu, J., Chen, J., Jiang, L. & Yin, X. 2014 Adsorption of mixed polycyclic aromatic hydrocarbons in surfactant solutions by activated carbon. *Journal of Industrial and Engineering Chemistry* **20** (2), 616–623.

Martins, C. I. M., Eding, E. H., Verdegem, M. C. J., Heinsbroek, L. T. N., Schneider, O., Blancheton, J. P., Orbcastel, E. R. & Verreth, J. A. J. 2010 New developments in recirculating aquaculture systems in Europe: a perspective on environmental sustainability. *Aquacultural Engineering* **43** (3), 83–93.

Munavalli, G. R. & Pise, N. S. 2012 Treatment of domestic wastewater by a hybrid natural system. *Journal of Water Reuse and Desalination* **2** (4), 194–203.

Nikolakis, V. 2005 Understanding interactions in zeolite colloidal suspensions: a review. *Current Opinion in Colloidal and Interface Science* **10**, 203–210.

Plugge, C. M. & Zoetendal, E. G. 2014 The family victivallaceae. In: *The Prokaryotes: Other Major Lineages of Bacteria and the Archaea*, E. Rosenberg (editor-in-chief) (E. F. DeLong, S. Lory, E. Stackebrandt & F. Thompson, eds), 4th edn. Springer, Heidelberg, pp. 1019–1021.

Sun, G., Liu, Y., Qiu, D., Yi, M., Li, X. & Li, Y. 2016 Effects of feeding rate and frequency on growth performance, digestion and nutrients balances of Atlantic salmon (*Salmo salar*) in recirculating aquaculture systems (RAS). *Aquaculture Research* **47**, 176–188.

Teresa, N. 2018 Water reuse and desalination in Spain – challenges and opportunities. *Journal of Water Reuse and Desalination* **8** (2), 153–168.

Vymazal, J. 2007 Removal of nutrients in various types of constructed wetlands. *Science of The Total Environment* **380** (1–3), 48–65.

Wang, L., Zhang, Z., Ding, A., Cheng, R., Zhang, M. & Wu, D. 2018 Modification of zeolite materials and their adsorption properties for the pollutants in aqueous solution. *Chemical Industry Engineering Progress* **37** (6), 2269–2281.

Wang, X., Zhu, H., Yan, B., Shutes, B., Banuelos, G. & Wen, H. 2020a Bioaugmented constructed wetlands for denitrification of saline wastewater: a boost for both microorganisms and plants. *Environment International* **138**, 105628.

Wang, H., Xu, J. & Sheng, L. 2020b Purification mechanism of sewage from constructed wetlands with zeolite substrates: a review. *Journal of Cleaner Production* **258**, 120760.

Zhao, Y., Park, H. D., Park, J. H., Zhang, F., Chen, C., Li, X., Zhao, D. & Zhao, F. 2016 Effect of different salinity adaptation on the performance and microbial community in a sequencing batch reactor. *Bioresource Technology* **216**, 808–816.

Zhou, Y. & Haynes, R. J. 2010 Sorption of heavy metals by inorganic and organic components of solid wastes: significance to use of wastes as low-cost adsorbents and immobilizing agents. *Critical Reviews in Environmental Science and Technology* **40**, 1–69.

First received 1 November 2020; accepted in revised form 7 January 2021. Available online 2 February 2021

© 2021 The Authors

Copper removal from semiconductor CMP wastewater in the presence of nano-SiO_2 through biosorption

Xiaoyu Wang, Gude Buer, Wei Fan, Lei Gao and Mingxin Huo

ABSTRACT

Copper-bearing wastewater from chemical mechanical planarization (CMP) is a typical semiconductor development byproduct. How to effectively treat Cu^{2+} in the CMP wastewater is a great concern in the microchip manufacturing industry. In this study, we investigated the potential for the microbial removal of Cu^{2+} by a multiple heavy metal-resistant bacterium *Cupriavidus gilardii* CR3. The environmental factors, including pH, nano-SiO_2, ionic strengths, and initial concentrations of Cu^{2+}, and adsorption times on the bioremoval of Cu^{2+} in CMP wastewater were optimized. Under optimal condition, the maximum biosorption capacity for Cu^{2+} was 18.25 mg g^{-1} and the bioremoval rate was 95.2%. The Freundlich model is described well for the biosorption of Cu^{2+} in CMP wastewater in the presence of nano-SiO_2 ($R^2 = 0.99$). The biosorption process obeyed the pseudo-second-order kinetic equation ($R^2 > 0.99$). In the column experiment, the advection–dispersion–retention model fitted the breakthrough curve of all experiments well ($R^2 > 0.95$). The attachment coefficient in the sand matrix coated by CR3 biofilm was 2.24–2.80 times as that in clean sand. Overall, *C. gilardii* CR3 is a promising candidate to remove Cu^{2+} from CMP wastewater. Nano-SiO_2 in CMP wastewater did not inhibit the bioremoval of Cu^{2+} but showed a slight promotion effect instead.

Key words | batch and packed column, biosorption, chemical mechanical planarization, metal-resistant bacteria

Xiaoyu Wang
Gude Buer
Wei Fan (corresponding author)
Lei Gao
Mingxin Huo
Engineering Lab for Water Pollution Control and Resources and Key Laboratory of Wetland Ecology and Vegetation Restoration, National Environmental Protection,
Northeast Normal University,
Changchun 130117,
China
E-mail: *fanw100@nenu.edu.cn*

HIGHLIGHTS

- Heavy metal-resistant bacterium *Cupriavidus gilardii* CR3 can effectively remove Cu^{2+} from the CMP wastewater in the presence of nano-SiO_2.
- Nano-SiO_2 under the common concentration in CMP wastewater does not affect the bioremoval of Cu^{2+}.
- The Freundlich model is better for describing the biosorption of Cu^{2+} CMP wastewater in the presence of Nano-SiO_2 relative to the Langmuir model.

INTRODUCTION

Increasing demand for advanced semiconductor products or microchips contributes to economic development in nations around the globe; however, it is also associated with a series of environmental issues. A major environmental problem is the large volumes of wastewater containing complicated contaminant composites generated from semiconductor manufacturing. Chemical mechanical planarization/polishing (CMP) is a vital process in semiconductor

This is an Open Access article distributed under the terms of the Creative Commons Attribution Licence (CC BY 4.0), which permits copying, adaptation and redistribution, provided the original work is properly cited (http://creativecommons.org/licenses/by/4.0/).

doi: 10.2166/wrd.2021.098

manufacturing for dielectrics and metal layer planarization. In CMP, a considerable quantity of chemicals and ultra-water is needed to remove the residues of nanoparticles and chemicals on the semiconductor devices and achieve dielectrics and metal layer planarization. Thus, the resulting CMP wastewater contains various contaminants, which accounts for 40% of the entire semiconductor production process wastewater (Lai & Lin 2004). Once treated appropriately, CMP wastewater can be reclaimed and pressures on water resources can be relieved in the semiconductor industry.

In CMP wastewater, cupric ions (Cu^{2+}) pose a great concern due to their persistent nature, accumulation tendency, and biological toxicity. Cu^{2+} in CMP wastewater originated from copper metallization, as well as the application of $CuSO_4$ as oxidizing agents during the CMP. The concentrations of Cu^{2+} in CMP wastewater generally are presented as parts per million although the concentration of Cu^{2+} fluctuates in different ranges, for example, 5–100 mg L^{-1} (Maketon 2007). The reported concentration of Cu^{2+} in CMP wastewater is much higher than the level of copper toxicity toward aquatic organisms at 5–10 parts per billion (ppb) (Chua *et al.* 1999). Accordingly, appropriate treatment of Cu^{2+} in CMP wastewater is necessary to meet the local discharge standards of pollutants for the semiconductor industry. In the districts of the developed semiconductor industry, the permissible range of Cu^{2+} is from 0.2 to 1.0 mg L^{-1} depending on the receiving waters.

Cu^{2+} in CMP wastewater is usually treated by the electrochemical method (Yang & Tsai 2008), ion exchange resin method (Maketon & Ogden 2009), and membrane filtration (Su *et al.* 2014). However, the high initial costs and additional chemicals hamper the application of physical–chemical methods. Alternatively, biosorption by growing cells can offer advantages for the removal of heavy metal ions at low concentrations (Malik 2004). In the case of Cu^{2+} treatment in CMP wastewater, biosorption of Cu^{2+} in CMP wastewater has been reported (Stanley & Ogden 2003; Mosier *et al.* 2015). Our previous study also investigated the characteristics of *C. gilardii* CR3 to treat Cu^{2+} in CMP wastewater (Yang *et al.* 2017). However, the synthesized CMP wastewater used in the previous studies focused on the bioremoval of Cu^{2+}. The CMP requires nanoparticles as abrasive particles, for example, nanoparticle SiO_2 (nano-SiO_2), which is the common nanoparticle in the CMP, usually ranging from 0.05 to 0.50 g L^{-1} (Yang & Tsai 2008). But the treatment of Cu^{2+} rarely varies based on the presence of nano-SiO_2 in CMP wastewater. Hence, it should be considered whether the nano-SiO_2 can affect the biosorption of Cu^{2+} when CMP wastewater is treated.

In this study, we investigated the bioremoval of Cu^{2+} by a copper-resistant bacterium *C. gilardii* CR3 in the presence of nano-SiO_2. In the batch adsorption system, the treatment conditions were optimized and the bioremoval mechanisms of Cu^{2+} were examined. Further, the transport and retention of Cu^{2+} was studied in a fixed-bed column packed with sand coated by strain CR3.

MATERIALS AND METHODS

Bacterium and synthetic wastewater

The bacterium *C. gilardii* CR3 used in this study was isolated in our previous study (Yang *et al.* 2017). The cryopreserved *C. gilardii* CR3 cells were inoculated into a 100 mL Luria–Bertani (LB) medium and cultured in a constant temperature shaker at 28 °C for 24 h. Then the resurrected bacterial cells in 1 mL were transferred into a new 100 mL LB media. The suspension was centrifuged at 10,000 r min^{-1} for 5 min to collect the living cells. The received cells by centrifugation were washed with phosphate-buffered saline three times and the resulting cell suspension was used for the batch and column experiments. The growth curve of the strain in the LB medium is shown in Supplementary Material, Figure S1.

To stimulate the real CMP wastewater containing Cu^{2+} (Cu-CMP wastewater), the synthetic Cu-CMP wastewater was prepared with $CuCl_2 \cdot H_2O$, nano-SiO_2 (50 nm), and deionized water.

Batch experiments

The effects of pH and nano-SiO_2 concentration, ionic strength, initial concentrations of Cu^{2+}, and contact time on the biosorption of Cu^{2+} by *C. gilardii* CR3 were studied. The concentration of Cu^{2+} used in batch experiments was 0.3 mM Cu^{2+} except for the test of Cu^{2+} initial concentrations. First, the pH (4.0, 5.0, and 6.0) and the concentration of nano-SiO_2 (0.1, 0.3, and 0.5) in CMP wastewater were optimized. Afterwards, the effects of ionic strength, initial concentrations of Cu^{2+}, and contact time on the biosorption of Cu^{2+} were conducted under the optimized conditions in the first step (pH = 5, nano-SiO_2 = 0.3 g L^{-1}). The ionic strength was measured at 0, 1.0, 1.5, 2.0, 2.5, 3.0, 4.0, and 5.0 mM with a $NaNO_3$ solution. The initial Cu^{2+} concentration was set to 10 levels, from 0.1 to 1 mM with intervals of 0.1. The contact times on Cu^{2+} biosorption were conducted from 0 to 180 min at 28 °C. In control experiments, the aqueous Cu^{2+} was replaced with the same volume of deionized water. The removal rate and biosorption capacity of strain CR3 on Cu^{2+} were calculated according to the following formula:

$$q_e = \frac{v(C_0 - C_e)}{m} \tag{1}$$

$$\text{Removal rate (\%)} = \frac{C_0 - C_e}{C_0} \times 100 \tag{2}$$

where q_e is the biosorption capacity of Cu^{2+} by the unit of biosorbents in mg g^{-1}, C_0 and C_e are the initial and equilibrium concentration, respectively, of Cu^{2+} in the solution in mg L^{-1}, m is the mass of the biosorbents in g, and v is the volume of Cu-CMP wastewater in the system in L. In this study, Langmuir and Freundlich models were used to describe the equilibrium process of biosorption in batch experiments. The models are expressed as follows:

$$\frac{C_e}{q_e} = \frac{1}{q_m b} + \frac{C_e}{q_m} \tag{3}$$

$$\ln q_e = \ln K_F + \frac{1}{n} \ln C_e \tag{4}$$

in which q_e is represented as the biosorption capacity of Cu^{2+} by the unit of biosorbent in mg g^{-1}, C_e is the equilibrium concentration of Cu^{2+} in the solution in mg L^{-1}, q_m is the theoretical monolayer biosorption capacity in mg g^{-1}, n and K_F are the Freundlich equilibrium constants, and b is the Langmuir equilibrium constant.

To study the biosorption of the system, the experimental results are fitted with pseudo-first-order kinetics and pseudo-second-order kinetics. The first- and second-order equations are expressed as follows:

$$\log(q_e - q_t) = \log q_e - \frac{k_1}{2.303} t \tag{5}$$

$$\frac{t}{q_t} = \frac{1}{k q_e^2} + \frac{1}{q_e} t \tag{6}$$

where q_t and q_e, respectively, represent the biosorption time t and the amount of Cu^{2+} adsorbed by the biosorbent when they reach biosorption equilibrium in mg g^{-1} and k_1 and k_2 are the first and second rate equation constants.

Scanning electron microscope, Fourier-transform infrared spectroscopy, and zeta potential

A scanning electron microscope (SEM) (XL-30 ESEM FEG) was used to investigate the quartz sand, nano-SiO_2, and *C. gilardii* CR3 with Extracellular Polymeric Substances (EPS) in this study.

Quartz sand was ultrasonically cleaned three times and incubated with strain CR3 for 24 h. The quartz sand, strain CR3, and sand–strain CR3 were prepared into a 1.0 g L^{-1} suspension. The Zeta potential of strain CR3, quartz sand, and sand–strain CR3 was measured before and after the biosorption of Cu^{2+} using a Zeta potentiometer (Malvern Zetasizer Nano ZS90) under the pH 5.0, 0.3 mM Cu^{2+}, and 0.3 g L^{-1} nano-SiO_2 condition.

The sand–strain CR3 biosorbent samples before and after biosorption of Cu^{2+} were freeze-dried and ground with KBr at a ratio of 1:100. The sand–strain CR3 biosorbent samples were analyzed using Fourier-transform infrared (FTIR) spectroscopy (SHIMADZU FTIR-8400S).

Transport column experiments

The Cu^{2+} transport experiments were conducted using an acrylic column under saturated flow conditions (Supplementary Material, Figure S2). The column was 2.2 cm in inner diameter and 11 cm in effective length. Quartz sand with a diameter of 0.7–1.0 mm was packed in the column as a porous medium, and the packing density was 1.39 g cm^{-3}. Screens were installed at both ends of the column to prevent the packing from flowing out and distributing the flow evenly. All column experiments were conducted in an up-flow mode using a peristaltic pump (Lead Fluid BT102S) at a flow rate of 4 mL min^{-1}. All experiments were performed at room temperature (19.5 ± 1.2 °C).

The transport column experiments were performed with pulsed injection of Cu^{2+} solution, while Cu^{2+} solution was injected continuously in the latter. Both quartz sand and strain CR3-coated quartz sand was used as the packed medium in both sets of column experiments. Cu^{2+} solution and the mixture of Cu^{2+} and nano-SiO_2 solutions were used as two different input conditions to identify the role of nanoparticles during the process. The strain CR3-coated sand column was prepared according to the following procedure. The quartz sand (58 g) was sterilized by an autoclave at 121 °C for 30 min. Afterwards, the sterilized sand and the strain CR3 (OD_{600} = 1.0) were mixed in a 250 mL flask and were cultured for 24 h on a shaker (28 °C, 120 r min^{-1}). The cells of strain CR3 were attached to the sand surface. Then the composite of sand and strain CR3 was filled into the columns. Next, LB media was injected into the column for 12 h to multiply the bacteria followed by 12-h diluted LB (20%) elution and 2-h DI water elution. This elution process aimed to remove all dead bacteria in the column.

In this column experiment, 20 pore volumes (PVs) of the background electrolyte solution of interest were first passed through the column to ensure that the column was fully equilibrated with this solution. Then a solution containing Cu^{2+} within the same background electrolyte composition was injected into the column for 3.3 PVs, followed by elution with background solution again. Before the Cu^{2+} transport experiments, the tracer (Br^-) breakthrough tests were performed to estimate the hydraulic dispersion coefficient. The concentration of the Br^- was monitored at 1.5 min intervals by ion chromatography (Metrohm, 881 Compact IC pro, Switzerland). Normalized effluent concentrations (C/C_0) were plotted against the number of PVs to obtain the breakthrough curves (BTCs).

Mathematical modeling for column experiments

The transport of Cu^{2+} within the saturated sand column was described by an advection–dispersion–retention (ADR) model considering dynamic blocking and depth-dependent straining. The governing equations can be written as:

$$\frac{\partial(nC)}{\partial t}+\rho\frac{\partial S}{\partial t}=nD\frac{\partial^2 C}{\partial \chi^2}-q\frac{\partial C}{\partial \chi} \tag{7}$$

$$\rho\frac{\partial S}{\partial t}=nk_{att}\psi C-k_{det}\rho S \tag{8}$$

$$\psi=1-\frac{S}{S_{max}} \tag{9}$$

where n is the porosity; C is the concentration of effluent Cu^{2+}; ρ is the bulk density of the porous media; S is the solid phase concentration adsorbed on the quartz sand; q is the flow rate; k_{att} and k_{det} are the first-order attachment coefficient and detachment coefficient, respectively; and ψ is a dimensionless function to account for the combined process of Langmuirian dynamics blocking and depth-dependent straining (Bradford & Bettahar 2006). S_{max} is the maximum solid phase particle concentration. The trace (Br^-) experiments were conducted before the Cu^{2+} penetration experiments to estimate the hydraulic dispersion coefficient (D). The above ADR model was fitted with the penetration curve to obtain k_{att}, k_{det}, and S_{max} parameters. The Cu^{2+} mass of water was calculated by the integral penetration curve BTCs based on the known Cu^{2+} mass input, and then the total adsorption mass of Cu^{2+} in the sand column was calculated according to the mass balance.

Measurement of Cu^{2+}

The Cu^{2+} concentration was measured by flame atomic absorption spectrometry (Varian 220FS, USA). All experiments were conducted in triplicate in this study. The data are given as mean values and the 95% confidence interval around the average value is indicated by error bars.

RESULTS AND DISCUSSION

Batch experiments – effects of pH, nano-SiO_2, ionic strength, initial Cu^{2+} concentration, and contact time

The pH value of the aqueous solution is an essential parameter that could affect the metallic ions biosorption (Alavi *et al.*

2015). The effects of pH value and nano-SiO_2 concentrations on the biosorption of Cu^{2+} by *C. gilardii* CR3 are shown in Figure 1(a) and Supplementary Material, Figure S3. Under all tested conditions of nano-SiO_2 (0, 0.1, 0.3, and 0.5 g L^{-1}), the bioremoval of Cu^{2+} reached a maximum at pH value of 5.0 compared with the results at pH value of 4.0 and 6.0. Thus, the optimized pH was 5.0. Under the optimized pH, the removal rate of Cu^{2+} was 95.4, 95.2, and 94.8% corresponding to the concentration of nano-SiO_2 at 0.1, 0.3, and 0.5 g L^{-1}, respectively, which were all higher than that in the absence of nano-SiO_2 (93.8%). Likewise, biosorption capacity for Cu^{2+} also exhibited the highest value at pH of 5 than that at pH 4 and 6 under different concentrations of nano-SiO_2 (Supplementary Material, Table S1). It is reported that the concentration of nanoparticle in the CMP usually ranges from 0.05 to 0.5 g L^{-1}, but in most cases, the nano-SiO_2 concentrations are about 0.3 g L^{-1} (Yang & Tsai 2006). Although the biosorption of Cu^{2+} from CMP wastewater has been investigated in previous studies (Stanley & Ogden 2003; Mosier *et al.* 2015; Yang *et al.* 2017), the effects of the presence of nano-SiO_2 on the removal of Cu^{2+} are not considered. In this study, bioremoval of Cu^{2+} and biosorption capacity for Cu^{2+} varied from 93.8%, 17.94 mg g^{-1} (nano-SiO_2 = 0 g L^{-1}) to 95.2%, 18.25 mg g^{-1} (nano-SiO_2 = 0.3 g L^{-1}) under the optimized pH of 5, indicating that the presence of nano-SiO_2 at actual concentration in CMP wastewater did not inhibit the Cu^{2+} removal but showed a slight promotion effect instead. According to our knowledge, this is the first study to investigate the effect of nano-SiO_2 on the bioremoval of Cu^{2+} in CMP wastewater.

Ionic strength is also an important factor affecting biosorption. The effect of different ionic strength for the biosorption of Cu^{2+} is shown in Figure 1(b). With the increase in ionic strength, the removal rate of Cu^{2+} decreased and reached the lowest (94.4%) at Na^+ concentration of 2.5 mM. The increase in ionic strength in solution will enhance the binding capacity of bacteria to Cu^{2+} (Darnall *et al.* 1986; Schiewer & Wong 2000). The competition of Na^+ and Cu^{2+} in solution

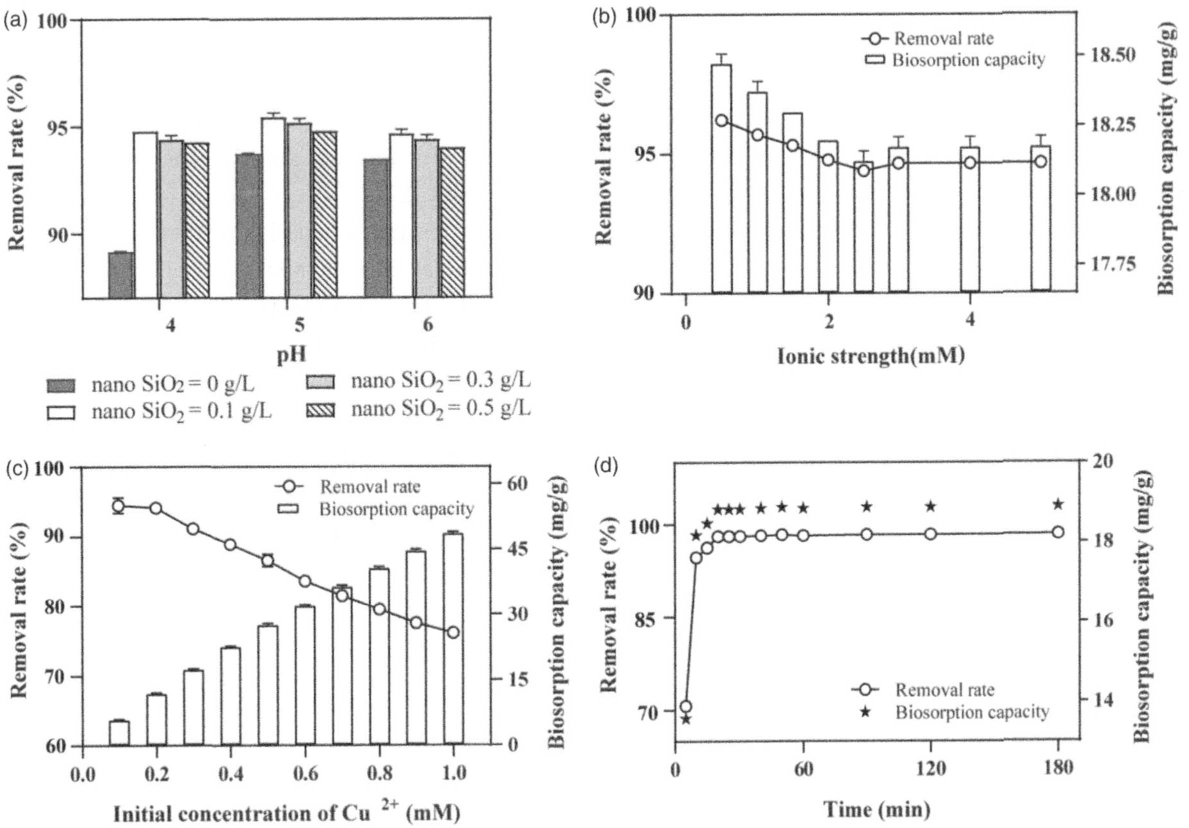

Figure 1 | Biosorption of Cu^{2+} by *C. gilardii* CR3 under static conditions: effect of pH and nano-SiO_2 (a), ionic strength (b), initial Cu^{2+} concentration (c), and contact time (d).

leading to Na^+ occupied a large number of biosorption sites on the surface of strain CR3. Hence, higher ionic strength decreased the bioremoval of Cu^{2+}.

The effect of the initial Cu^{2+} concentration on the bioremoval of Cu^{2+} by *C. gilardii* CR3 is shown in Figure 1(c). The removal rate of Cu^{2+} decreased with the increase in Cu^{2+} concentration (0.1–1.0 mM). The lowest removal rate of Cu^{2+} reached 91.2% at the common Cu^{2+} concentration (0.3 mM). When the Cu^{2+} concentration in CMP wastewater was 1 mM, the lowest removal rate of Cu^{2+} was still over 76.1% at 1 mM of Cu^{2+} concentration; the maximum biosorption capacity of Cu^{2+} was 48.7 mg g^{-1}. *Aureispira* sp. can remove approximately 90% of Cu^{2+} in a Cu^{2+} solution of 25–50 mg L^{-1} (Hasyimah *et al.* 2020). At the 1 mM concentration of Cu^{2+}, the maximum biosorption capacity of *Acidianus manzaensis* is 45.1 mmol kg^{-1} (Li *et al.* 2020). Hence, *C. gilardii* CR3 is a promising bacterial candidate for the treatment of Cu-CMP or other wastewater containing Cu^{2+} at a low concentration.

The effect of contact times for the biosorption of Cu^{2+} by the *C. gilardii* CR3 is shown in Figure 1(d). The removal rate of Cu^{2+} by the *C. gilardii* CR3 reached a maximum of 98% within the first 20 min. The removal rate of Cu^{2+} was almost stable from 20 to 180 min; the equilibrium biosorption capacity was 18.88 mg g^{-1}. It is known that the process includes two steps and the first step was rapid biosorption. First, a large number of biosorption sites on the surface of bacteria can be used to quickly adsorb Cu^{2+} (Liu *et al.* 2014). In the second step, biosorption equilibrium is reached and the biosorption sites are saturated, with the removal rate unchanged (Torab-Mostaedi *et al.* 2013). Our results were in agreement with this pattern.

Biosorption-isotherms and kinetics

Table 1 shows the results of relevant parameters using the Langmuir and Freundlich models. The Langmuir model fits well the related experimental data of Cu^{2+} initial concentration ($R^2 = 0.98$) (Supplementary Material, Figure S4). The predicted maximum biosorption capacity of *C. gilardii* CR3 was 59.62 mg g^{-1}, which was very close to the experimental biosorption capacity value (48.70 mg g^{-1}). For the Freundlich model, the fitting coefficient value is 0.99. In our previous study, the Langmuir model ($R^2 = 0.99$) was better than the Freundlich model ($R^2 = 0.49$) (Yang *et al.* 2017). Although Langmuir and Freundlich isotherms are usually used to simulate the biosorption in single ion solutions, they are slightly different. Specifically, the Langmuir model usually is used to describe the homogeneous monolayer biosorption, whereas the Freundlich model is used to describe heterogeneous biosorption (Veit *et al.* 2005). Our results indicated that the Freundlich model is better for describing the biosorption of Cu^{2+} CMP wastewater in the presence of nano-SiO_2.

Biosorption kinetics can show the binding speed of metal ions on the surface of the biosorbent (Michalak *et al.* 2013). Kinetic model fitting helps us to understand the reaction pathway and biosorption reaction mechanism in depth (Ho & McKay 2000). The experimental results in Figure 1(d) are fitted with quasi-first-order kinetics and pseudo-second-order kinetics models. The fitting results are shown in Table 2.

The fitting results of quasi-first-order kinetics and pseudo-second-order kinetics model are shown in Table 2. The fitting result of the quasi-first-order kinetic model was poor ($R^2 = 0.53$), thus the resulting q_e value (0.271 mg g^{-1}) is lower than the value of equilibrium biosorption capacity (18.88 mg g^{-1}). The pseudo-second-order kinetic model obeyed the experimental data ($R^2 > 0.99$). Meanwhile, the q_e value obtained by the pseudo-second-order kinetic is 18.97 mg g^{-1}, which is in good agreement with the equilibrium biosorption capacity.

Table 1 | Adsorption isotherms of *C. gilardii* CR3 for Cu^{2+}

Model	Parameter value		Correlation coefficient (R^2)	R_L	Fitting equation
Langmuir	$q_m = 59.620$	$b = 0.218$	0.989	0.067	$q_e = \frac{12.9746C_e}{1 + 0.2176C_e}$
Freundlich	$n = 2.120$	$K_F = 13.476$	0.998		$q_e = 13.4757C_e^{0.4714}$

Table 2 | Adsorption kinetics of *C. gilardii* CR3 for Cu^{2+}

Model	*K*	R^2	q_e (mg g^{-1})	Fitting equation
Quasi-first-order kinetics	0.015	0.5300	0.271	$\ln(q_e - q_t) = -1.3054 - 0.0151t$
Pseudo-second-order kinetics	0.102	0.9999	18.970	$\frac{t}{q_t} = \frac{t}{18.97} + 0.0273$

Analysis of SEM, zeta potential, and FTIR

The SEM images of *C. gilardii* CR3 before and post Cu^{2+} binding are shown in Figure 2. Compared to the sample without Cu^{2+} (Figure 2(a)), a large number of irregularly drawn EPS was covered or wrapped on the surface of bacterial cells (Figure 2(b)). The genome of *C. gilardii* CR3 includes EPS genes and a T3SS system responsible for secreting EPS; Cu^{2+} can lead to significant up-regulation of 11 genes of T3SS in *C. gilardii* CR3 (Huang *et al.* 2019a, 2019b). Hence, it can be inferred that Cu^{2+} stimulates the secretion of EPS. EPS is known as possessing a great capability for binding various metal ions (Comte *et al.* 2008). Additionally, EPS is beneficial to protect growing cells to avoid the toxic effect of heavy metal ions (Iyer *et al.* 2005). In sum, the increased EPS improved the ability of the bacterium cell surface to bind to Cu^{2+}, which further increased the biosorption of Cu^{2+}. Here, we provide a confocal laser scanning microscopy image of the EPS produced by the strain *C. gilardii* CR3 in Supplementary Material, Figure S5. The presence of protein and polysaccharides in EPS was indicated by a multiple fluorescence staining method (Cui *et al.* 2019).

The Zeta potential of strain CR3, quartz sand, and the composite of quartz sand-strain CR3 were measured before and after the biosorption of Cu^{2+} (Table 3). Among the three sorbents, the potential value of composite of quartz sand–strain CR3 was the most negative (−55.4 mV), followed by strain CR3 (−22.5 mV) and quartz sand (−45.9 mV) before biosorption of Cu^{2+}, which indicated that composite biosorbent carried more negative charges than that of strain CR3 and quartz sand. Likewise, the potential value showed a similar trend with the before biosorption of Cu^{2+}. Zeta potential represents the surface charge of the biosorbent and reflects the electrochemical properties of the biosorbent surface (van der Mei & Busscher 2001). Thus, the results show that the composite of quartz sand–strain CR3 is favorable to bind more Cu^{2+} as compared to strain CR3 and quartz sand because the composite possesses a stronger biosorption capacity for Cu^{2+}.

The results of the FTIR analysis showed that the peak shape of the composite of quartz sand–strain CR3 remained unchanged before and after the biosorption of Cu^{2+} (Figure 3). In other words, Cu^{2+} did not destroy the structure of the composite. In the composite of quartz sand–strain CR3, the major sorption peaks were as follows: 3,491 cm^{-1} corresponding to -NH and -OH stretching; 2,921 cm^{-1} corresponding to $-CH_2$ symmetrical or asymmetrical stretching; 1,082 cm^{-1} corresponding to C-OH/C–O–C/C–C/P = O; and the remaining absorption peak corresponds to the SiO_2 peak position in quartz sand. The specific functional groups on the surface of

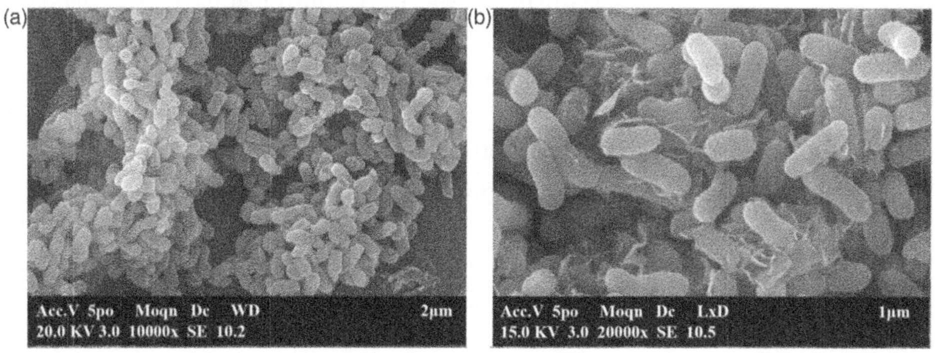

Figure 2 | SEM images of *C. gilardii* CR3 before (a) and after (b) Cu^{2+} binding.

Table 3 | Zeta potential of *C. gilardii* CR3, quartz sand, and the composite of quartz sand–strain CR3 before and after binding 0.3 mM Cu^{2+} at pH 5 and 0.3 g L^{-1} nano-SiO_2

	Zeta potential (mV)		
	Before biosorption	After biosorption	Absolute value
Strain CR3	−22.5	−18.5	4.0
Quartz sand	−45.9	−14.7	31.2
Sand–strain CR3	−55.4	−20.1	35.3

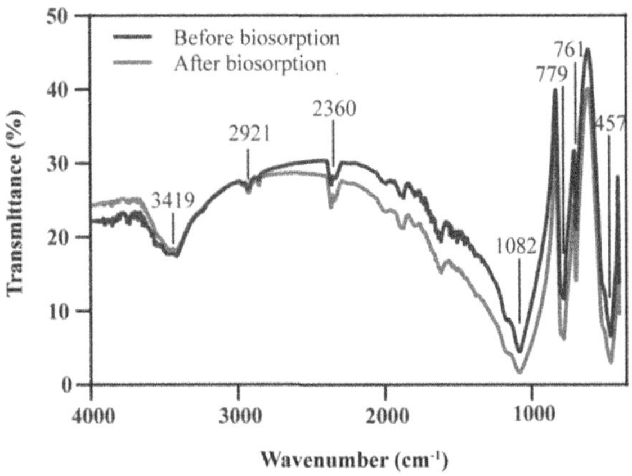

Figure 3 | The Fourier-transform infrared spectroscopy of the composite of quartz sand–strain CR3 before and after the biosorption of Cu^{2+} under the optimized condition (pH 5, 0.3 g L^{-1} nano-SiO_2).

the biosorbent play a key role in the biosorption of heavy metals (Bai & Abraham 2005). The main functional groups involved in the Cu^{2+} biosorption of the composite of quartz sand-strain CR3 were hydroxyl, carboxyl, amino, and phosphate groups, which played an important role in the binding with metal ions (Huang *et al.* 2013; Xia *et al.* 2015). The absorption peaks and shift indeed occurred before and after biosorption of Cu^{2+} although the variations of the absorption peaks were not obvious, which provided evidence that the composite of quartz sand–strain CR3 bound Cu^{2+} from the side of the functional groups on the sorbent.

Observed and modeled BTCs of the bioremoval of Cu^{2+} in columns

The observed and model-fitted BTCs of the tracer (Br^-) and Cu^{2+} in all columns are shown in Figure 4. The results of all tracer tests show very little difference with each other, and the maximum breakthrough concentrations, $(C/C_0)_{max}$, in all experiments were close to 1. This indicated good reproducibility of the experimental procedure (Fan *et al.* 2015). The hydrodynamic dispersion coefficient D was estimated as 0.28 ± 0.07 cm^2 min^{-1} ($R^2 > 0.96$). This fitted value of D was then used to model the transport of Cu^{2+}, and the best-fit transport parameters in the model are listed in Table 4 (the numerical model described all the experimental BTCs well, $R^2 > 0.90$). The mass percentages of Cu^{2+} recovered in the effluent, retained in the column, and total recovery ratio is summarized in Table 4. The recovery ratio ranges from 95.28 to 104.06%. The mass balance indicated that the operation and measurement were reliable.

As shown in Figure 4, the coexisting nano-SiO_2 could facilitate the transport of Cu^{2+}, since the retention ratio of Cu^{2+} was 29.73% in clean sand without nano-SiO_2 and it decreased to 23.90% when nano-SiO_2 was introduced. Nano-SiO_2 worked as the Cu^{2+}-carrier and promoted Cu^{2+} transport, which was further validated in the strain CR3-coated column (Figure 4(b) and 4(c)). The presence of strain CR3 biofilm enhanced Cu^{2+} deposition both in the presence or absence of nano-SiO_2. The retention ratios of Cu^{2+} in strain CR3-coated sand columns (Figure 4(b) and 4(d)) were 1.24–1.37 times those in the clean sand columns (Figure 4(a) and 4(c)).

The variations of k_{det} and S_{max} in modeling verified these trends quantitatively. Higher values of k_{att} mean more intensive deposition of Cu^{2+} in the column. The k_{att} in the presence of nano-SiO_2 was only 0.174 of that in the absence of nano-SiO_2 in the clean sand columns, while this k_{att} was 0.565 in strain CR3-coated sand. After the sand was coated by the *C. gilardii* CR3 biofilm, the k_{att} increased 2.24–2.80 times the k_{att} in clean sand. The k_{det} was nearly 3–5 orders of magnitude lower than k_{att}, which confirmed that the remobilization of previously retained Cu^{2+} could be inconsiderable and the slight differences of k_{det} under these scenarios could be ignored. Overall, the presence of nano-SiO_2 could enhance Cu^{2+} transport and thus may increase spreading risk. By contrast, *C. gilardii* CR3 biofilm in the sand matrix can be a barrier to Cu^{2+} breakthrough.

The aforementioned facilitated bioremoval of Cu^{2+} by strain CR3 was confirmed by additional continuous flow column experiments, which was described in Section S1 in Supplementary Material. We employed two kinds of CMP

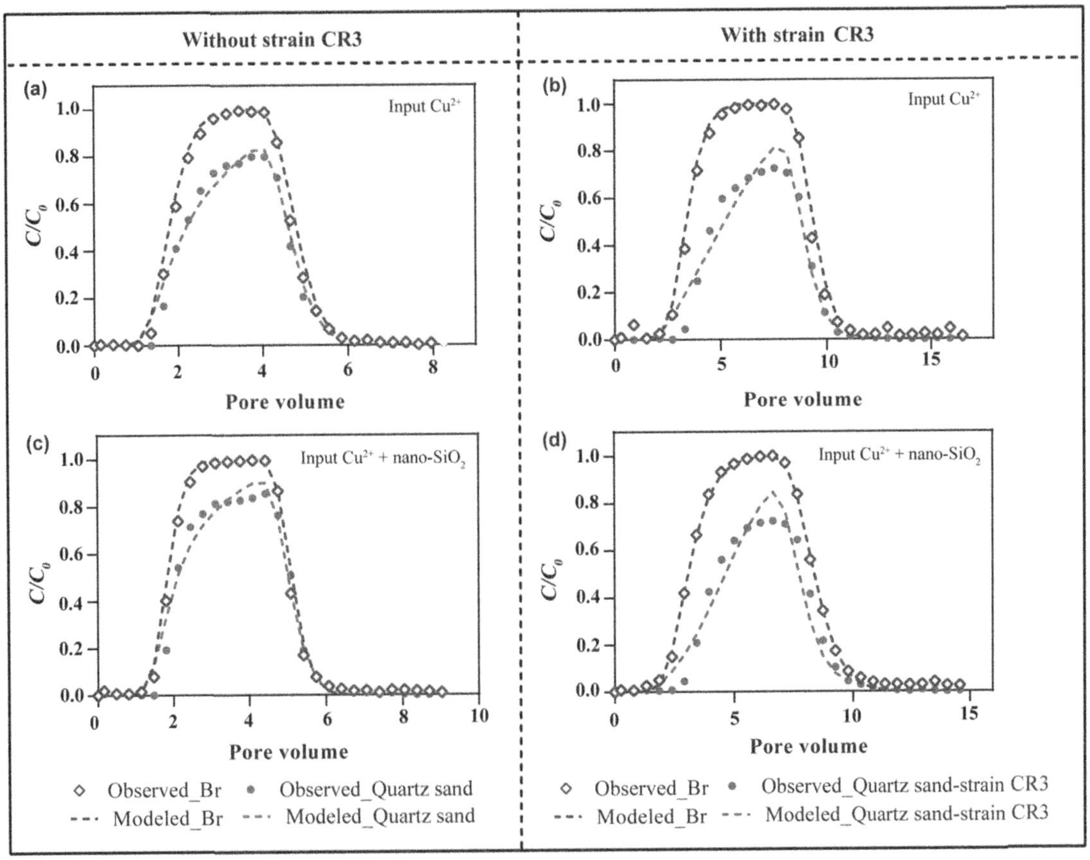

Figure 4 | Observed and modeled breakthrough curves of the tracer (Br^-) and Cu^{2+} in quartz sand columns under Cu^{2+} single solution control (a); Cu^{2+} single solution with strain CR3 (b); Cu^{2+} and nano-SiO_2 mixture solution control (c); Cu^{2+} and nano-SiO_2 mixture solution with strain CR3 (d).

Table 4 | Summary of experimental conditions and model results of Cu^{2+} transport

	Mass balance (%)							
Exp. no.	**Effluent**	**Retained**	**Total**	$(C/C_0)_{max}$	K_{att} (min^{-1})	K_{det} (min^{-1})	S_{max} (mg g^{-1})	R^2
a	72.19	29.73	101.92	0.8	0.215	0.001	0.176	0.987
b	59.10	37.13	96.23	0.72	0.818	0.001	0.256	0.967
c	80.16	23.90	104.06	0.85	0.174	0.004	0.132	0.984
d	62.50	32.78	95.28	0.72	0.565	0.005	0.259	0.957

wastewaters to be injected into the columns continuously: one only contained Cu^{2+} and one contained Cu^{2+} and nano-SiO_2 together. The results showed that the relative effluent concentration of Cu^{2+} (C/C_0) was higher in quartz sand without strain CR3 compared with that in CR3-coated sand matrix (Supplementary Material, Figure S6) for both of the two CMP wastewater, while the average C/C_0 at $t = 60$ min was 0.90 for the former and 0.81 for the latter. Note that the biosorption in these breakthrough columns was weaker than those in batch adsorption system; it can be attributed to the shorter hydraulic retention time in the columns (less than 12 min). Nevertheless, it still could be seen that the Cu^{2+} transportation was retarded by CR3 biofilm. By contrast, the differences between different C/C_0 curves in the same medium with/o nano-SiO_2 were less than 2.20% (Supplementary Material, Figure S6).

The coexistence of nano-SiO_2 did not inhibit this process, which was in keeping with the results of the batch adsorption experiments.

Overall, bacterial removal for heavy metal ions from wastewater is considered as an environmental friendly and cost-effective method, which is effective especially for ions at low concentrations. It is known that three mechanisms, including surface sorption, bioaccumulation, and biotransformation, are responsible for the bioremoval of metal ions (Li & Tao 2013). Among these, surface sorption occurring on the cell surface is the first step and metabolism-independent, and therefore accounts for a major part for heavy metal removal. The surface sorption of heavy metal ions depends on various factors, such as the nature of the bacterial species (the isoelectric point, the composition and structure of bacterial cell walls), heavy metal species, and pH value of the aqueous solution. According to the FTIR results and characteristics of bacterium *C. gilardii* CR3, we summarized the mechanisms of surface sorption Cu^{2+} in Figure 5. In the aqueous solution bearing copper, there are multiple ions, such as H^+, H_3O^+, OH^-, Cu^{2+}, and $Cu(OH)^+$. When the pH is lower to 2.2, i.e. the isoelectric point of strain CR3 (Yang *et al.* 2017), H^+, H_3O^+, and OH^- will compete the binding sites on the cell surface with Cu^{2+} and $Cu(OH)^+$, which means lower bioremoval of Cu^{2+}. If the pH value is over 5.0, some Cu^{2+} will present as $Cu(OH)_2$ according to the solubility product constant (2.2×10^{-20}). Since the aim of this study is to estimate the capability of bacterium to remove Cu^{2+}, higher pH condition (>5.0) is not considered although the insolubility of copper hydroxide can lead to small extra Cu^{2+} removal. When pH value ranges between 2.2 and 5.0, Cu^{2+} and $Cu(OH)^+$ are more preferable in the aqueous solution. As a result, these copper species will interact with the function groups (Figure 3), for example -SH, -OH, and -COOH, on the surface of microbial cells

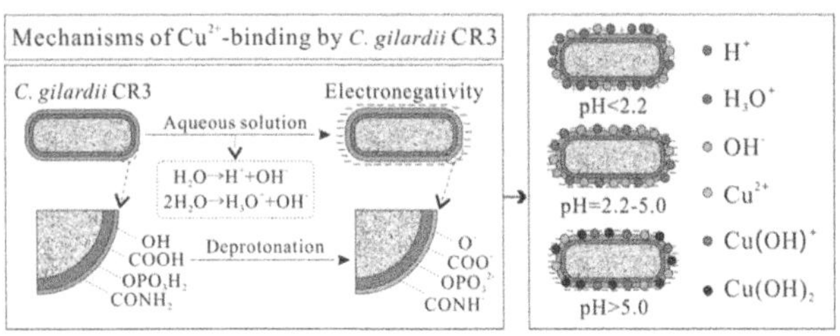

Figure 5 | Conceptual model for the Cu^{2+} binding onto the surface of *C. gilardii* CR3.

and increasing the binding of Cu^{2+}. In addition to the surface sorption, bioaccumulation and biotransformation also contribute to some extent to the removal of Cu^{2+}. Once the binding on the cell sites are fully occupied, bioaccumulation will contribute to the bioremoval of Cu^{2+} (Ayangbenro & Babalola 2017). At the same time, the intracellular Cu^{2+} might be effluxed out to keep the level of intracellular Cu^{2+} below the toxicity threshold (Machalová *et al.* 2015). It should be noted that the synthesized CMP wastewater used in the above-mentioned studies contained Cu^{2+} without nano-SiO_2. According to our knowledge, the finding in our study is the first to indicate that coexisting nano-SiO_2 does not inhibit the biosorption Cu^{2+} by strain CR3.

CONCLUSIONS

In this study, bioremoval of Cu^{2+} from CMP wastewater in the presence of nano-SiO_2 was investigated through batch adsorption experiments and transport column experiments using a typical copper-resistant bacterium *C. gilardii* CR3. The Freundlich isotherm is suitable for describing the biosorption process in Cu^{2+}-nano-SiO_2-CR3 ternary system. The biosorption kinetic data fit well with the pseudo-second-order equation. Overall, the maximum biosorption capacity for Cu^{2+} was 18.25 mg g^{-1} and the bioremoval rate was as high as 95.2%, displaying high removal efficacy of strain CR3. In the transport column experiments, the BTCs fit well with the ADR model. The attachment coefficient in the sand matrix coated by CR3 biofilm was 2.24–2.80 times as that in clean sand, confirming that CR3 biofilm can be a barrier to retard Cu^{2+} transport. In particular, in all experiments, the presence of nano-SiO_2 did not inhibit the bioremoval of Cu^{2+} from CMP wastewater but showed a slight promotion effect instead. Therefore, *C. gilardii* CR3 is a promising microbe biosorbent to remove Cu^{2+} from CMP wastewater even in the presence of nano-SiO_2.

ACKNOWLEDGEMENT

This work was funded by the National Natural Science Foundation of China (no. 51678122, no. 52070037, and no. 51978135).

DATA AVAILABILITY STATEMENT

All relevant data are included in the paper or its Supplementary Information.

REFERENCES

Alavi, S. A., Zilouei, H. & Asadinezhad, A. 2015 *Otostegia persica* biomass as a new biosorbent for the removal of lead from aqueous solutions. *International Journal of Environmental Science and Technology* **12** (2), 489–498. doi:10.1007/s13762-014-0705-x.

Ayangbenro, A. & Babalola, O. 2017 A new strategy for heavy metal polluted environments: a review of microbial biosorbents. *International Journal of Environmental Research and Public Health* **14** (1), 94. doi:10.3390/ijerph14010094.

Bai, S. R. & Abraham, T. E. 2005 Continuous adsorption and recovery of Cr(VI) in different types of reactors. *Biotechnology Progress* **21** (6), 1692–1699. doi:10.1021/bp050196c.

Bradford, S. A. & Bettahar, M. 2006 Concentration dependent transport of colloids in saturated porous media. *Journal of Contaminant Hydrology* **82** (1–2), 99–117. doi:10.1016/j.jconhyd.2005.09.006.

Chua, H., Yu, P., Sin, S. & Cheung, M. 1999 Sub-lethal effects of heavy metals on activated sludge microorganisms. *Chemosphere* **39** (15), 2681–2692. doi:10.1016/S0045-6535(99)00203-9.

Comte, S., Guibaud, G. & Baudu, M. 2008 Biosorption properties of extracellular polymeric substances (EPS) towards Cd, Cu and Pb for different pH values. *Journal of Hazardous Materials* **151** (1), 185–193. doi:10.1016/j.jhazmat.2007.05.070.

Cui, X., Chen, C., Liu, Y., Zhou, D. & Liu, M. 2019 Exogenous refractory protein enhances biofilm formation by altering the quorum sensing system: a potential hazard of soluble microbial proteins from WWTP effluent. *Science of the Total Environment* **667**, 384–389. doi:10.1016/j.scitotenv.2019.02.370.

Darnall, D. W., Greene, B., Henzl, M. T., Hosea, J. M., McPherson, R. A., Sneddon, J. & Alexander, M. D. 1986 Selective recovery of gold and other metal ions from an algal biomass. *Environmental Science & Technology* **20** (2), 206–208. doi:10.1021/es00144a018.

Fan, W., Jiang, X. H., Yang, W., Geng, Z., Huo, M. X., Liu, Z. M. & Zhou, H. 2015 Transport of graphene oxide in saturated porous media: effect of cation composition in mixed Na–Ca electrolyte systems. *Science of the Total Environment* **511**, 509–515. doi:10.1016/j.scitotenv.2014.12.099.

Hasyimah, N. A. R., Furusawa, G. & Amirul, A. A. 2020 Biosorption of a dye and heavy metals using dead cells of filamentous bacterium, *Aureispira* sp. CCB-QB1. *International Journal of Environmental Science and Technology* **18** (6), 1–10. doi:10.1007/s13762-020-02918-3.

Ho, Y. & McKay, G. 2000 The kinetics of sorption of divalent metal ions onto sphagnum moss peat. *Water Research* **34** (3), 735–742. doi:10.1016/S0043-1354(99)00232-8.

Huang, F., Dang, Z., Guo, C., Lu, G., Gu, R. R., Liu, H. & Zhang, H. 2013 Biosorption of Cd(II) by live and dead cells of *Bacillus cereus* RC-1 isolated from cadmium-contaminated soil. *Colloids and Surfaces B: Biointerfaces* **107**, 11–18. doi:10.1016/j.colsurfb.2013.01.062.

Huang, N., Mao, J., Hu, M., Wang, X. & Huo, M. 2019a Responses to copper stress in the metal-resistant bacterium *Cupriavidus gilardii* CR3: a whole-transcriptome analysis. *Journal of Basic Microbiology* **59** (5), 446–457. doi:10.1002/jobm.201800693.

Huang, N., Mao, J., Zhao, Y., Hu, M. & Wang, X. 2019b Multiple transcriptional mechanisms collectively mediate copper resistance in *Cupriavidus gilardii* CR3. *Environmental Science & Technology* **53** (8), 4609–4618. doi:10.1021/acs.est.8b06787.

Iyer, A., Mody, K. & Jha, B. 2005 Biosorption of heavy metals by a marine bacterium. *Marine Pollution Bulletin* **50** (3), 340–343. doi:10.1016/j.marpolbul.2004.11.012.

Lai, C. L. & Lin, S. H. 2004 Treatment of chemical mechanical polishing wastewater by electrocoagulation: system performances and sludge settling characteristics. *Chemosphere* **54** (3), 235–242. doi:10.1016/j.chemosphere.2003.08.014.

Li, P. & Tao, H. 2013 Cell surface engineering of microorganisms towards adsorption of heavy metals. *Critical Reviews in Microbiology* **41** (2), 140–149. doi:10.3109/1040841X.2013.813898.

Li, M., Huang, Y., Yang, Y., Wang, H., Hu, L., Zhong, H. & He, Z. 2020 Heavy metal ions removed from imitating acid mine drainages with a thermoacidophilic archaea: *Acidianus manzaensis* YN25. *Ecotoxicology and Environmental Safety* **190**, 110084. doi:10.1016/j.ecoenv.2019.110084.

Liu, Y., Cui, G., Luo, C., Zhang, L., Guo, Y. & Yan, S. 2014 Synthesis, characterization and application of amino-functionalized multi-walled carbon nanotubes for effective fast removal of methyl orange from aqueous solution. *RSC Advances* **4** (98), 55162–55172. doi:10.1039/C4RA10047F.

Machalová, L., Pipíška, M., Trajteľová, Z. & Horník, M. 2015 Comparison of Cd^{2+} biosorption and bioaccumulation by bacteria – a radiometric study. *Nova Biotechnologica et Chimica* **14** (2), 158–175. doi:10.1515/nbec-2015-0024.

Maketon, W. 2007 *Treatment of Cu-CMP Waste Streams Containing Copper(II) Using Polyethyleneimine (PEI).* PhD Thesis. University of Arizona, Arizona, USA.

Maketon, W. & Ogden, K. 2009 Synergistic effects of citric acid and polyethyleneimine to remove copper from aqueous solutions. *Chemosphere* **75** (2), 206–211. doi:10.1016/j.chemosphere.2008.12.005.

Malik, A. 2004 Metal bioremediation through growing cells. *Environment International* **30** (2), 261–278. doi:10.1016/j.envint.2003.08.001.

Michalak, I., Chojnacka, K. & Witek-Krowiak, A. 2013 State of the art for the biosorption process – a review. *Applied Biochemistry and Biotechnology* **170** (6), 1389–1416. doi:10.1007/s12010-013-0269-0.

Mosier, A. P., Behnke, J., Jin, E. T. & Cady, N. C. 2015 Microbial biofilms for the removal of Cu^{2+} from CMP wastewater. *Journal of Environmental Management* **160**, 67–72. doi:10.1016/j.jenvman.2015.05.016.

Schiewer, S. & Wong, M. 2000 Ionic strength effects in biosorption of metals by marine algae. *Chemosphere* **41** (1–2), 271–282. doi:10.1016/S0045-6535(99)00421-X.

Stanley, L. C. & Ogden, K. L. 2003 Biosorption of copper (II) from chemical mechanical planarization wastewaters. *Journal of Environmental Management* **69** (3), 289–297. doi:10.1016/j.jenvman.2003.09.009.

Su, Y., Lin, W., Hou, C. & Den, W. 2014 Performance of integrated membrane filtration and electrodialysis processes for copper recovery from wafer polishing wastewater. *Journal of Water Process Engineering* **4**, 149–158. doi:10.1016/j.jwpe.2014.09.012.

Torab-Mostaedi, M., Asadollahzadeh, M., Hemmati, A. & Khosravi, A. 2013 Equilibrium, kinetic, and thermodynamic studies for biosorption of cadmium and nickel on grapefruit peel. *Journal of the Taiwan Institute of Chemical Engineers* **44** (2), 295–302. doi:10.1016/j.jtice.2012.11.001.

van der Mei, H. & Busscher, H. 2001 Electrophoretic mobility distributions of single-strain microbial populations. *Applied and Environmental Microbiology* **67** (2), 491–494. doi:10.1128/AEM.67.2.491-494.2001.

Veit, M. T., Tavares, C. R. G., Gomes-da-Costa, S. M. & Guedes, T. A. 2005 Adsorption isotherms of copper(II) for two species of dead fungi biomasses. *Process Biochemistry* **40** (10), 3303–3308. doi:10.1016/j.procbio.2005.03.029.

Xia, L., Xu, X., Zhu, W., Huang, Q. & Chen, W. 2015 A comparative study on the biosorption of Cd^{2+} onto *Paecilomyces lilacinus* XLA and *Mucoromycote* sp. XLC. *International Journal of Molecular Sciences* **16** (7), 15670–15687. doi:10.3390/ijms160715670.

Yang, G. C. C. & Tsai, C. 2006 Performance evaluation of a simultaneous electrocoagulation and electrofiltration module for the treatment of Cu-CMP and oxide-CMP wastewaters. *Journal of Membrane Science* **286** (1–2), 36–44. doi:10.1016/j.memsci.2006.09.007.

Yang, G. C. C. & Tsai, C. 2008 Preparation of carbon fibers/carbon/alumina tubular composite membranes and their applications in treating Cu-CMP wastewater by a novel electrochemical process. *Journal of Membrane Science* **321** (2), 232–239. doi:10.1016/j.memsci.2008.04.060.

Yang, Y., Hu, M., Zhou, D., Fan, W., Wang, X. & Huo, M. 2017 Bioremoval of Cu^{2+} from CMP wastewater by a novel copper-resistant bacterium *Cupriavidus gilardii* CR3: characteristics and mechanisms. *RSC Advances* **7** (30), 18793–18802. doi:10.1039/C7RA01163F.

First received 30 October 2020; accepted in revised form 23 December 2020. Available online 25 January 2021

© 2021 The Authors

Influence of light quality on *Chlorella* growth, photosynthetic pigments and high-valued products accumulation in coastal saline-alkali leachate

Xiao-ya Liu, Yu Hong IWA and Wen-ping Gu

ABSTRACT

Using saline-alkali leachate to cultivate microalgae is an effective way to realize the utilization of wastewater and alleviate the shortage of water resources. Light source is usually used as an optimized parameter to further improve the cultivation efficiency of microalgae. In this work, the influence of light qualities on the growth and high-valued substances accumulation of *Chlorella* sp. HQ in coastal saline-alkali leachate were investigated. The specific growth rate of *Chlorella* in coastal saline-alkali leachate was 0.27–0.60 d^{-1}. At the end of cultivation, the algal density under blue light reached $8.71 \pm 0.15 \times 10^7$ cells·mL^{-1}, which was significantly higher than the other light groups. The lipid content in the biomass was 29.31–62.95%, and the highest lipid content and TAGs content were obtained under red light and blue-white mixed light, respectively. Percentages of total chlorophylls (0.81–1.70%) and carotenoids (0.08–0.25%) were obtained in the final biomass of the coastal saline-alkali leachate. In addition, the contents of photosynthetic pigments and three high-valued products under mixed light were higher than those of monochromatic light, and the protein, total sugar and starch content under blue-red mixed light was 1.52–3.76 times, 1.54–3.68 times and 1.06–3.35 times of monochromatic blue light and red light, respectively.

Key words | *Chlorella*, high-valued products, light quality, pigments, saline-alkali leachate

Xiao-ya Liu
Yu Hong IWA (corresponding author)
Wen-ping Gu
Beijing Key Lab for Source Control Technology of Water Pollution, College of Environmental Science and Engineering,
Beijing Forestry University,
Beijing 100083,
China
E-mail: *yuhong829908@gmail.com*

HIGHLIGHTS

- *Chlorella* in coastal saline-alkali leachate exhibited better growth curve under blue light.
- The highest total lipid content per unit biomass of *Chlorella* was obtained under red light.
- Mixed LEDs light can improve the photosynthetic ability of *Chlorella* in coastal saline-alkali leachate.
- High-valued substances content of *Chlorella* under mixed LEDs light was higher than that of monochromatic light.

This is an Open Access article distributed under the terms of the Creative Commons Attribution Licence (CC BY 4.0), which permits copying, adaptation and redistribution, provided the original work is properly cited (http://creativecommons.org/licenses/by/4.0/).

doi: 10.2166/wrd.2021.088

GRAPHICAL ABSTRACT

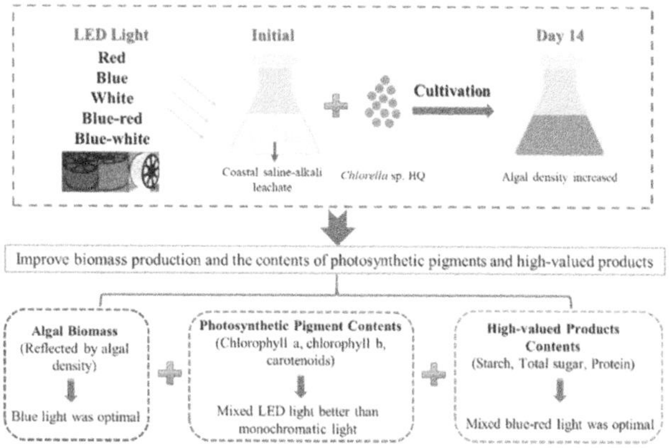

INTRODUCTION

Soil salinization is a widespread environmental problem worldwide, and there are approximately 954 million hectares of saline-alkali land in the world (Yue *et al.* 2020). Coastal saline-alkali soils, have a high content of exchangeable sodium, which is easy to cause soil consolidation, and the high salinity in the soil prevents plant roots from absorbing water and obtaining nutrients (Tang *et al.* 2020). In addition, as saline-alkali land is an important reserve land resource for food production, the restoration of saline-alkali land is essential to global food demand. Nowadays, the improvement of saline-alkali land often adopts various methods such as physical restoration, chemical modification and bioremediation, and the common practiced method in the world is the combination of irrigation and drainage, which removes the salt in the soil through leaching (Wang *et al.* 2018; Meena *et al.* 2019). Although this method can improve the soil quality, it also produces a large amount of wastewater with high salinity. The wastewater contains inorganic salts, which is an abundant source of energy to be recovered and reused, but if it is directly discharged, it may cause groundwater pollution and deterioration of water quality (Fang *et al.* 2019). At present, the global water consumption is gradually increasing, and a large part of the population still lives in water-scarce areas (Zhu & Dou 2018; Archer *et al.* 2019). Given the above, the resource utilization of saline-alkali leachate is essential to alleviate the shortage of water resources and maintain the sustainable development of the social economy. Our previous studies have shown that the coastal saline-alkali leachate can be used to cultivate three oleaginous microalgae at low cost to realize the conversion of wastewater resources (Liu *et al.* 2020).

As a global emerging industry, microalgae cultivation has been extensively researched and marketed in recent years due to its enormous economic and social potential (Mata *et al.* 2010). Microalgae are highly efficient photosynthetic cell factories, which contain high amount of high-valued substances, such as proteins, pigments, carbohydrates and lipids. Among the possible applications of the microalgae, they can be used in the food, feed, life sciences and renewable fuels, for example carbohydrates and fatty acids can be converted into alcohol and biodiesel (Chew *et al.* 2017; Chandra *et al.* 2019). However, the cost of large-scale production of microalgae and related biological products is still high, which has brought challenges to the development of microalgae biotechnology. Current researches are aimed at increasing the biomass of microalgae, improving the contents of lipid and high-valued substances, or reducing the cost of microalgae cultivation (Chen *et al.* 2009). How to improve the cultivation efficiency of microalgae in saline-alkali leachate and the content of high-valued substances is of great significance for reducing the production cost of microalgae-based biological products.

It is known that environmental factors such as light, nutrients, temperature and CO_2 concentration can significantly affect the algal growth and lipid accumulation including content and composition (Ananthi *et al.* 2021). Light, as the energy source of microalgae photosynthesis, is one of the important factors affecting the growth of algae and the accumulation of active substances. In indoor microalgae cultivation, artificial light sources such as fluorescent light are commonly used as energy sources, which undoubtedly increase the power consumption and may exacerbate the greenhouse phenomenon (Sun *et al.* 2017). Light emitting diode (LED) has been widely used in microalgae cultivation due to its advantages of narrowing specific wavelengths with low power consumption, small chip size, and long duration (Kim *et al.* 2014). It has been reported that the microalgal photosystem II and photosystem I can be enhanced or induced by red and blue light wavelengths, respectively (Ravelonandro *et al.* 2008). In addition, since pigments such as chlorophyll a and chlorophyll b are sensitive to light of different wavelengths, it would be pertinent to select LEDs to regulate the distribution of various products in microalgae (Schulze *et al.* 2014). Shu *et al.* (2012) found that light quality has a significant effect on algae cell growth and product formation, and this effect is dependent on the algae species. Abiusi *et al.* (2014) investigated the influence of light quality on algae cell size, growth, productivity, photosynthetic efficiency of *Tetraselmis suecica* F&M-M33, and found that red light can be profitably used for the production of this marine microalga for aquaculture.

Therefore, in order to further improve the cultivation efficiency of *Chlorella* in the coastal saline-alkali leachate and the content of high-valued substances, the effects of five different light sources (including monochromatic blue, red, white LED light and blue-red mixed light, blue-white mixed light) on the growth of *Chlorella* and the accumulation of lipid, pigments, protein, total sugar and starch were explored. The findings can provide a theoretical guidance on how to accurately use light quality to make *Chlorella* in coastal saline-alkali leachate obtain higher content of high-valued substances. This will not only provide a resource utilization method for the coastal saline-alkali leachate, but also reduce the production cost of microalgae-based biological products.

MATERIALS AND METHODS

Microorganism

The microalga *Chlorella* sp. HQ (No. GCMCC7601) was used in the experiment, which was isolated in our previous study and has been kept in China General Microbiological Culture Collection Center. The strain was routinely cultivated in an axenic Selenite Enrichment (SE) medium, which contained 250 $mg{\cdot}L^{-1}$ of $NaNO_3$, 75 $mg{\cdot}L^{-1}$ of $K_2HPO_4{\cdot}3H_2O$, 75 $mg{\cdot}L^{-1}$ of $MgSO_4{\cdot}7H_2O$, 25 $mg{\cdot}L^{-1}$ of $CaCl_2{\cdot}2H_2O$, 175 $mg{\cdot}L^{-1}$ of KH_2PO_4, 25 $mg{\cdot}L^{-1}$ of NaCl, 5 $mg{\cdot}L^{-1}$ of $FeCl_3{\cdot}6H_2O$, 0.81 $mg{\cdot}L^{-1}$ of $FeCl_3$, 10 $mg{\cdot}L^{-1}$ of Na_2EDTA, 2.86 $mg{\cdot}L^{-1}$ of H_3BO_3, 1.81 $mg{\cdot}L^{-1}$ of $MnCl_2{\cdot}4H_2O$, 0.22 $mg{\cdot}L^{-1}$ of $ZnSO_4{\cdot}H_2O$, 0.079 $mg{\cdot}L^{-1}$ of $CuSO_4{\cdot}5H_2O$, and 0.039 $mg{\cdot}L^{-1}$ of $(NH_4)_6Mo_7O_{24}{\cdot}4H_2O$. All chemicals were of analytical grade. Prior to the experiments, cultures were maintained in Erlenmeyer flask at 25 ± 1 °C, 4800 lux incident light intensity in a 14:10 h light/dark cycle and the flasks were shaken for three times a day.

Cultivation setup

The coastal saline-alkali soil samples used in this study were taken from Dongying, Shandong Province, China. The detailed preparation method of the coastal saline-alkali leachate was described by Liu *et al.* (2020). The obtained leachate was filtered through 0.45-μm filtration membrane and sterilized at 121 °C for 30 min, and was stored after cooling to room temperature for later use.

The experiments were performed in a climate chamber equipped with LEDs as the light source. *Chlorella* sp. HQ was cultivated in 250 mL Erlenmeyer flasks with 180 mL sterile coastal saline-alkali leachate and initial cell density of 2×10^5 $cells{\cdot}mL^{-1}$, and the temperature was maintained at 25 ± 1 °C. The illumination was supplied at an average light intensity of 4500 lux and a photoperiod of 14:10 h light/dark. Based on light intensity, two mixed LED light treatments (red:blue = 1:2; blue:white = 1:2) and three monochrome LED light treatments (monochrome red; blue; and white

LED) were set up in this study. In order to avoid the influence of the light source on the other experimental groups, each treatment group was separated by aluminum foil.

In the experiment, the algal density was measured every other day, and the biomass, lipids, TAGs, total sugar, starch, protein, and photosynthetic pigment content of the microalgae were measured at the end of cultivation. The Erlenmeyer flask was manually shaken three times a day to ensure that the algae cells and the culture medium were thoroughly mixed, and to reduce experimental errors.

Analytical methods

Microalgal growth and mathematic model analysis

Algal density was determined using a Neubauer hemocytometer. To determine the yield of biomass, a 20 mL algae solution sample was filtered through a pre-weighed 0.45-μm membrane. The membrane was washed twice with distilled water, dried at 110 °C for 24 h until completely dehydrated and weighed again, then the biomass yield was calculated.

A generalized Logistic model was fitted to the measured data (cell density) with Origin software.

$$N_t = \frac{K}{1+e^{a-rt}} \quad (1)$$

$$\frac{dN}{dt} = rN\left(\frac{K-N}{K}\right) \quad (2)$$

$$R_{max} = \frac{rK}{4} \quad (3)$$

In the above formula, N_t (cells·mL^{-1}) is cell density at cultivation time t (d). K, a and r represent the maximum cell density during the whole cultivation period, the relative position from the origin and algae cell intrinsic growth rate, respectively. We determined the parameters (K, a, r) of each light treatment group through non-linear regression, and R_{max} (maximum population growth rate, in which case N equals to half of K) were calculated (Equation (3)).

Lipid and triacylglycerols determination

The lipid content was determined based on the method of modified chloroform-methanol extraction (Bligh & Dyer 1959). Using a high-speed refrigerated centrifuge (CR22G, HITACHI, Japan), 40 mL of algae liquid was centrifuged at 12,000 rpm, 4 °C for 10 min, and concentrated to 0.8 mL. For extraction, 2 mL of chloroform, 2 mL of methanol, and 1 mL of distilled water were added to the above concentrated algae solution in this order, and mixed well. After extraction, centrifuged again for 10 min (4,000 rpm, 4 °C), then transferred the bottom chloroform layer to the pre-weighed dry glass tubes, blown it to a constant weight with a nitrogen blower. Finally, the microalgal lipid was calculated by gravimetric method. In addition, after the measurement of lipid was completed, 0.4 mL of isopropyl alcohol was added to the glass tube, and the triacylglycerols (TAGs) was determined by the enzyme colorimetric kit method (A110-1 GPO-PAP M Enzyme Method Single Reagent Type) (Li *et al.* 2010).

Photosynthetic pigments determination

The photosynthetic pigments analysis followed the modified procedure as used from Kirk & Allen (1965). A 5 mL sample was centrifuged at 8,000 rpm for 5 min. The pellets were re-suspended in 5 mL of 80% acetone solution at 4 °C for 24 h in darkness. After centrifugation at 4 °C, 4,000 rpm for 15 min, the obtained supernatant was measured for absorbance at 480 nm, 645 nm, and 663 nm, respectively. The 80% acetone solution was used as the blank. The photosynthetic pigments content was calculated as shown below:

$$\text{Chlorophyll a (mg}\cdot\text{L}^{-1}) = 12.7 \times OD_{663} - 2.69 \times OD_{645} \quad (4)$$

$$\text{Chlorophyll b (mg}\cdot\text{L}^{-1}) = 22.9 \times OD_{645} - 2.69 \times OD_{663} \quad (5)$$

$$\text{Carotenoids (mg}\cdot\text{L}^{-1}) = OD_{480} + (0.114 \times OD_{663}) - (0.638 \times OD_{645}) \quad (6)$$

where OD_λ is optical density at wavelength λ (nm).

Protein, total sugar and starch determination

Total sugar, starch and protein were determined using modified standard methods previously described by Liu *et al.* (2019).

Statistics

All experiments were carried out in triplicates, and the data were the mean values ± standard deviation from three independent experiments. All statistical analyses were conducted using SPSS software. One-way ANOVA analysis and Duncan's multiple range test were performed to assess the significant differences among the light source treatments ($p < 0.05$).

RESULTS AND DISCUSSION

Algal growth performance in coastal saline-alkali leachate under different light sources

In order to improve the growth efficiency of *Chlorella* in coastal saline-alkali leachate, LED lights with different light qualities were used as the light source for the growth of *Chlorella*. The growth curves of *Chlorella* sp. HQ in the coastal saline-alkali leachate under different light sources are shown in Figure 1. The five treatments exposed to blue, red, white, blue-red and blue-white mixed light, *Chlorella* sp. HQ maintained high speed growth, and the cell density exceeded 1.00×10^7 cells·mL^{-1} after 5 days of cultivation. After the end of cultivation, it was found that the algal density under blue light was the highest, reaching $8.71 \pm 0.15 \times 10^7$ cells·mL^{-1}, which was about 2–3 times that of the other treatment groups. In addition, *Chlorella* reached the log phase earlier under blue light than the other treatment groups, which indicated that the growth of *Chlorella* was significantly affected by light quality. Meanwhile, the biomass yield of *Chlorella* in the coastal saline-alkali leachate was determined at the end of cultivation under five different light sources (Figure 2). The results showed that biomass yield of *Chlorella* grown under monochromatic blue and white light were significantly higher than that of other groups ($p < 0.05$). And, the algal biomass obtained by *Chlorella* using white light in coastal saline-alkali leachate was 1.76 times higher than that using red light, which indicated that choosing a light source with a specific light quality can significantly increase the biomass of microalgae. At present, there are currently studies on increasing the biomass of microalgae by adjusting other parameters. Gao *et al.* (2019) achieved the purpose of increasing the biomass of *Chlorella* sp. G-9 by adjusting the initial organic carbon-nitrogen ratio of simulated wastewater, and the results showed the algal biomass productivities obtained in wastewater with TOC/TN ratio of 1 and 3 were 1.2 and 1.5 times higher than the value in photoautotrophic condition (TOC/TN = 0).

Based on the the experimental data of microalgae cultivation exhibited in Figure 1, a Logistic model was used to describe the light sources effects on the growth of *Chlorella*

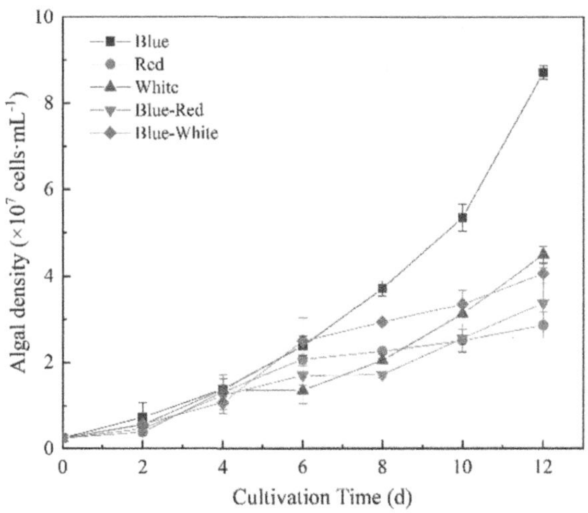

Figure 1 | Growth curves of *Chlorella* sp. HQ in coastal saline-alkali leachate under different light sources.

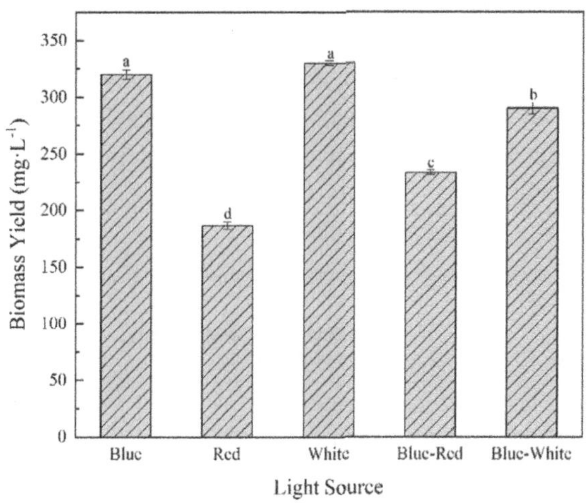

Figure 2 | The biomass yield of *Chlorella* in coastal saline-alkali leachate under different light sources. Different letters above each column indicate significant differences between values ($p < 0.05$).

sp. HQ. Through the model fitting, the fitting parameters of K, r and R_{max} are listed in Table 1. The results showed that using red light as the light source, the specific growth rate of *Chlorella* sp. HQ in coastal saline-alkali leachate was the highest, reaching 0.60 d^{-1}. Maaitah *et al.* (2020) used olive oil washing wastewater to provide a low-cost substrate for algal biomass production and found that the maximum specific growth rate of *C. pyrenoidosa* was 0.4872 d^{-1}, which was lower than the maximum specific growth rate obtained in this experiment. Compared with r and K, R_{max} is more suitable for predicting the growth potential of microalgae (Campos *et al.* 2014). In this study, the R_{max} values of *Chlorella* in the monochromatic blue, white light, and blue-white mixed light groups were higher, which was consistent with the law of algal density at the end of cultivation. In contrast, the blue-red mixed light group gave the lowest growth performances in view of the R_{max}. This showed that *Chlorella* sp. HQ was more suitable to use blue light as a light source to obtain higher algal biomass in the coastal saline-alkali leachate.

Several studies have shown that the growth ability of microalgae under monochromatic blue or red light is similar or enhanced compared to white light (Mouget *et al.* 2004; Abiusi *et al.* 2014). Herein, monochromatic blue light was the optimal light source for *Chlorella* sp. HQ growth because it achieved the highest cell density and the highest value of R_{max}. Similarly, Atta *et al.* (2013) found that *Chlorella vulgaris* grew faster under blue light compared to white light with the light intensity of 200 μmol photons $m^{-2}{\cdot}s^{-1}$. In another study, *Nannochloropsis* sp. and *Tetraselmis* sp. were cultured for 14 d under blue, red, red-blue LED light and white fluorescent light, based on the results of optical density measurements, it was found that microalgae exhibited better growth curve under blue light (Teo *et al.* 2014). Researchers have reported that blue light not only participates in photosynthesis and energy activation, but also helps regulate gene transcription and enzyme activation (Ruyters 1984).

However, many studies have obtained different results, that is, red light is more suitable for the growth of microalgae. This may be due to the different responses of different species to light, which is reflected in product formation and cell growth of microalgae (Shu *et al.* 2012). Wang *et al.* (2007) found that the biomass yield and specific growth rate of *Spirulina platensis* under red light were higher than those of blue light. Xu *et al.* (2013) found that under different C/N ratios, the amount of algal dry weight exposed by the red, white, and yellow light groups was similar, much higher than that of the purple and blue light groups.

Photosynthetic pigment contents of *Chlorella* in coastal saline-alkali leachate under different light sources

The generation of photosynthetic pigments is related to the growth of microalgae. Generally, the percentage of pigments in the algal biomass that is finally harvested from the culture medium is constant (Maaitah *et al.* 2020). In this study, the effect of five different light sources on pigment contents of *Chlorella* sp. HQ in the coastal saline-alkali leachate are presented in Figure 3. The total chlorophylls content and total carotenoids content of *Chlorella* sp. HQ in the coastal saline-alkali leachate were 0.81–1.70% and 0.08–0.25%, respectively. It has been reported that the relative proportion of photosynthetic pigments changes with the spectral quality of light (Rochet *et al.* 1986). The photosynthetic pigments content of *Chlorella* in the coastal

Table 1 | The fitting parameters of the logistic model for *Chlorella* in coastal saline-alkali leachate under different light sources

Light sources	Maximum population density K (10^7 cells·mL^{-1})	Intrinsic growth rates r (d^{-1})	Maximum population growth rates R_{max} (10^7 cells·mL^{-1}·d^{-1})	R^2
Blue light	8.86 ± 0.97	0.36 ± 0.02	0.80 ± 0.10	0.9984
Red light	2.69 ± 0.14	0.60 ± 0.12	0.41 ± 0.08	0.9716
White light	11.54 ± 2.30	0.27 ± 0.02	0.78 ± 0.16	0.9993
Blue-Red light	3.90 ± 0.46	0.32 ± 0.02	0.31 ± 0.04	0.9403
Blue-White light	4.04 ± 0.33	0.48 ± 0.09	0.49 ± 0.09	0.9720

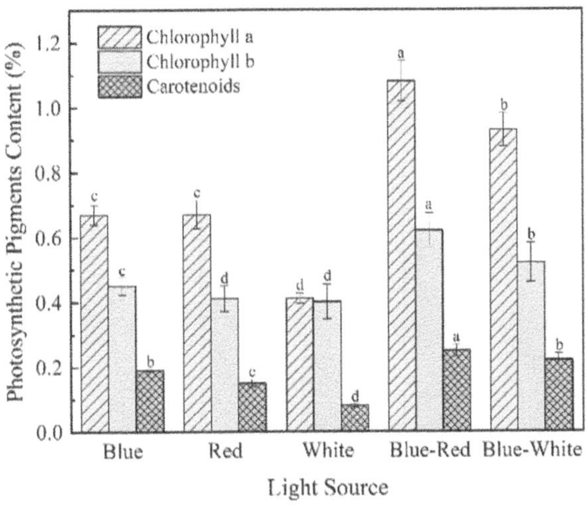

Figure 3 | The photosynthetic pigments content of *Chlorella* in coastal saline-alkali leachate under different light sources at the end of cultivation. Different letters above each column indicate significant differences among different light sources in pigments content ($p < 0.05$).

saline-alkali leachate under the blue-red mixed light was significantly higher than that of monochromatic blue and red light ($p < 0.05$). The chlorophyll a content of *Chlorella* in the blue-white mixed light group was 1.38 times and 2.27 times that of the monochromatic blue and white light groups, respectively. Similarly, the content of chlorophyll b and carotenoids in the blue-white mixed light group was also higher than that in the monochromatic blue and white light groups. The above results showed that compared with the monochromatic light group, the mixed light group was more conducive to the accumulation of pigments of *Chlorella* sp. HQ in the coastal saline-alkali leachate. Pereiraa & Oterob (2019) reported that blue light can increase the total carotenoid content in *Dunaliella salina*, and when it was mixed with red light, due to the synergistic effect, the carotenoid content was significantly higher than the other two monochromatic light experimental groups. The explanation on the higher production of chlorophyll and carotenoid under mixed LED light treatments (blue-red and blue-white mixed light) can be related to complementary chromatic adaptation. According to Emerson and Brody's theory, complementary chromatic adaption is the process of adjusting relative content of major light harvesting pigments which have similar absorption spectra patterns to that of the incident light. And complementary chromatics adaptation can improve the algae cells photosynthetic ability (Rochet *et al.* 1986; Mouget *et al.* 2004).

In addition, the chlorophyll b and carotenoids content in the blue light group were higher than those in the red light group, while the content of chlorophyll a was not significantly different ($p < 0.05$). The reason may be that chlorophyll a in microalgae can absorb blue and red light, while other auxiliary pigments (such as carotenoids) absorb blue light (Kandilian *et al.* 2013). Similarly, Vadiveloo *et al.* (2015) found that the content of chlorophyll a in *Nannochloropsis* sp. exposed the blue light was higher than that of red light. Tamburic *et al.* (2014) found that the microalgae had a higher photosynthetic rate when exposed to monochromatic blue light than to monochromatic red light, which due to that blue light was more easily absorbed by microalgae than red light and was more efficiently transmitted to photosystem II under light-saturated illumination. Therefore, the photosynthetic efficiency of *Chlorella* under blue light was higher than red light, and higher biomass productivity can be obtained. In this work, we obtained similar results, that is, the biomass yield of *Chlorella* in the coastal saline-alkali leachate under blue light was higher than that of red light (Figure 2).

Comparison of high-valued products accumulation of *Chlorella* in coastal saline-alkali leachate under different light sources

Figure 4 demonstrates the total lipid per algal biomass and TAGs content per total lipids yield of *Chlorella* sp. HQ in

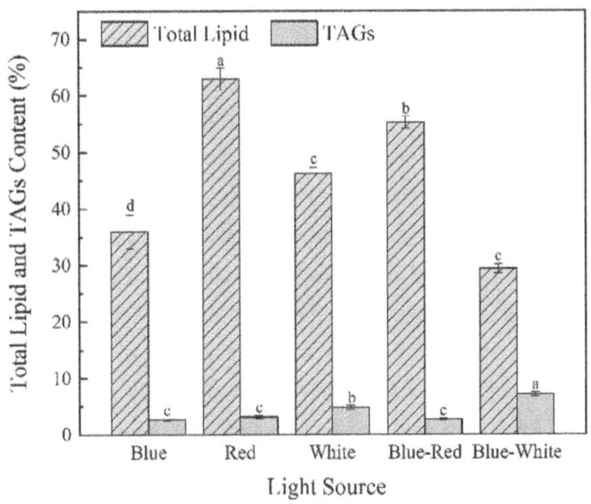

Figure 4 | The total lipids content per algal biomass and TAGs content per total lipids yield of *Chlorella* in coastal saline-alkali leachate under different light sources at the end of cultivation. Different letters above each column indicate significant differences among different light sources in total lipid content and TAGs content ($p < 0.05$).

the coastal saline-alkali leachate with different light sources at the end of cultivation. The results showed that the lipid content of *Chlorella* in the coastal saline-alkali leachate was 29.31–62.95%, and the value under red light was the highest, which was significantly higher than the blue light group and the blue-red mixed light group ($p < 0.05$). Maaitah *et al.* (2020) found that using different dilutions of olive oil washing wastewater as the culture medium of *Chlorella pyrenoidosa*, the lipid content of *Chlorella* under undiluted conditions was the highest, reaching 51.5%, which was lower than the highest lipid content of *Chlorella* sp. HQ in the coastal saline-alkali leachate.

The TAGs content per lipid yield of *Chlorella* was the highest under blue-white mixed light, reaching 7.10 ± 0.35%, which were 2.73 and 1.48 times that of monochromatic blue and white light, respectively. The accumulation of lipids in algal biomass is often associated with a decrease in biomass productivity (Duarte *et al.* 2019). In this experiment, the total lipid content per unit biomass under red light was significantly higher than the blue light and blue-red mixed light, while the biomass under red light was the lowest (Figures 2 and 4). This indicated that red light was conductive to lipid accumulation but was not good to the growth of biomass of *Chlorella* in coastal saline-alkali leachate.

In addition, in the previous study, the fluorescent lamp was used as light source, the protein, total sugar and starch content of *Chlorella* sp. HQ in coastal saline-alkali leachate were higher than those of SE medium, indicating that the accumulation of high-valued substances can be improved under salinity stress (Liu *et al.* 2020). In this study, under blue light conditions where *Chlorella* sp. HQ can obtain higher biomass in saline-alkali leachate, the protein and total sugar content of *Chlorella* was 50.09% and 24.73%, respectively. Rashid *et al.* (2019) found that *Ettlia* sp. and *Chlorella* sp. HS-2 was co-cultured in BG11, and its protein and carbohydrate content were 44 and 33%, respectively. This indicated that the coastal saline-alkali leachate can obtain a higher content of high-value substances under the adjustment of suitable light quality. As shown in Figures 5 and 6, the accumulation characteristics of high-value substances of *Chlorella* under different light qualities have been further explored. The results showed that at the end of experiment, the protein, total sugar and starch contents in blue-red mixed light cultures were higher than those in monochromatic red and blue light, which were $64.71 \pm 1.63 \times 10^{-7}$ μg·cell^{-1}, $46.92 \pm 1.05 \times 10^{-7}$ μg·cell^{-1} and $10.20 \pm 0.23 \times 10^{-7}$ μg·cell^{-1}, respectively. In addition, statistically significant differences were observed among blue-white mixed light, monochromatic blue, and white light groups, the protein, total sugar and starch content in blue-white mixed light cultures remaining higher than the monochromatic blue and white light groups ($p < 0.05$). This indicated that mixed light treatments (blue-red and blue-white) were more appropriate for microalgal protein, total sugar and starch accumulation than monochromatic LED light groups (monochromatic red, blue, and white) in the coastal saline-alkali leachate.

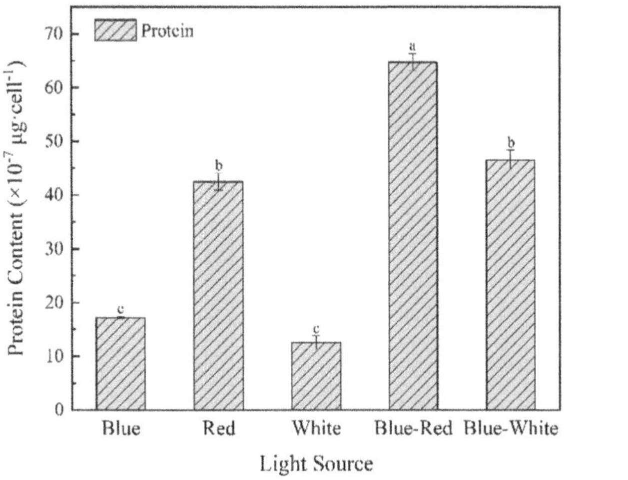

Figure 5 | The protein content of *Chlorella* in coastal saline-alkali leachate under different light sources at the end of cultivation. Different letters above each column indicate significant differences between values ($p < 0.05$).

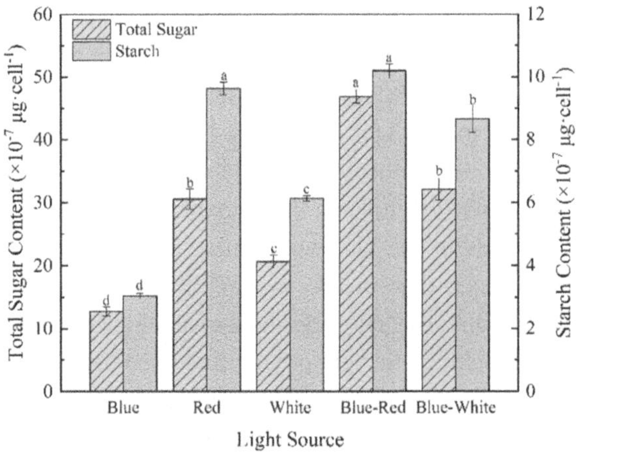

Figure 6 | The total sugar and starch content of *Chlorella* in coastal saline-alkali leachate under different light sources at the end of cultivation. Different letters above each column indicate significant differences in total sugar content and starch content among different light sources ($p < 0.05$).

The total sugar and starch content of *Chlorella* in the coastal saline-alkali leachate under blue light lower than those of other light groups, which may be due to the fact that blue light can promote the production and activity of respiratory enzymes in microalgae. The increase in the concentration and activity of respiratory enzymes accelerates the rate of decomposition of carbohydrates, resulting in a reduction in carbohydrate content (Kowallik 1982; Ruyters 1984). Similarly, Vadiveloo *et al.* (2015) found that the carbohydrate content of *Nannochloropsis* sp. under blue light was lower than that of other light treatments. However, several studies have shown that under different light qualities, the light-harvesting pigments and membranes of microalgae are changed due to various types of algae, resulting in different aspects responses of microalgae to light spectra (Saavedra & Voltolina 1994).

Besides, the effect of light quality on the microalgae growth is a complex process, which is not only related to photosynthesis pathways, but also stimulates photoreceptors, affects microalgae cell division, enzyme activation, gene transcription, and so on (Oldenhof *et al.* 2006; Abiusi *et al.* 2014). For example, in this paper, the total lipid content of *Chlorella* sp. HQ in the coastal saline-alkali leachate under red light was significantly higher than that of monochromatic blue and white light, while another study found that *Nannochloropsis* sp. had significantly higher lipid content under blue light than that of red light and white light (Vadiveloo *et al.* 2015). The reason may be that the activities of carbonic anhydrase and ribulose biphosphate carboxylase/oxygenase which play a vital role in the regulation of the microalgal carbon cycle are enhanced under blue light (Eskins *et al.* 1991).

In general, by optimizing the light quality of the light source required for culturing *Chlorella* in saline-alkali leachate, the light conditions can be controlled more accurately, and the high yield of algal biomass and the maximum utilization of coastal saline-alkali leachate resources can be achieved.

CONCLUSION

Based on the experimental results obtained in this study, optimizing light quality can harvest higher algal biomass from the coastal saline-alkali leachate. Compared with monochromatic lights, blue-red mixed light and blue-white mixed light were more conducive to the accumulation of photosynthetic pigment of *Chlorella*, and complementary chromatics adaptation can improve the algae cells photosynthetic ability. While the highest total lipid content per unit biomass was achieved under monochromatic red light. In addition, mixed LED light (blue-red light and blue-white light) can be used to improve the production of high-valued products. Therefore, if obtaining more biomass and high-valued substances is the desired goal, then a two-stage control of light quality can be considered, such as using blue light first and then using blue-red mixed light as the light source, so as to maximize the benefits of saline-alkali leachate resources utilization.

ACKNOWLEDGEMENTS

The authors would like to thank Prof. Hu Hong-Ying for the experimental facilities. This study was supported by National Natural Science Foundation (No. 52071030) and the Fundamental Research Funds for the Central Universities (No. 2017ZY43).

DATA AVAILABILITY STATEMENT

All relevant data are included in the paper or its Supplementary Information.

REFERENCES

Abiusi, F., Sampietro, G., Marturano, G., Biondi, N., Rodolfi, L., D'Ottavio, M. & Tredici, M. R. 2014 Growth, photosynthetic efficiency, and biochemical composition of *Tetraselmis suecica* F&M-M33 grown with LEDs of different colors. *Biotechnology and Bioengineering* **111** (5), 956–964. https://doi.org/10.1002/bit.25014.

Ananthi, V., Brindhadevi, K., Pugazhendhi, A. & Arun, A. 2021 Impact of abiotic factors on biodiesel production by microalgae. *Fuel* **284**. https://doi.org/10.1016/j.fuel.2020.118962.

Archer, J., Luffman, I., Joyner, T. & Nandi, A. 2019 Identifying untapped potential: a geospatial analysis of Florida and California's 2009 recycled water production. *Journal of*

Water Reuse and Desalination **9** (2), 173–192. https://doi.org/10.2166/wrd.2018.012.

Atta, M., Idris, A., Bukhari, A. & Wahidin, S. 2013 Intensity of blue LED light: a potential stimulus for biomass and lipid content in fresh water microalgae *Chlorella vulgaris*. *Bioresource Technology* **148**, 373–378. https://doi.org/10.1016/j.biortech.2013.08.162.

Bligh, E. G. & Dyer, W. J. 1959 A rapid method of total lipid extraction and purification. *Canadian Journal of Biochemistry and Physiology* **37**, 911–917. https://doi.org/10.1139/o59-099.

Campos, H., Boeing, W. J., Dungan, B. N. & Schaub, T. 2014 Cultivating the marine microalga *Nannochloropsis salina* under various nitrogen sources: effect on biovolume yields, lipid content and composition, and invasive organisms. *Biomass and Bioenergy* **66**, 301–307. https://doi.org/10.1016/j.biombioe.2014.04.005.

Chandra, R., Iqbal, H. M. N., Vishal, G., Lee, H. S. & Nagra, S. 2019 Algal biorefinery: a sustainable approach to valorize algal-based biomass towards multiple product recovery. *Bioresource Technology* **278**, 346–359. https://doi.org/10.1016/j.biortech.2019.01.104.

Chen, P., Min, M., Chen, Y., Liang, W. & Ruan, R. 2009 Review of biological and engineering aspects of algae to fuels approach. *International Journal of Agricultural & Biological Engineering* **2** (4), 1–30. https://doi.org/10.3965/j.issn.1934-6344.2009.04.001-030.

Chew, K. W., Yap, J. Y., Show, P. L., Suan, N. H., Juan, J. C., Ling, T. C., Lee, D. & Chang, J. 2017 Microalgae biorefinery: high value products perspectives. *Bioresource Technology* **229**, 53–62. https://doi.org/10.1016/j.biortech.2017.01.006.

Duarte, J. H., de Souza, C. O., Druzian, J. I. & Costa, J. A. V. 2019 Light emitting diodes applied in *Synechococcus nidulans* cultures: effect on growth, pigments production and lipid profiles. *Bioresource Technology* **280**, 511–514. https://doi.org/10.1016/j.biortech.2019.02.009.

Eskins, K., Jiang, C. Z. & Shibles, R. 1991 Light-quality and irradiance effects on pigments, light-harvesting proteins and Rubisco activity in a chlorophyll- and light-harvesting-deficient soybean mutant. *Physiologia Plantarum* **83** (1), 47–53. https://doi.org/10.1111/j.1399-3054.1991.tb01280.x.

Fang, S. B., Tu, W. R., Mu, L., Sun, Z. L., Hu, Q. Y. & Yang, Y. 2019 Saline alkali water desalination project in Southern Xinjiang of China: a review of desalination planning, desalination schemes and economic analysis. *Renewable and Sustainable Energy Reviews* **113**. https://doi.org/10.1016/j.rser.2019.109268.

Gao, F., Yang, H. L., Li, C., Peng, Y. Y., Lu, M. M., Jin, W. H., Bao, J. J. & Guo, Y. M. 2019 Effect of organic carbon to nitrogen ratio in wastewater on growth, nutrient uptake and lipid accumulation of a mixotrophic microalgae *Chlorella* sp. *Bioresource Technology* **282**, 118–124. https://doi.org/10.1016/j.biortech.2019.03.011.

Kandilian, R., Lee, E. & Pilon, L. 2013 Radiation and optical properties of *Nannochloropsis oculata* grown under different irradiances and spectra. *Bioresource Technology* **137**, 63–73. https://doi.org/10.1016/j.biortech.2013.03.058.

Kim, D. G., Lee, C., Park, S. & Choi, Y. 2014 Manipulation of light wavelength at appropriate growth stage to enhance biomass productivity and fatty acid methyl ester yield using *Chlorella vulgaris*. *Bioresource Technology* **159**, 240–248. https://doi.org/10.1016/j.biortech.2014.02.078.

Kirk, J. T. O. & Allen, R. L. 1965 Dependence of chloroplast pigment synthesis on protein synthesis: effect of actidione. *Biochemical & Biophysical Research Communications* **21** (6), 523–530. https://doi.org/10.1016/0006-291X(65)90516-4.

Kowallik, W. 1982 Blue light effects on respiration. *Annual Review of Plant Physiology* **33**, 51–72. https://doi.org/10.1146/annurev.pp.33.060182.000411.

Li, X., Hu, H. Y., Yang, J. & Wu, Y. H. 2010 Enhancement effect of ethyl-2-methyl acetoacetate on triacylglycerols production by a freshwater microalga, *Scenedesmus* sp. LX1. *Bioresource Technology* **101** (24), 9819–9821. https://doi.org/10.1016/j.biortech.2010.07.103.

Liu, X., Hong, Y., He, Y. & Liu, Y. 2019 Growth and high-valued products accumulation characteristics of microalgae in saline-alkali leachate from Inner Mongolia. *Environmental Science and Pollution Research* **26**, 36985–36992. https://doi.org/10.1007/s11356-019-06842-z.

Liu, X., Hong, Y., He, Y. & Gu, W. 2020 Comparison of oleaginous microalgal growth and lipid accumulation in saline-alkali leachate: a case from Shandong Province. *Desalination and Water Treatment* **187**, 390–398. https://doi.org/10.5004/dwt.2020.25477.

Maaitah, M., Hodaifa, G., Malvis, A. & Sánchez, S. 2020 Kinetic growth and biochemical composition variability of *Chlorella pyrenoidosa* in olive oil washing wastewater cultures enriched with urban wastewater. *Journal of Water Process Engineering* **35**. https://doi.org/10.1016/j.jwpe.2020.101197.

Mata, T. M., Martins, A. A. & Caetano, N. S. 2010 Microalgae for biodiesel production and other applications: a review. *Renewable and Sustainable Energy Reviews* **14** (1), 217–232. https://doi.org/10.1016/j.rser.2009.07.020.

Meena, M. D., Yadav, R. K., Narjary, B., Yadav, G., Jat, H. S., Sheoran, P., Meena, M. K., Antil, R. S., Meena, B. L., Singh, H. V., Meena, V. S., Rai, P. K., Ghosh, A. & Moharana, P. C. 2019 Municipal solid waste (MSW): strategies to improve salt affected soil sustainability: a review. *Waste Management* **84**, 38–53. https://doi.org/10.1016/j.wasman.2018.11.020.

Mouget, J., Rosa, P. & Tremblin, G. 2004 Acclimation of *Haslea ostrearia* to light of different spectral qualities–confirmation of 'chromatic adaptation' in diatoms. *Journal of Photochemistry and Photobiology B: Biology* **75**, 1–11. https://doi.org/10.1016/j.jphotobiol.2004.04.002.

Oldenhof, H., Zachleder, V. & Van Den Ende, H. 2006 Blue- and red-light regulation of the cell cycle in *Chlamydomonas reinhardtii* (Chlorophyta). *European Journal of Phycology* **41** (3), 313–320. https://doi.org/10.1080/09670260600699920.

Pereiraa, S. & Oterob, A. 2019 Effect of light quality on carotenogenic and non-carotenogenic species of the genus

Dunaliella under nitrogen deficiency. *Algal Research* **44**. https://doi.org/10.1016/j.algal.2019.101725.

Rashid, N., Ryu, A. J., Jeong, K. G., Lee, B. & Chang, Y. K. 2019 Co-cultivation of two freshwater microalgae species to improve biomass productivity and biodiesel production. *Energy Conversion and Management* **196**, 640–648. https://doi.org/10.1016/j.enconman.2019.05.106.

Ravelonandro, P. H., Ratianarivo, D. H., Joannis Cassan, C., Isambert, A. & Raherimandimby, M. 2008 Influence of light quality and intensity in the cultivation of *Spirulina platensis* from Toliara (Madagascar) in a closed system. *Journal of Chemical Technology & Biotechnology* **83** (6), 842–848. https://doi.org/10.1002/jctb.1878.

Rochet, M., Legendre, L. & Demers, S. 1986 Photosynthetic and pigment responses of sea-ice microalgae to changes in light intensity and quality. *Journal of Experimental Marine Biology & Ecology* **101** (3), 211–226. https://doi.org/10.1016/0022-0981(86)90264-9.

Ruyters, G. 1984 Effects of blue light on enzymes. Blue light effects in biological system. *Proceedings in Life Sciences* 283–301. https://doi.org/10.1007/978-3-642-69767-8_32.

Saavedra, M. D. P. S. & Voltolina, D. 1994 The chemical composition of *Chaetoceros* sp. (Bacillariophyceae) under different light conditions. *Comparative Biochemistry and Physiology Part B: Comparative Biochemistry* **107** (1), 39–44. https://doi.org/10.1016/0305-0491(94)90222-4.

Schulze, P. S. C., Barreira, L. A., Pereira, H. G. C., Perales, J. A. & Varela, J. C. S. 2014 Light emitting diodes (LEDs) applied to microalgal production. *Trends in Biotechnology* **32** (8), 422–430. https://doi.org/10.1016/j.tibtech.2014.06.001.

Shu, C., Tsai, C., Liao, W., Chen, K. & Huang, H. 2012 Effects of light quality on the accumulation of oil in a mixed culture of *Chlorella* sp. and *Saccharomyces cerevisiae*. *Journal of Chemical Technology & Biotechnology* **87** (5), 601–607. https://doi.org/10.1002/jctb.2750.

Sun, B., Fan, X., Ye, H., Fan, J., Qian, C., van Driel, W. & Zhang, G. 2017 A novel lifetime prediction for integrated LED lamps by electronic-thermal simulation. *Reliability Engineering & System Safety* **163**, 14–21. https://doi.org/10.1016/j.ress.2017.01.017.

Tamburic, B., Szabó, M., Tran, N. T., Larkum, A. W. D., Suggett, D. J. & Ralph, P. J. 2014 Action spectra of oxygen production and chlorophyll a fluorescence in the green microalga *Nannochloropsis oculata*. *Bioresource Technology* **169**, 320–327. https://doi.org/10.1016/j.biortech.2014.07.008.

Tang, J. W., Zhang, S. D., Zhang, X. T., Chen, J. H., He, X. Y. & Zhang, Q. Z. 2020 Effects of pyrolysis temperature on soil-plant-microbe responses to *Solidago canadensis* L.-derived biochar in coastal saline-alkali soil. *Science of the Total Environment* **731**. https://doi.org/10.1016/j.scitotenv.2020.138938.

Teo, C. L., Atta, M., Bukhari, A., Taisir, M., Yusuf, A. M. & Idris, A. 2014 Enhancing growth and lipid production of marine microalgae for biodiesel production via the use of different LED wavelengths. *Bioresource Technology* **162**, 38–44. https://doi.org/10.1016/j.biortech.2014.03.113.

Vadiveloo, A., Moheimani, N. R., Cosgrove, J. J., Bahri, P. A. & Parlevliet, D. 2015 Effect of different light spectra on the growth and productivity of acclimated *Nannochloropsis* sp. (Eustigmatophyceae). *Algal Research* **8**, 121–127. https://doi.org/10.1016/j.algal.2015.02.001.

Wang, C., Fu, C. & Liu, Y. 2007 Effects of using light-emitting diodes on the cultivation of *Spirulina platensis*. *Biochemical Engineering Journal* **37** (1), 21–25. https://doi.org/10.1016/j.bej.2007.03.004.

Wang, R., Wan, S., Sun, J. & Xiao, H. 2018 Soil salinity, sodicity and cotton yield parameters under different drip irrigation regimes during saline wasteland reclamation. *Agricultural Water Management* **209**, 20–31. https://doi.org/10.1016/j.agwat.2018.07.004.

Xu, B., Cheng, P., Yan, C., Pei, H. & Hu, W. 2013 The effect of varying LED light sources and influent carbon/nitrogen ratios on treatment of synthetic sanitary sewage using *Chlorella vulgaris*. *World Journal of Microbiology & Biotechnology* **29** (7), 1289–1300. https://doi.org/10.1007/s11274-013-1292-6.

Yue, Y., Shao, T., Long, X., He, T., Gao, X., Zhou, Z., Liu, Z. & Rengel, Z. 2020 Microbiome structure and function in rhizosphere of Jerusalem artichoke grown in saline land. *Science of the Total Environment* **724**. https://doi.org/10.1016/j.scitotenv.2020.138259.

Zhu, Z. & Dou, J. 2018 Current status of reclaimed water in China: an overview. *Journal of Water Reuse and Desalination* **8** (3), 293–307. https://doi.org/10.2166/wrd.2018.070.

First received 3 April 2020; accepted in revised form 2 September 2020. Available online 25 January 2021

© 2021 The Authors

Seasonal variations of pollutants removal and microbial activity in integrated constructed wetland–microbial fuel cell systems

Xiaoou Wang and Yimei Tian

ABSTRACT

This study investigated the seasonal variations of pollutants removal and microbial activity in constructed wetland–microbial fuel cell systems (CW–MFCs). The results showed that the atmospheric temperature significantly influenced the bioelectricity generation and removal of organics and nitrogen in CW–MFCs by primarily influencing the microbial enzymatic activity. The electricity output of CW-MFCs was extremely low below 5 °C, and reached the maximum above 25 °C. The organics and nitrogen removal of closed-circuit CW–MFC reached the highest in summer and autumn, followed by spring, and decreased by an average of 10.5% COD, 14.2% NH_3-N and 10.7% TN in winter, demonstrating smaller seasonal fluctuations compared to open-circuit CW–MFC in which the difference between summer and winter was 13.4% COD, 15.1% NH_3-N and 15.1% TN. Even at low temperatures, the MFC current could enhance the enzymatic activity and stabilize the growth of microorganisms on the electrodes, moreover, the closed circuit operation can promote the bacteria diversity on CW–MFC anodes as well as the abundance of electrogens on CW–MFC anodes and cathodes, and thus reduce the adverse effect of cooling on organics and nitrogen removal in CWs. However, neither MFC nor temperature had a significant influence on phosphorus removal in CW–MFCs.

Key words | constructed wetland, microbial activity, microbial fuel cell, nitrogen, organics, seasonal variation

Xiaoou Wang
Key Laboratory of Clean Energy Utilization and Pollutant Control in Tianjin, School of Energy and Environmental Engineering,
Hebei University of Technology,
Tianjin,
China

Yimei Tian (corresponding author)
School of Environmental Science and Engineering,
Tianjin University,
Tianjin,
China
E-mail: *tian43_b305@126.com*

HIGHLIGHTS

- Temperature significantly influenced CW-MFCs by primarily influencing microbial enzymatic activity.
- Closed-circuit CW-MFC showed smaller seasonal variations of organics and nitrogen removal than open-circuit CW-MFC.
- Closed circuit mode promoted bacteria diversity and electrogens' abundance on CW-MFC anodes, even at low temperatures.
- MFC reduced adverse effects of cooling on organics and nitrogen removal in CWs.

This is an Open Access article distributed under the terms of the Creative Commons Attribution Licence (CC BY 4.0), which permits copying, adaptation and redistribution, provided the original work is properly cited (http://creativecommons.org/licenses/by/4.0/).

doi: 10.2166/wrd.2021.094

INTRODUCTION

Constructed wetlands (CWs) are an attractive ecological restoration technology, and have been applied in treating various wastewaters worldwide (Wu *et al.* 2011; Wang *et al.* 2017; Cao *et al.* 2019). Pollutants removal in CWs is accomplished through a variety of physical, chemical and biological processes, including sedimentation, interception, precipitation, filtration, adsorption, absorption, volatilization, plant uptake, and microbial degradation (Saeed & Sun 2012). Due to the dependency of biological and biochemical processes on temperature, many studies have reported poorer performance of pollutants removal in CWs under low temperatures. For example, Akratos & Tsihrintzis (2007) reported that the average removal rates of ammonia (NH_3-N) and total Kjeldahl nitrogen (TKN) in a pilot-scale horizontal-flow CW (HFCW) were 37.9 and 58.5% at temperatures of <15 °C, and 69.1 and 73.9% at temperatures of >15 °C, respectively. Song *et al.* (2006) reported that for a full-scale CW with a total area of 80 ha in Shandong, China, the average removal of 5-day biochemical oxygen demand (BOD_5), chemical oxygen demand (COD), NH_3-N and total phosphorus (TP) was lowest in winter (67.8, 59.4, 32.4 and 28.9%, respectively), and highest in summer (74.1, 66.3, 54.5 and 35.0%, respectively). Sani *et al.* (2013) evaluated the seasonal performance of a vertical-flow CW (VFCW), and significantly higher COD, NH_3-N, and nitrate (NO_3^--N) removals were recorded in summer. Saeed & Sun (2012) reported that temperatures between 16.5 and 32 °C favor nitrification in CWs, and nitrification barely occurs at temperatures of ≤5–6 and ≥40 °C; the suitable temperature for denitrification is 20–25 °C, and, at temperatures of <5 °C, denitrification proceeds very slowly.

Some measures have been taken to mitigate the adverse impacts of low temperature on CWs. For example, Ouellet-Plamondon *et al.* (2006) reported that in a reed CW artificial aeration improved the removal of TKN, NH_3-N and COD in winter by approximately 2.2, 29.4 and 7.5%, respectively. Wu *et al.* (2011) reported that a 0.4 m sawdust layer cover on the CW could keep the temperature inside the wetland constantly above 6 °C, even when the atmospheric temperature dropped to −8 °C during winter, which provided effective system thermal insulation and maintained high pollutants removal (95.0% BOD_5, 84.6% NH_3-N, and 88.2% TP) in the freezing winter period. Kadlec & Wallace (2009) reported that during the winter icing period, wetlands could be operated under a frozen layer by adjusting the water level, and a relatively high water temperature would be maintained inside the wetlands. Wang *et al.* (2017) summarized that hybrid CWs consisting of various types of CWs arranged in series possess higher treatment performance than a single CW in a cold climate.

The interest in integrated constructed wetland–microbial fuel cell (CW–MFC) systems has increased due to their ability to produce electricity and enhance the wastewater treatment efficiency (Doherty *et al.* 2015; Guadarrama-Pérez *et al.* 2019). Fang *et al.* (2013) found that an MFC could improve the decolorization rate and COD removal of a CW by 15 and 12.7%, respectively. Wang *et al.* (2016) reported that the MFC significantly promoted the relative abundance of beta-*Proteobacteria*, nitrobacteria, and denitrifying bacteria in the CW, and thus increased the average COD and NO_3^--N removal by 8.3 and 40.2%, respectively. Srivastava *et al.* (2015) reported that the MFC increased COD removal in a CW by 27–49%. Xu *et al.* (2018) reported that the rates of nitrification and denitrification increased by approximately 82% in a three-biocathode CW–MFC. Yu *et al.* (2020) found that the constructed wetland–microbial electrolysis cell (CW-MES) system successfully enhanced NH_3-N removal at low temperatures (5.6–7.9 °C), while CW-MFC did not exhibit a positive effect. Overall, few of the reported studies have referred to the seasonal variations of CW–MFCs, and the performance of CW–MFCs at low temperature should be further explored.

This study investigated the seasonal variations of pollutants removal and microbial activity in CW–MFCs, aiming to: (1) quantitatively analyze the decreasing degree of CW–MFCs performance caused by low temperature; (2) investigate whether MFC could still help in enhancing the pollutants removal in CWs at low temperature; and (3) investigate the variations of enzymatic activity and microbial community in CW–MFCs under different seasons, with the expectation of offering a reference for enhancing the treatment efficiency of CWs in a cold climate.

MATERIALS AND METHODS

Experimental setup

Two parallel integrated up-flow CW–MFC reactors (closed-circuit CW–MFCc and open-circuit CW–MFCo) were built in the open air. As shown in Figure 1, the wetland reactors were made of Perspex glass, and were 0.7 m long, 0.6 m wide, and 1.0 m high. The substrates (provided by Hebei Yanxi Mineral Processing Factory, China) had four layers: the bottom supporting layer was 20–40 mm gravel (0.15 m deep), topped with 10–30 mm lava (0.30 m deep), and then filled with 5–10 mm gravel (0.05 m deep), and 1–3 mm sand (0.05 m deep). The porosity of the substrates was approximately 0.48. There was a free water surface approximately 0.20 m deep in the reactors. There was a water collection pipe 0.15 m above the substrates' surface through which the effluent was discharged to an effluent water area with a dimension of $0.1 \times 0.6 \times 1.0$ m (length × width × height).

The cathode and anode were graphite plates and self-made carbon fiber brushes, respectively. Four graphite plates ($100 \times 100 \times 8$ mm, provided by Tianjin Dongguang Huijin Co., Ltd, China) were evenly placed on the substrates' surface. The carbon fiber brush was 0.7 m long, 0.4 m of which contained the carbon fiber (0.1 m long, provided by Beijing Chuanjing Technology Development Co., Ltd, China). Five carbon fiber brushes were positioned vertically at the four corners and middle of the substrates, and the bottom of the brush was 0.05 m away from the bottom of the substrates. All the graphite plates and carbon fiber brushes were connected by titanium wires as a whole cathode and anode, respectively. The cathode and anode were connected across a variable external resistor (0–9999.9 Ω) by insulated copper wires.

Canna indica was selected as the wetland plants. After planting, the reactors were submerged in tap water immediately for the plants and microbes to develop. KH_2PO_4, $MgSO_4{\cdot}7H_2O$, NH_4Cl, and $CaCl_2{\cdot}2H_2O$ (in 15 mg/L tap water) were added to the tap water to provide nutrients for plants. After one month, the MFC circuit of both reactors was connected and the external resistors were set to 1,000 Ω to start up the CW–MFCs. During the startup, the inoculum (supernatant (1:1) of aerobic and anaerobic sludge) and carbon source (500 mg/L sodium acetate) were mixed at a ratio of 1:1, and the mixture was continuously pumped into the reactors from the bottom at 0.073 m^3/d. The inoculation supernatant was collected from a municipal wastewater treatment plant in Tianjin, China. Half of the effluent was returned to the reactors to enhance the enrichment of electrogens on electrodes. Additionally, graphite plates were submerged in the mixed solution for 24 hours before starting up the reactors, so as to accelerate the development of microbes on cathodes. Reproducible maximum voltages of 0.73–0.74 V arrived after 10 days of operation, indicating the successful startup of CW–MFCs. Then the external resistance of reactor CW–MFCc was set to 250 Ω, and the circuit of reactor CW–MFCo was disconnected (as control). The influents were then continuously dosed into the reactors at 0.049 m^3/d to begin the experiment, and the effluent reflux ratio was 50% (reflux flow rate was 0.024 m^3/d). The average hydraulic retention time in the reactors was 1.5 days. The influent and reflux flow rate was controlled by a liquid flowmeter installed at the inlet pipe and reflux pipe, respectively. Dissolved oxygen (DO) concentration in the cathode area remained at approximately 1.5 mg/L through continuous aeration by an air compressor. Approximately 1 week after the dosing, the output voltages of CW–MFCc stabilized and sampling began.

Synthetic domestic wastewater was used as influent, and was prepared using analytically pure CH_3COONa, $C_2H_5NO_2$, NH_4Cl, KNO_3, $NaNO_2$, KH_2PO_4, $Na_5P_3O_{10}$, $MgSO_4{\cdot}7H_2O$, and$CaCl_2{\cdot}2H_2O$. The major characteristics of the influents are listed in Table 1.

The whole monitoring period was from September 2016 to August 2017, covering the four seasons, and the CW–MFC reactors stopped operation during the winter icing period (Dec 13, 2016–Feb 13, 2017) because the water in the reactors froze. The 5-day average temperature of spring, summer, autumn and winter is 10–22, ≥22, 22–10 and ≤10 °C, respectively.

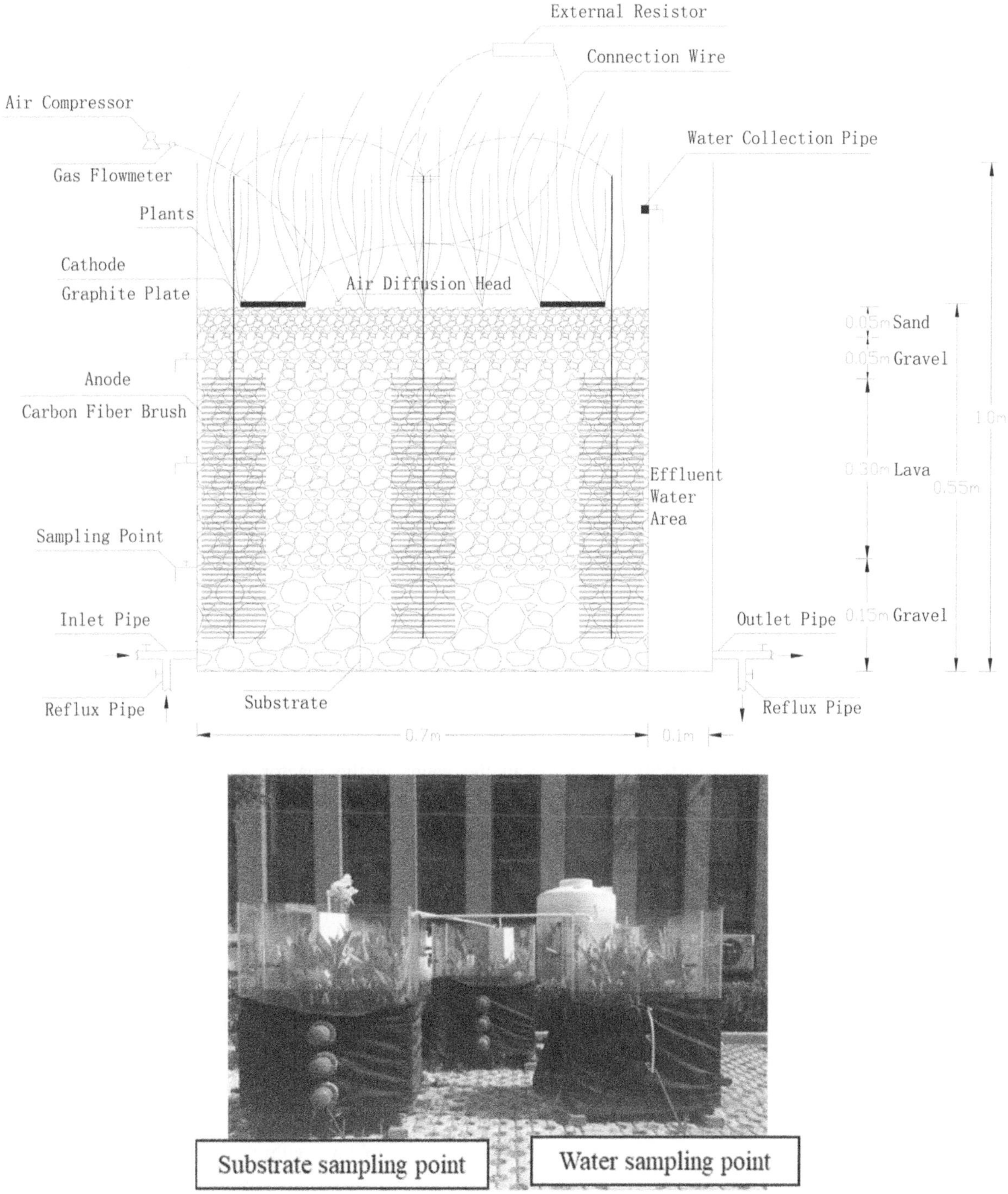

Figure 1 | Schematic diagram of the constructed wetland–microbial fuel cell (CW–MFC) system (above) and a picture of the CW–MFC reactors (below).

Water sampling and analysis

Water samples were collected between 9.00 and 10.00 am every 3 days from the effluent, and were immediately analyzed in the lab for COD, NH_3-N, TN, PO_4^{3-}-P, and TP using a Digital Reactor Block 200 and a HACH DR 2800 spectrophotometer, according to the standard procedure provided by HACH Company, USA. Specifically, COD

Table 1 | Major characteristics of the synthetic influents (mg/L)

pH	COD	NH_3-N	NO_2^--N	NO_3^--N	Org-N	TN	PO_4^{3-}-P	$P_3O_{10}^{5-}$-P	TP
7.0 ± 0.10	200	25	0.1	0.5	15	40.6	4	1	5

NO_2^--N, nitrite nitrogen; Org-N, organic nitrogen; TN, total nitrogen; PO_4^{3-}-P, phosphate; $P_3O_{10}^{5-}$-P, polyphosphate; TP, total phosphorus.

was measured by the quick digestion spectrophotometry method, NH_3-N was measured by the salicylate method, TN was measured by the persulfate digestion method, PO_4^{3-}-P was measured by the molybdovanadate method, and TP was measured by the molybdovanadate method with acid persulfate digestion. The wastewater DO and temperature within the CW–MFCs were measured every 30 minutes by an online DO detector (provided by Hangzhou Sinomeasure Automation Technology Co., Ltd, China). The atmospheric temperature was monitored using a temperature recorder (provided by Hangzhou Sinomeasure Automation Technology Co., Ltd, China), and the data were collected every 30 minutes. The pollutants removal rate (R) was calculated as follows:

$$R = \frac{(C_i - C_e)}{C_i} \times 100\% \tag{1}$$

where C_i and C_e are the mean influent and effluent concentration (mg/L), respectively.

Bioelectricity measurement and analysis

The output voltage (U) was collected using a multi-channel data logger (Model CT-4008-5v10 mA-164, Shenzhen Neware Electronics Co., Ltd, China) and recorded by a computer at intervals of 30 min. The output current (I) was calculated by Ohm's law. The volumetric power density (P_d) was calculated as follows:

$$P_d = \frac{P}{V} = \frac{U^2}{VR_{ex}} \tag{2}$$

where P is the power (mW), V is the total volume of the CW–MFC reactor (m^3), U is the output voltage (V), and R_{ex} is the external resistance (Ω).

Microbial sampling and analysis

Microbial sampling was conducted once during spring, summer, autumn and winter, respectively, between 9.00 and 10.00 am. The anode samples were collected by cutting a small amount of carbon fiber at a depth of 10–15 cm from the substrate surface, and samples from the five carbon fiber brushes were uniformly mixed as one sample. The cathode samples were collected from the biomass attached on the graphite plates, and samples from the four graphite plates were uniformly mixed as one sample. In order to minimize the disturbance of sampling to the cathodes, only 1.0 × 1.0 cm biomass was gently scraped away from the graphite plate, accounting for 1% of the total area of the plate. The substrate (lava) samples were collected at a depth of 10–25 and 25–40 cm, and were uniformly mixed as one sample. During substrate sampling, the influents dosing was stopped and the water in the reactor was drained, and then the substrate sampling holes (10 cm in diameter, Figure 1) were opened to collect substrate samples. Once the sampling was completed, the influents dosing was restarted. All microbial sampling was conducted between 9.00 and 10.00 am.

The enzymatic activity was determined by the analysis of dehydrogenase and catalase activity. Samples were freeze-dried first, and then ground and screened through a 16-mesh sieve for determination of enzymatic activity. Dehydrogenase activity (DHA) was measured with the triphenyl-tetrazolium chloride (TTC) method, for which 1 g (dry basis) of sample was cultured using 1 mL TTC solution (5 g/L) and 0.4 mL glucose solution (0.1 M) in an incubator at 37 °C for 12 h, and the resulting formazan was extracted with toluene and then measured by spectrophotometric quantification at 480 nm. DHA was expressed as μL (H)/g, 12 h, 37 °C for the substrate and μL (H)/cm^2, 12 h, 37 °C for the anodes and cathodes. Catalase activity (CA) was determined by the titration of residual H_2O_2 (3%) with

$KMnO_4$ (0.02 M), and was expressed as mL (0.02 M $KMnO_4$)/g, 1 h for the substrate and mL (0.02 M $KMnO_4$)/cm^2, 1 h for the anodes and cathodes.

The microbial diversity was determined by the Illumina sequencing analysis of the microbial community. Samples were submitted to Novogene Technology Co., Ltd (Beijing, China) to perform DNA extraction, PCR amplification and high throughput sequencing analysis. The specific procedures are provided in the supplementary material.

Data analysis

One-way analysis of variance (ANOVA) was conducted to detect significant differences in the treatment efficiency between the two CW–MFC reactors, followed by a Duncan post hoc test ($P < 0.05$). All of the statistical analyses were conducted using SPSS and Origin software.

RESULTS AND DISCUSSION

As illustrated in Figure 2, during the whole monitoring period (Sep 2016–Aug 2017, except for Dec 13, 2016–Feb 13, 2017), the recorded average daily minimum and maximum atmospheric temperatures were -6 to 27 and 0–36 °C, respectively. Compared to the atmospheric temperature, the wastewater temperature within the CW–MFCs was higher by 1–2 °C in winter and lower by 2–4 °C in summer. Additionally, in both closed-circuit CW–MFCc and open-circuit CW–MFCo, the anode area inside the reactors (0–45 cm from the bottom) remained anoxic/anaerobic (DO < 0.5 mg/L), while the cathode area was under aerobic conditions (DO > 1.0 mg/L) because of aeration.

Bioelectricity generation of the CW–MFC reactor

Figure 3 shows the bioelectricity generation of CW–MFCc with atmospheric temperature. The electricity output of CW–MFCc was quite low when the temperature was

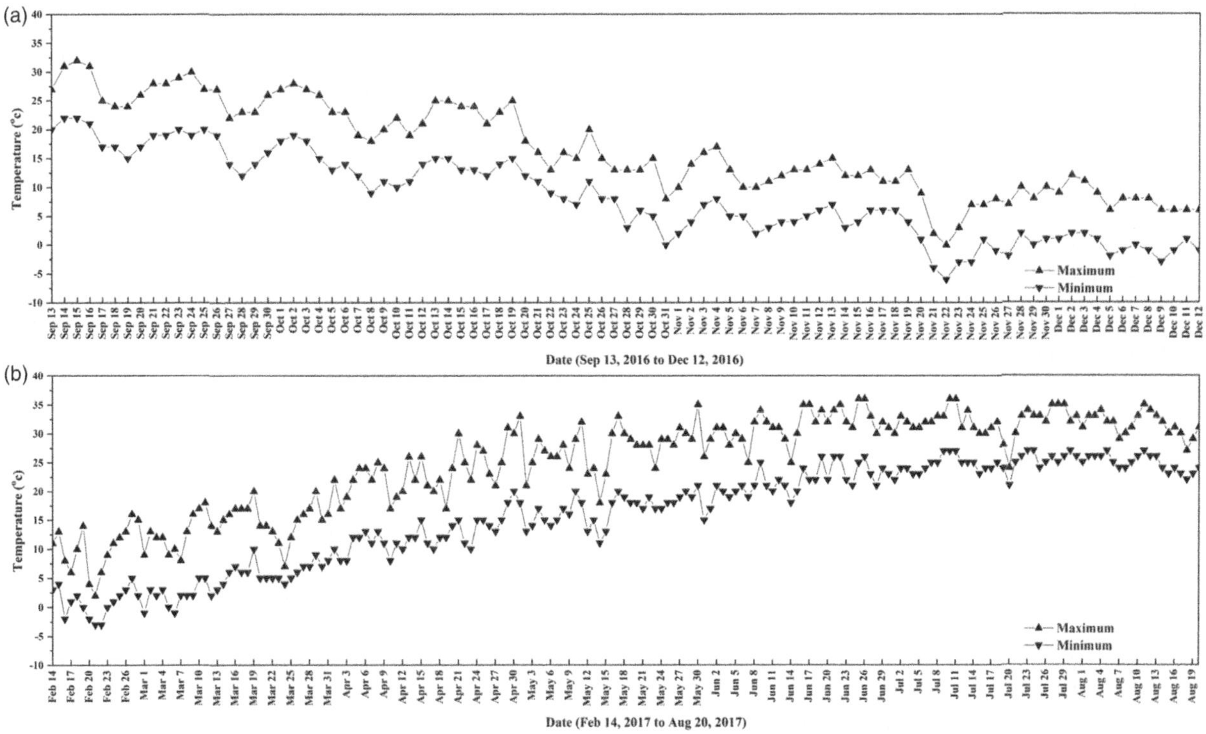

Figure 2 | Atmospheric temperature during the whole monitoring period (Sep 2016–Aug 2017, except for Dec 13, 2016–Feb 13, 2017).

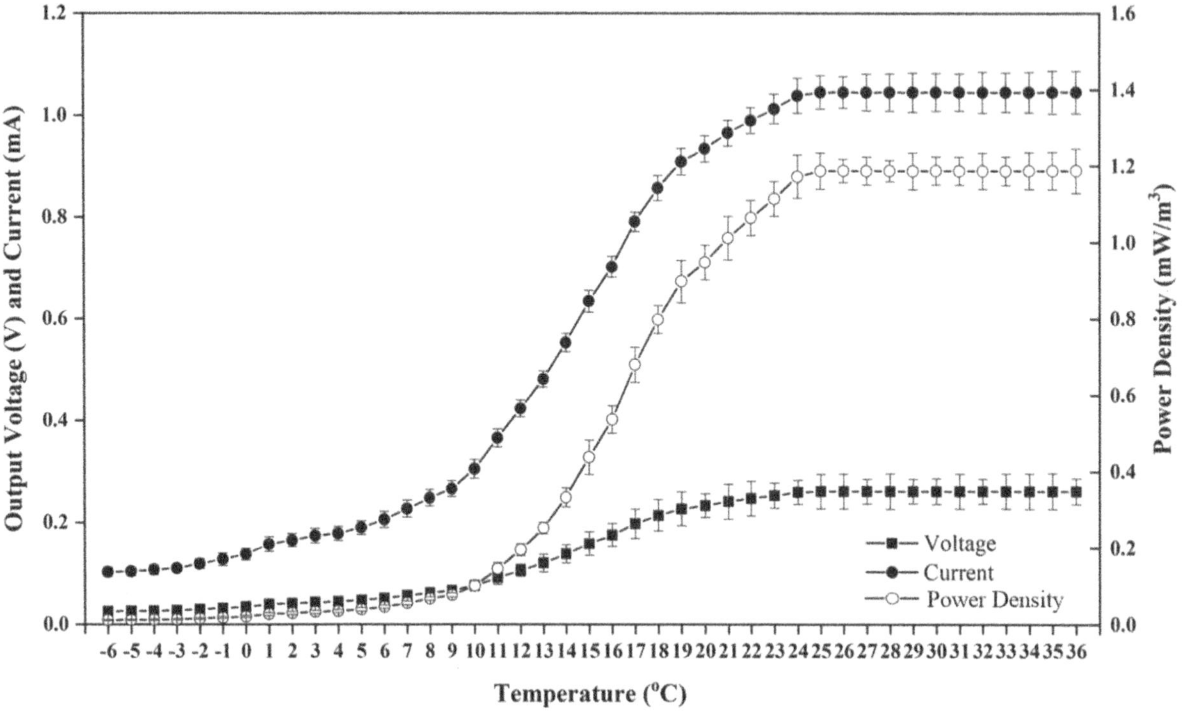

Figure 3 | Bioelectricity generation of the closed-circuit constructed wetland–microbial fuel cell system at various atmospheric temperatures.

$\leq$5 °C ($U < 0.0500$ V, $I < 0.20$ mA, $P_d < 0.04$ mW/m^3), indicating the weak activity of electrogens at temperatures below 5 °C. When the temperature increased from 6 to 10 °C, the electricity output of CW–MFCc began to increase slowly, indicating that the activity of electrogens gradually recovered as the temperature increased. When the temperature exceeded 10 °C, the electricity output of CW–MFCc began to rapidly increase with the increasing temperature. Especially when the temperature was between 14 and 19 °C, the output voltage increased by approximately 0.013–0.022 V as the temperature increased by 1 °C, and the corresponding output current and power density increased by approximately 0.05–0.09 mA and 0.10–0.14 mW/m^3, respectively. As the temperature exceeded 19 °C, the increase in electricity output gradually slowed down with the increasing temperature. When the temperature was above 25 °C, the electricity output of CW–MFCc stopped increasing and remained stable ($U \approx 0.26$ V, $I \approx$ 1.05 mA, $P_d \approx 0.04$ mW/m^3), demonstrating that the activity of electrogens had reached the maximum at $\geq$25 °C.

Temperature is a key factor influencing the MFC performance. Within a certain temperature range the bioelectricity generation in MFCs will increase as the temperature increases (Adelaja *et al.* 2015). This is because increasing the temperature can enhance the electrochemical activity of electrogens (Michie *et al.* 2011) as well as the rate of ion electromigration (Behera *et al.* 2011), which both could reduce the internal resistance of the MFC, thereby increasing its output power. However, microbial proteins will denature at excessively high temperatures, which is detrimental to the MFC performance. Adelaja *et al.* (2015) found that the biodegradation rates of a petroleum hydrocarbon mix (i.e. phenanthrene and benzene) and maximum power density in MFCs were both two times higher at 40 °C (97.10% and 1.15 mW/m^2 anode, respectively) than those at 30 °C, but were four times lower when the operating temperature was raised to 50 °C. It has been reported that the most suitable temperature for the electrochemical activity of anode biofilms is 30–45 °C (Michie *et al.* 2011).

Overall, the atmospheric temperature significantly influenced the electricity generation of CW-MFCs by influencing the electrogens' activity. The electricity output of the CW-MFC system was extremely low when the temperature was ≤5 °C, and reached the maximum at ≥25 °C.

Pollutants removal of the CW–MFC reactor

Pollutants removal in the CW–MFC systems under different atmospheric temperature is provided in Figure 4. The 5-day average temperature varied between 3 and 30 °C. The pollutants removal stabilized at the 5-day average temperature of ≥11 °C in closed-circuit CW–MFCc (84.5–91.0% COD, 83.9–90.8% NH_3-N and 84.0–92.9% TN) and ≥14 °C in open-circuit CW–MFCo (74.5–82.0% COD, 79.5–85.7% NH_3-N and 75.9–81.6% TN), showing significant difference ($P < 0.05$) between CW–MFCc and CW–MFCo. This demonstrated that incorporating MFC could help strengthen the resistance of CWs to lower temperature in terms of organics and nitrogen removal. Considering the dependency of organics and nitrogen bio-reaction kinetics on temperature, it is unsurprising to find the decreases in the removal of COD, NH_3-N and TN at lower temperatures. The maximum difference in the average removal rate under higher (≥11 °C for CW–MFCc and ≥14 °C for CW–MFCo) and lower (3–6 °C) 5-day average temperature was approximately 12.2% COD, 16.8% NH_3-N, 12.6% TN for CW–MFCc, and 15.6% COD, 19.9% NH_3-N, 16.2% TN for CW–MFCo. Comparing CW–MFCc with CW–MFCo, it can be found that the decreasing degree of organics and nitrogen removal under lower temperatures was smaller in CW–MFCc. However, phosphorus removal in both CW–MFCc (85.2–90.2% PO_4^{3-}-P and 84.4–89.0% TP) and CW–MFCo (84.2–89.2% PO_4^{3-}-P and 83.4–87.4% TP) barely changed with temperature, demonstrating more stable removal than that of organics and nitrogen. Moreover, the removal rates of COD, NH_3-N and TN in CW–MFCc were significantly ($P < 0.05$) higher than that in CW–MFCo by 8.0–15.0%, 2.7–5.9% and 10.2–17.8%, respectively, while there was negligible increase of average removal rates for PO_4^{3-}-P and TP in CW–MFCc (0.81 and 0.87%, respectively).

Pollutants removal in the CW–MFC systems under different seasons is provided in Figure 5. The average removal rates in closed-circuit CW–MFCc under spring, summer, autumn and winter were, respectively, higher than that in open-circuit CW–MFCo by approximately 11.2, 9.4, 10.0 and 12.4% for COD, 4.8, 4.8, 5.1 and 5.7% for NH_3-N, and 12.42, 11.19, 11.60 and 15.6% for TN. It can be found that the gap between CW–MFCc and CW–MFCo in terms of organics and nitrogen removal widened in winter. For closed-circuit CW–MFCc, the organics and nitrogen removal in summer did not significantly ($P > 0.05$) differ from that in autumn, but was significantly ($P < 0.05$) higher than that in spring by an average of 3.2% COD, 4.2% NH_3-N, 3.5% TN, and that in winter by an average of 10.5% COD, 14.2% NH_3-N, 10.7% TN. In the meantime, the average removal rate of COD, NH_3-N and TN in open-circuit CW–MFCo demonstrated a larger difference between summer and winter, which was 13.4, 15.1 and 15.1%, respectively. This further proved that incorporating MFC could still help improve the organics and nitrogen removal in CWs during winter. No significant ($P > 0.05$) difference was observed in phosphorus removal in both CW–MFCc (88.0–88.5% PO_4^{3-}-P and 86.5–87.0% TP) and CW–MFCo (87.1–87.6% PO_4^{3-}-P and 85.5–86.0% TP) under the four seasons.

In CWs, biodegradation is considered as the dominant process responsible for organics removal, and microbial nitrification and denitrification is the primary method of nitrogen removal (Saeed & Sun 2012). Therefore, it is widely acknowledged that temperature significantly influenced the organics and nitrogen removal in CWs by primarily affecting the microbial activities (Yan & Xu 2014). MFC can enhance the anaerobic degradation of organics in the anodes, and thus improve the organics removal of CW–MFCs (Doherty *et al.* 2015). Nitrite and nitrate can be used as electron acceptors in the cathode of an MFC for reducing nitrogen in wastewater while producing bioelectricity (Puig *et al.* 2011). NH_3-N can be used as one of the main substrates for electricity generation and thus its removal was enhanced (Yu *et al.* 2020). Despite the weak bioelectricity generation of the CW–MFC under lower temperature (Figure 3), the presence of MFC reduced the adverse effect of cooling on the organics and nitrogen removal in CWs to a certain extent. As for the similar removal of organics and nitrogen in the CW–MFCs during summer and autumn, this may be because there were only 34 days in the autumn of this study, and 26 of the days had a 5-day average temperature of more than 15 °C.

Many studies have found that the physio-chemical processes of substrates were mainly responsible for P removal

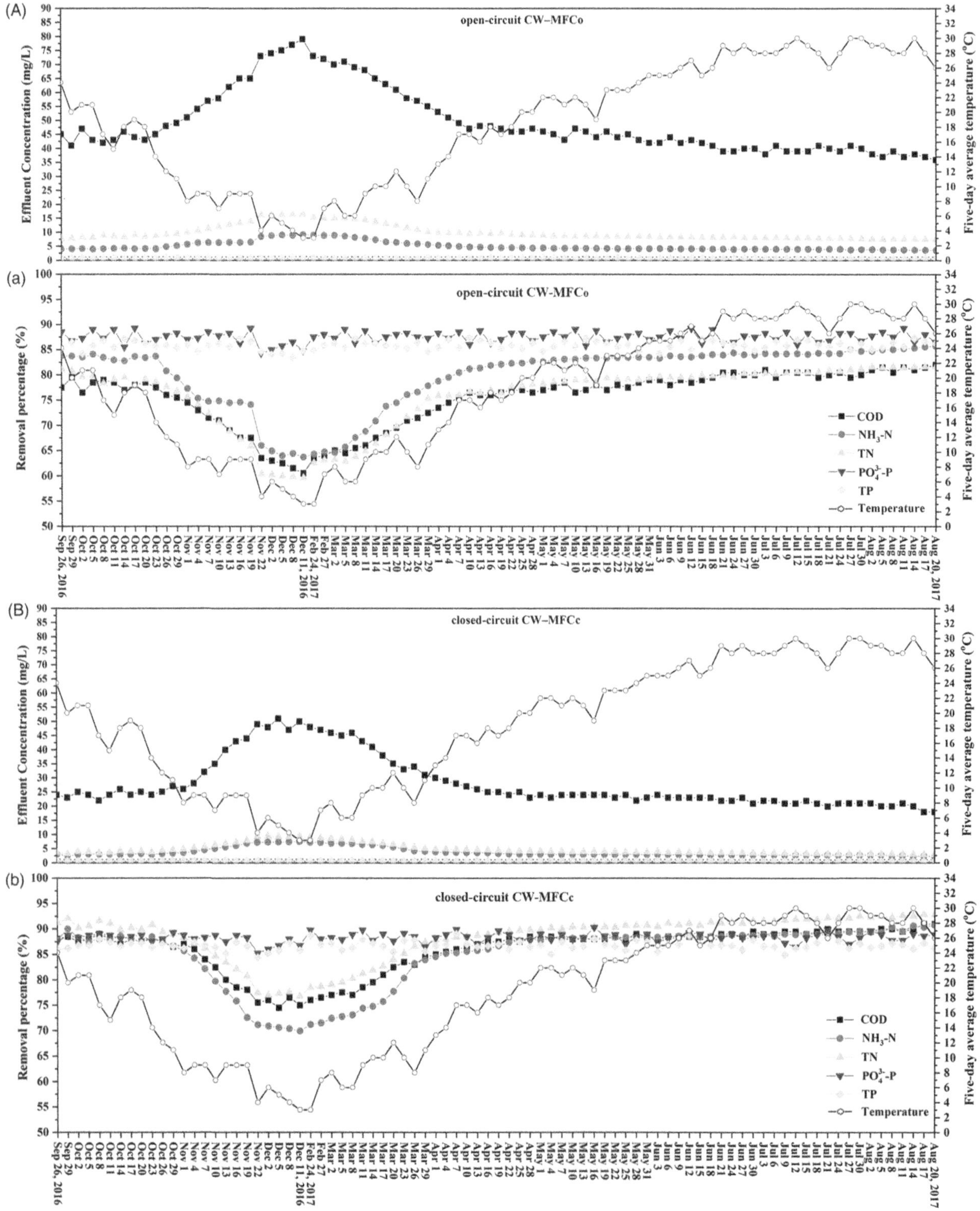

Figure 4 | Pollutants removal of the constructed wetland–microbial fuel cell systems at various atmospheric temperatures.

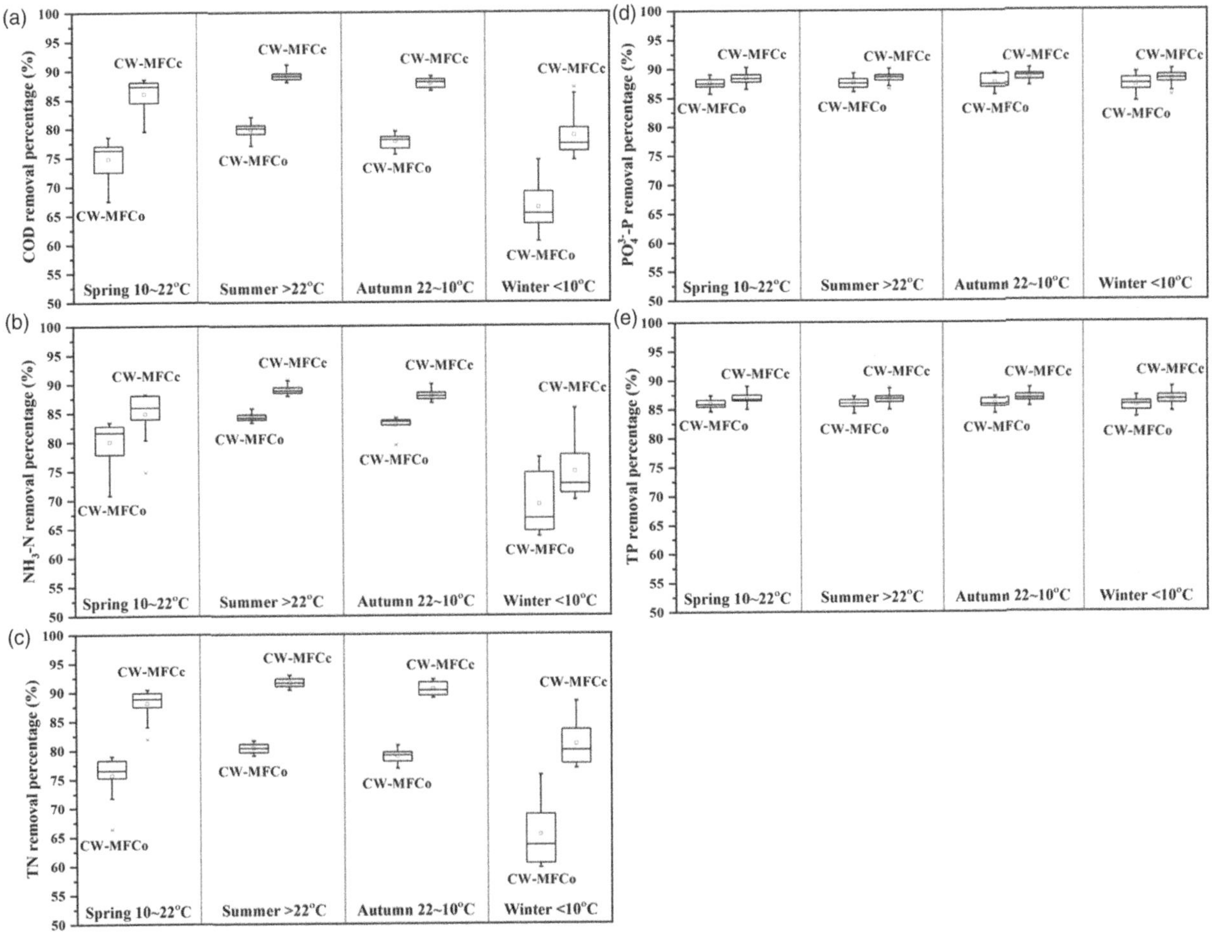

Figure 5 | Pollutants removal of the constructed wetland–microbial fuel cell systems under different seasons. Error bars are ±1 standard deviation.

in CWs, including sedimentation, filtration, interception, adsorption, absorption, precipitation, ion exchange, and complexation reactions (Lan *et al.* 2018). From the results of this study, it can be summarized that the primary method of P removal in CW–MFCs was also the physio-chemical processes of substrates.

Overall, the MFC not only significantly ($P < 0.05$) improved the organics and nitrogen removal but also mitigated the negative effects of lower temperature on organics and nitrogen removal in CWs. However, both MFC and temperature had no significant influence on phosphorus removal in CW–MFCs.

Enzymatic activity

As an intermediate carrier of hydrogen, DHA can reflect the microbial oxidative capability during the organics degradation, therefore it is often used to measure microbial activity (Barrena *et al.* 2008). As shown in Figure 6(a) and 6(b), DHA of the anodes was higher than that of the cathodes by approximately 3.2–4.6 times in closed-circuit CW–MFCc and 3.3–5.2 times in open-circuit CW–MFCo, respectively. This may be owing to the much higher COD concentration in the wetland interior, since the influents were dosed into the reactors from the bottom. Catalase is an oxidoreductase mainly associated with the activity of aerobic microorganisms. As shown in Figure 6(a) and 6(b), CA of the cathodes was significantly ($P < 0.05$) higher than that of the anodes by approximately 23.7–35.6% in closed-circuit CW–MFCc and 16.7–33.1% in open-circuit CW–MFCo, respectively, which resulted from the aerobic environment (DO > 1.0 mg/L) in the cathode area. Compared to open-circuit CW–MFCo, enzymatic activity of the anodes and cathodes in closed-circuit CW–MFCc under

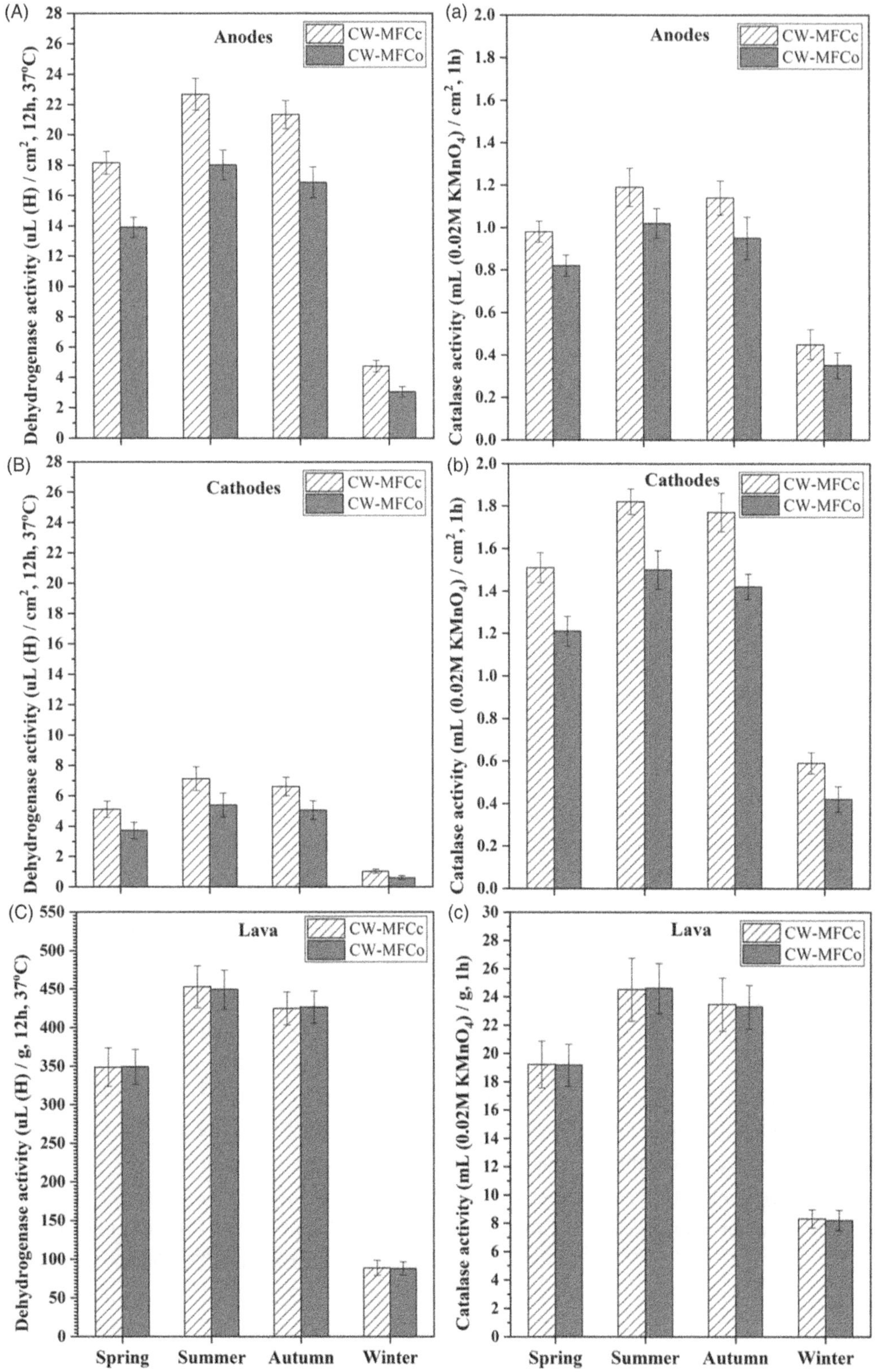

Figure 6 | Enzymatic activity of dehydrogenase and catalase in the constructed wetland–microbial fuel cell systems under different seasons. Error bars are ±1 standard deviation.

the four seasons was significantly ($P < 0.05$) higher by 25.9–55.4% and 30.6–72.9% for DHA, and 16.7–28.6% and 21.3–40.5% for CA, respectively, while DHA and CA of the substrate were similar under the same season. This proved that the MFC current could enhance the growth of microorganisms on the electrodes. It was obvious that DHA and CA of CW–MFCs reached the highest in summer and autumn, followed by spring, and the lowest in winter, which was consistent with the seasonal variations of organics and nitrogen removal (Figure 5) as well as the variations of electricity generation with temperature (Figure 3). In closed-circuit CW–MFCc, enzymatic activity of the anodes, cathodes and substrate during winter, respectively, significantly ($P < 0.05$) dropped by 79.1, 85.7 and 80.4% for DHA, and 62.2, 67.6 and 66.1% for CA, when compared to summer. As has been noted, the wastewater temperature within the CW–MFCs was higher than the atmospheric temperature by 1–2 °C in winter. Since the anodes were installed inside the substrate, while the cathodes were placed on the substrate's surface, the lower temperature resulted in the larger decrease of the enzymatic activity in the cathode area. Moreover, the decrease of enzymatic activity in the anodes and cathodes during winter was smaller in closed-circuit CW–MFCc than that in open-circuit CW–MFCo. This demonstrated that the MFC current could also stabilize the growth of microorganisms on the electrodes at lower temperatures to a certain extent.

Microbial diversity

Table 2 shows the observed microbial species and alpha diversity indices of anodes, cathodes and lava in open-circuit CW–MFCo and closed-circuit CW–MFCc under the four seasons. The Shannon and Simpson index was used to identify community diversity, the Chao1 and ACE estimator was used to identify community richness, and the Good's coverage was used to characterize sequencing depth. It was obvious that the microbial species and diversity in both CW–MFCo and CW–MFCc declined in winter when compared to summer and autumn. For the anodes, the observed species and alpha diversity indices in CW–MFCc were much higher than in CW–MFCo, especially in winter. This result indicated that the closed circuit mode promoted the bacteria diversity on MFC anodes, which was consistent with the study by Li *et al.* (2019). For the cathodes, it was only in summer that the bacterial community diversity in CW–MFCc was higher than that in CW–MFCo. This may be because the effect of temperature on microbes was stronger than the stimulation of MFC, since the cathodes were placed on the substrate's surface where they were directly exposed to the exterior atmosphere. For lava, the bacteria diversity in CW–MFCc and CW–MFCo was similar, which demonstrated that the enhancement of bacteria diversity by MFC was limited to the electrode surface. This was because the electricity generation in CW–MFCs was by direct transfer of electrons to the electrode by electrogens growing on the electrode (Liu *et al.* 2005).

The reported electrochemically active bacteria (EAB) in CW–MFCs includes *Proteobacteria*, *Firmicutes*, *Bacteroidetes* and *Acidobacteria*, among which *Proteobacteria* and *Firmicutes* are the main ones (Wang *et al.* 2016; Li *et al.* 2019). As shown in Figure 7, the bacterial community of the anodes, cathodes and lava samples at the phyla level was primarily composed of 10 phyla, and the dominant microbes were phyla *Proteobacteria*. For anodes, the closed circuit mode significantly promoted the abundance of *Firmicutes* and *Bacteroidetes*. The result that phyla *Bacteroidetes* was enriched on an MFC anode was consistent with previous studies (De *et al.* 2010; Li *et al.* 2019). For cathodes, *Proteobacteria* and *Firmicutes* were enriched in closed-circuit CW–MFCc, while there was approximately 7.5–14.7% *Cyanobacteria* on the cathodes of open-circuit CW–MFCo. As shown in Figure 8(b), the typical EAB at the genera level in closed-circuit CW–MFCc mainly included *Pseudomonas*, *Rhodoferax*, *Clostridium*, *Escherichia*, *Enterobacter*, *Shewanella*, *Desulfovibrio* and *Geobacter*. In CW–MFCc, the proportion of the eight genera EAB was approximately 16.7–17.5% on anodes and 10.1–10.5% on cathodes in spring and winter, while it decreased to 13.5–13.8% on anodes and 8.2–9.0% on cathodes in summer and autumn. In open-circuit CW–MFCo, the eight genera EAB were less than 3% on both anodes and cathodes (Figure 8(a)). This further proved that the closed circuit operation contributed to the accumulation of EAB on MFC electrodes at low temperatures in winter.

Overall, in closed-circuit CW-MFCc, the microbial community composition at the phyla level on anodes and cathodes was similar under the four seasons. The closed

Table 2 | Microbial diversity indexes in the constructed wetland–microbial fuel cell systems under different seasons

		Observed-species	Shannon	Simpson	Chao1	ACE	Good's coverage	PD_whole tree
Open-circuit CW–MFCo								
Anode	Spring	928	4.421	0.713	1,219.572	1,246.234	0.990	66.801
	Summer	1,538	7.339	0.964	2,013.814	2,067.515	0.984	97.745
	Autumn	1,272	5.875	0.852	1,737.153	1,736.001	0.986	86.046
	Winter	734	3.694	0.648	932.985	973.350	0.993	57.067
Cathode	Spring	1,215	7.373	0.985	1,749.989	1,780.930	0.986	84.120
	Summer	1,382	7.798	0.986	1,796.299	1,858.841	0.986	91.862
	Autumn	1,220	7.421	0.986	1,733.890	1,747.039	0.986	84.788
	Winter	1,172	7.102	0.979	1,814.247	1,782.391	0.986	81.463
Lava	Spring	1,209	6.619	0.945	1,632.460	1,687.157	0.987	86.788
	Summer	1,578	7.319	0.956	2,132.064	2,073.974	0.985	111.167
	Autumn	1,485	7.726	0.984	1,652.602	1,767.321	0.989	98.682
	Winter	917	5.279	0.838	1,168.487	1,182.301	0.991	71.241
Closed-circuit CW–MFCc								
Anode	Spring	1,440	7.555	0.974	1,625.902	1,738.800	0.989	94.789
	Summer	1,880	8.950	0.995	2,029.463	2,090.419	0.990	121.148
	Autumn	1,442	7.633	0.984	1,853.157	1,925.595	0.986	93.599
	Winter	1,408	7.687	0.978	1,799.267	1,796.147	0.987	97.101
Cathode	Spring	1,138	6.012	0.935	1,601.580	1,649.927	0.987	77.823
	Summer	1,553	8.049	0.990	1,810.973	1,932.238	0.987	100.648
	Autumn	1,262	6.771	0.964	1,747.763	1,890.075	0.985	85.998
	Winter	1,035	5.853	0.932	1,231.711	1,329.222	0.991	72.464
Lava	Spring	979	6.193	0.956	1,278.640	1,312.090	0.990	71.804
	Summer	1,487	6.445	0.903	1,994.040	2,013.345	0.984	102.439
	Autumn	1,429	7.928	0.987	1,639.218	1,679.065	0.990	99.017
	Winter	860	4.487	0.730	1,077.600	1,106.129	0.992	63.445

circuit mode can promote the bacteria diversity on CW–MFC anodes as well as the abundance of EAB on CW–MFC anodes and cathodes, even at low temperatures, and thus enhance the organics and nitrogen removal of CW–MFCs in winter.

CONCLUSIONS

The atmospheric temperature significantly influenced the bioelectricity generation and removal of organics and nitrogen in CW–MFCs by primarily influencing the microbial enzymatic activity. The electricity output of CW-MFCs was extremely low below 5 °C, and reached the maximum above 25 °C. The organics and nitrogen removal of closed-circuit CW–MFC reached the highest in summer and autumn, followed by spring, and decreased by an average of 10.5% COD, 14.2% NH_3-N and 10.7% TN in winter, demonstrating smaller seasonal fluctuations compared to open-circuit CW–MFC in which the difference between summer and winter was 13.4% COD, 15.1% NH_3-N and 15.1% TN. Even at low temperatures, the MFC current could enhance the enzymatic activity and stabilize the growth of microorganisms on the electrodes. Moreover, the closed circuit mode can promote the bacteria diversity on CW–MFC anodes as well as the abundance of EAB on

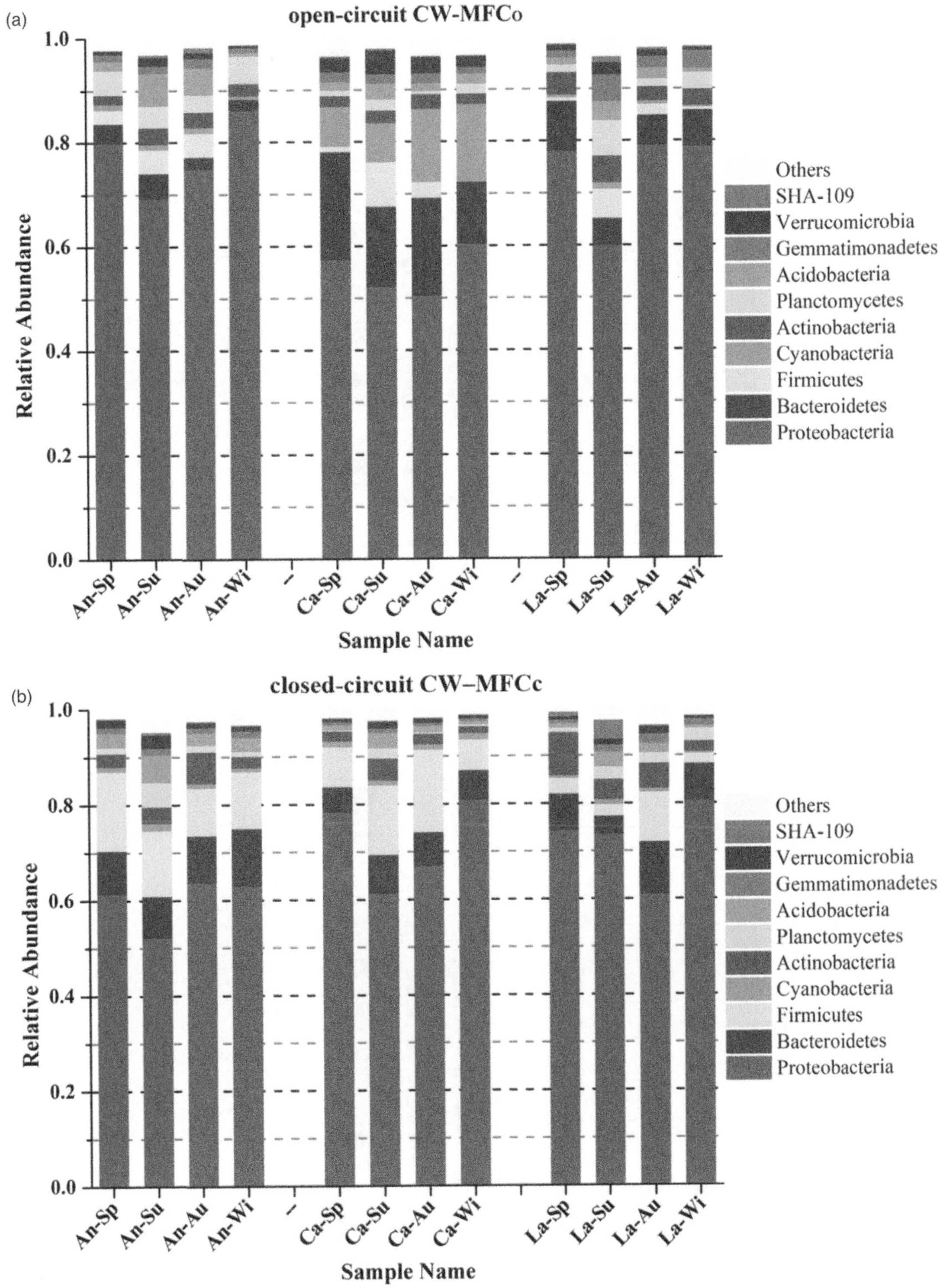

Figure 7 | Relative abundance plot at phyla level based on sequencing results of 16s rRNA. An: Anode; Ca: Cathode; La: Lava; Sp: Spring; Su: Summer; Au: Autumn; Wi: Winter.

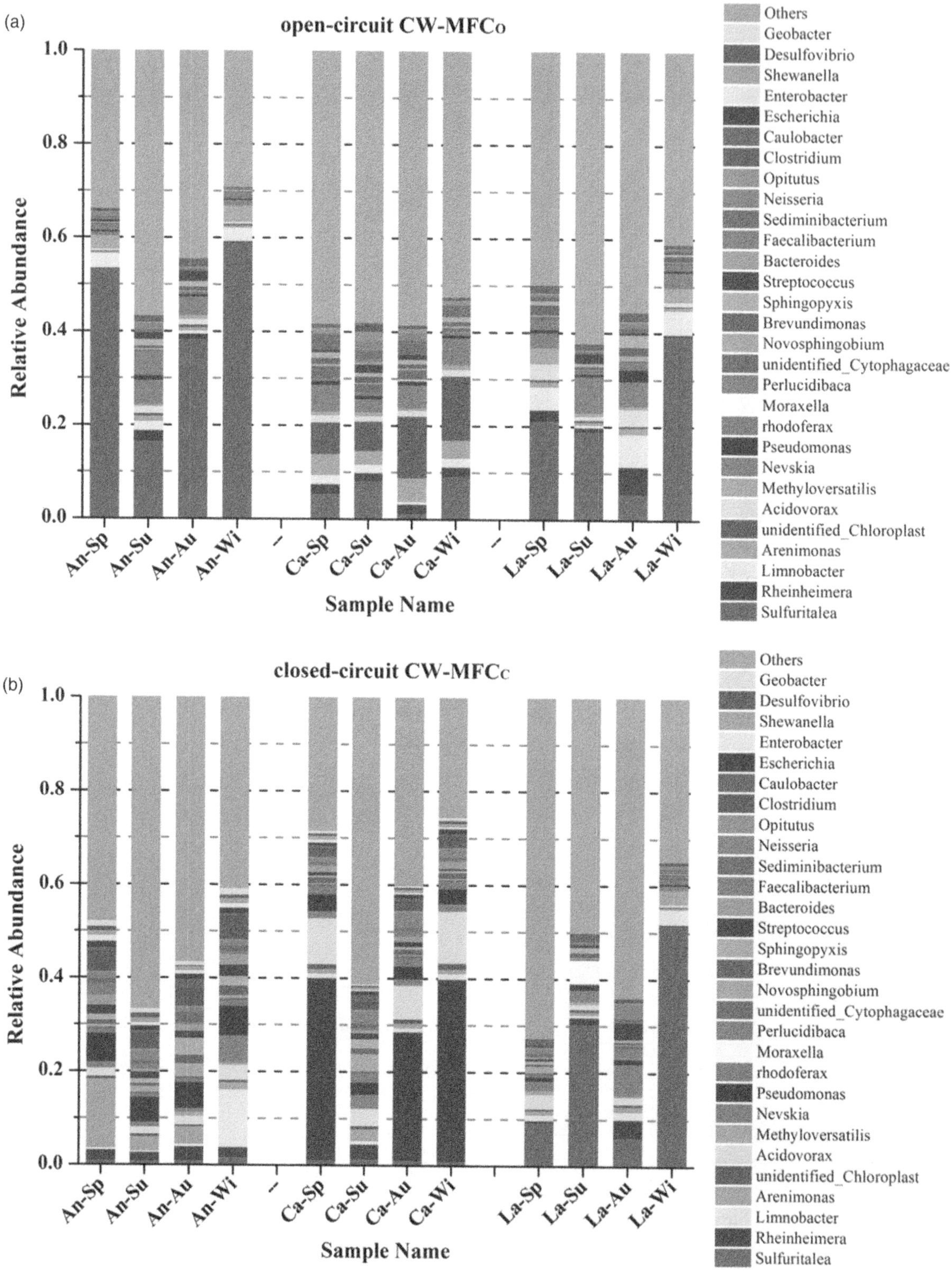

Figure 8 | Relative abundance plot at genera level based on sequencing results of 16s rRNA. An: Anode; Ca: Cathode; La: Lava; Sp: Spring; Su: Summer; Au: Autumn; Wi: Winter.

CW–MFC anodes and cathodes, and thus reduce the adverse effect of cooling on the organics and nitrogen removal in CWs. However, both MFC and temperature did not significantly influence the phosphorus removal in CW–MFCs.

ACKNOWLEDGEMENTS

This work was supported by The Key Special Program on the S&T for the Pollution Control and Treatment of Water Bodies of China (2014ZX07203-009).

DATA AVAILABILITY STATEMENT

All relevant data are included in the paper or its Supplementary Information.

REFERENCES

Adelaja, O., Keshavarz, T. & Kyazze, G. 2015 The effect of salinity, redox mediators and temperature on anaerobic biodegradation of petroleum hydrocarbons in microbial fuel cells. *Journal of Hazardous Materials* **283**, 211–217.

Akratos, C. S. & Tsihrintzis, V. A. 2007 Effect of temperature, HRT, vegetation and porous media on removal efficiency of pilot-scale horizontal subsurface flow constructed wetlands. *Ecological Engineering* **29** (2), 173–191.

Barrena, R., Vázquez, F. & Sanchez, A. 2008 Dehydrogenase activity as a method for monitoring the composting process. *Bioresource Technology* **99**, 905–908.

Behera, M., Murthy, S. S. R. & Ghangrekar, M. M. 2011 Effect of operating temperature on performance of microbial fuel cell. *Water Science and Technology* **64** (4), 917–922.

Cao, S. W., Jing, Z. Q., Yuan, P., Wang, Y. & Wang, Y. 2019 Performance of constructed wetlands with different substrates for the treated effluent from the municipal sewage plant. *Journal of Water Reuse and Desalination* **9** (4), 452–462.

De, S. L., Cabezas, A., Marzorati, M., Friedrich, M. W., Boon, N. & Verstraete, W. 2010 Microbial community analysis of anodes from sediment microbial fuel cells powered by rhizodeposits of living rice plants. *Applied & Environmental Microbiology* **76**, 2–8.

Doherty, L., Zhao, Y. Q., Zhao, X. H., Hu, Y. S., Hao, X. D., Xu, L. & Liu, R. B. 2015 A review of a recently emerged technology: constructed wetland-Microbial fuel cells. *Water Research* **85**, 38–45.

Fang, Z., Song, H. L., Cang, N. & Li, X. N. 2013 Performance of microbial fuel cell coupled constructed wetland system for decolorization of azo dye and bioelectricity generation. *Bioresource Technology* **144** (6), 165–171.

Guadarrama-Pérez, O., Gutiérrez-Macías, T., García-Sánchez, L., Guadarrama-Pérez, V. H. & Estrada-Arriaga, E. B. 2019 Recent advances in constructed wetland–microbial fuel cells for simultaneous bioelectricity production and wastewater treatment: a review. *International Journal of Energy Research* **43**, 5106–5127.

Kadlec, R. H. & Wallace, S. D. 2009 *Treatment Wetlands*, 2nd edn. CRC Press, Taylor and Francis Group, Boca Raton, FL, USA.

Lan, W., Zhang, J., Hu, Z., Ji, M. D., Zhang, X. W., Zhang, J. D., Li, F. Z. & Yao, G. Q. 2018 Phosphorus removal enhancement of magnesium modified constructed wetland microcosm and its mechanism study. *Chemical Engineering Journal* **335**, 209–214.

Li, H., Zhang, S., Yang, X. L., Yang, Y. L., Xu, H., Li, X. N. & Song, H. L. 2019 Enhanced degradation of bisphenol A and ibuprofen by an up-flow microbial fuel cell-coupled constructed wetland and analysis of bacterial community structure. *Chemosphere* **217**, 599–608.

Liu, H., Cheng, S. A. & Logan, B. E. 2005 Production of electricity from acetate or butyrate using a single-chamber microbial fuel cell. *Environmental Science & Technology* **39** (2), 658–662.

Michie, I. S., Kim, J. R., Dinsdale, R. M., Guwy, A. J. & Premier, G. C. 2011 Operational temperature regulates anodic biofilm growth and the development of electrogenic activity. *Applied Microbiology and Biotechnology* **92**, 419–430.

Ouellet-Plamondon, C., Chazarenc, F., Comeau, Y. & Brisson, J. 2006 Artificial aeration to increase pollutant removal efficiency of constructed wetlands in cold climate. *Ecological Engineering* **27**, 258–264.

Puig, S., Serra, M., Vilar-Sanz, A., Cabré, M., Bañeras, L., Colprim, J. & Balaguer, M. D. 2011 Autotrophic nitrite removal in the cathode of microbial fuel cells. *Bioresource Technology* **102** (6), 4462–4467.

Saeed, T. & Sun, G. Z. 2012 A review on nitrogen and organics removal mechanisms in subsurface flow constructed wetlands: dependency on environmental parameters, operating conditions and supporting media. *Journal of Environmental Management* **112**, 429–448.

Sani, A., Scholz, M. & Bouillon, L. 2013 Seasonal assessment of experimental vertical-flow constructed wetlands treating domestic wastewater. *Bioresource Technology* **147**, 585–596.

Song, Z. W., Zheng, Z. P., Li, J., Sun, X. F., Han, X. Y., Wang, W. & Xu, M. 2006 Seasonal and annual performance of a full-scale constructed wetland system for sewage treatment in China. *Ecological Engineering* **26** (3), 272–282.

Srivastava, P., Yadav, A. K. & Mishra, B. K. 2015 The effects of microbial fuel cell integration into constructed wetland on the performance of constructed wetland. *Bioresource Technology* **195**, 223–230.

Wang, J. F., Song, X. S., Wang, Y. H., Abayneh, B., Li, Y. H., Yan, D. H. & Bai, J. H. 2016 Nitrate removal and bioenergy production in constructed wetland coupled with microbial fuel cell: establishment of electrochemically active bacteria community on anode. *Bioresource Technology* **221**, 358–365.

Wang, M., Zhang, D. Q., Dong, J. W. & Tan, S. K. 2017 Constructed wetlands for wastewater treatment in cold climate – A review. *Journal of Environmental Sciences* **57**, 293–311.

Wu, S., Austin, D., Liu, L. & Dong, R. 2011 Performance of integrated household constructed wetland for domestic wastewater treatment in rural areas. *Ecological Engineering* **37** (6), 948–954.

Xu, L., Zhao, Y. Q., Wang, X. D. & Yu, W. Z. 2018 Applying multiple bio-cathodes in constructed wetland-microbial fuel cell for promoting energy production and bioelectrical derived nitrification-denitrification process. *Chemical Engineering Journal* **344**, 105–113.

Yan, Y. J. & Xu, J. C. 2014 Improving winter performance of constructed wetlands for wastewater treatment in northern China: a review. *Wetlands* **34**, 243–253.

Yu, B., Liu, C. L., Wang, S. Y., Wang, W. D., Zhao, S. Y. & Zhu, G. B. 2020 Applying constructed wetland-microbial electrochemical system to enhance NH+ 4 removal at low temperature. *Science of the Total Environment* **724**, 138017.

First received 16 October 2020; accepted in revised form 17 January 2021. Available online 2 March 2021

IWA Publishing's authorised EU representative for General Product Safety Regulations is Diane D'Arras, 15 rue Duret, 75116 Paris, France, e-mail: safety@iwap.co.uk.

Printed and bound by CPI Group (UK) Ltd, Croydon, CR0 4YY
12/05/2026
02108892-0002